TODAY'S TECHNICIAN ™

Classroom Manual for
Manual Transmissions and Transaxles

Fourth Edition

■

Jack Erjavec

DELMAR
CENGAGE Learning™

Australia   Canada   Mexico   Singapore   Spain   United Kingdom   United States

Today's Technician: Manual Transmissions and Transaxles Classroom Manual 4E

Jack Erjavec

Vice President, Technology and Trades SBU:

Alar Elken

Editorial Director:

Sandy Clark

Senior Acquisitions Editor:

David Boelio

Development Editor:

Matthew Thouin

Marketing Director:

David Garza

Channel Manager:

Bill Lawrensen

Marketing Coordinator:

Mark Pierro

Production Director:

Mary Ellen Black

Production Editor:

Toni Hansen

Art/Design Specialist:

Cheri Plasse

Technology Project Manager:

Kevin Smith

Technology Project Specialist:

Linda Verde

Editorial Assistant:

Andrea Domkowski

For product information and technology assistance, contact us at
Cengage Learning Customer & Sales Support, 1-800-354-9706
For permission to use material from this text or product, submit all requests online at **cengage.com/permissions**
Further permissions questions can be emailed to
permissionrequest@cengage.com

ExamView® and ExamView Pro® are registered trademarks of FSCreations, Inc. Windows is a registered trademark of the Microsoft Corporation used herein under license. Macintosh and Power Macintosh are registered trademarks of Apple Computer, Inc. Used herein under license.

© 2007 Cengage Learning. All Rights Reserved. Cengage Learning WebTutor™ is a trademark of Cengage Learning.

Library of Congress Control Number: 2005015956

ISBN-13: 978-1-4018-7753-8

ISBN-10: 1-4018-7753-2

Delmar Cengage Learning
5 Maxwell Drive
Clifton Park, NY 12065-2919
USA

Cengage Learning products are represented in Canada by Nelson Education, Ltd.

For your lifelong learning solutions, visit **delmar.cengage.com**

Visit our corporate website at **www.cengage.com**

Notice to the Reader

Publisher does not warrant or guarantee any of the products described herein or perform any independent analysis in connection with any of the product information contained herein. Publisher does not assume, and expressly disclaims, any obligation to obtain and include information other than that provided to it by the manufacturer. The reader is expressly warned to consider and adopt all safety precautions that might be indicated by the activities described herein and to avoid all potential hazards. By following the instructions contained herein, the reader willingly assumes all risks in connection with such instructions. The publisher makes no representations or warranties of any kind, including but not limited to, the warranties of fitness for particular purpose or merchantability, nor are any such representations implied with respect to the material set forth herein, and the publisher takes no responsibility with respect to such material. The publisher shall not be liable for any special, consequential, or exemplary damages resulting, in whole or part, from the readers' use of, or reliance upon, this material.

Printed in Canada
4 5 6 7 11 10 09 08

CONTENTS

PREFACE

Thanks to the support the Today's Technician™ Series has received from those who teach automotive technology, Delmar Cengage Learning, the leader in automotive related textbooks, is able to live up to its promise to provide new editions of the series regularly. We have listened and responded to our critics and our fans and present this new updated and revised fourth edition. By revising our series regularly, we can and will respond to changes in the industry, changes in technology, changes in the certification process, and to the ever-changing needs of those who teach automotive technology.

The Today's Technician™ Series, by Delmar Cengage Learning, features textbooks that cover all mechanical and electrical systems of automobiles and light trucks (whereas the Heavy-duty Trucks portion of the series does the same for Heavy-duty vehicles). Principally, the individual titles correspond to the main areas of ASE (National Institute for Automotive Service Excellence) certification. Additional titles include remedial skills and theories common to all of the certification areas and advanced or specific subject areas that reflect the latest technological trends. Each text is divided into two volumes: a Classroom Manual and a Shop Manual.

Unlike yesterday's mechanic, the technician of today and for the future must know the underlying theory of all automotive systems and be able to service and maintain those systems. Dividing the material into two volumes provides the reader with the information needed to begin a successful career as an automotive technician without interrupting the learning process by mixing cognitive and performance learning objectives into one volume.

The design of Delmar Cengage Learning's Today's Technician™ series was based on features that are known to promote improved student learning. The design was further enhanced by a careful study of survey results, in which the respondents were asked to value particular features. Some of these features can be found in other textbooks, whereas others are unique to this series.

Each Classroom Manual contains the principles of operation for each system and subsystem. The Classroom Manual also contains discussions on design variations of key components used by the different vehicle manufacturers. This volume is organized to build on basic facts and theories. The primary objective of this volume is to allow the reader to gain an understanding of how each system and subsystem operates. This understanding is necessary to diagnose the complex automobiles of today and tomorrow. Although the basics contained in the Classroom Manual provide the knowledge needed for diagnostics, diagnostic procedures appear only in the Shop Manual. An understanding of the basics is also a requirement for competence in the skill areas covered in the Shop Manual.

A spiral bound Shop Manual covers the "how-to's." This volume includes step-by-step instructions for diagnostic and repair procedures. Photo Sequences are used to illustrate some of the common service procedures. Other common procedures are listed and are accompanied with line drawings and photos that allow the reader to visualize and conceptualize the finest details of the procedure. This volume also contains the reasons for performing the procedures as well as when that particular service is appropriate.

The two volumes are designed to be used together and are arranged in corresponding chapters. Not only are the chapters in the volumes linked together, but also the contents of the chapters are linked. This linking of content is evidenced by marginal callouts that refer the reader to the chapter and page that the same topic is addressed in the other volume. This feature is valuable to instructors. Without this feature, users of other two-volume textbooks must search the index or table of contents to locate supporting information in the other volume. This is not only cumbersome but also creates additional work for an instructor when planning the presentation of material and when making reading assignments. It is also valuable to the students; with page references, they also know exactly where to look for supportive information.

Both volumes contain clear and thoughtfully selected illustrations, many of which are original drawings or photos specially prepared for inclusion in this series. This means that the art is a vital part of each textbook and not merely inserted to increase the numbers of illustrations.

The page layout, used in the series, is designed to include information that would otherwise break up the flow of information presented to the reader. The main body of the text includes all of the "need-to-know" information and illustrations. In the wide side margins of each page are many of the special features of the series. Items that are truly "nice-to-know" information include simple examples of concepts just introduced in the text, explanations or definitions of terms that will not be defined in the glossary, examples of common trade jargon used to describe a part or operation, and exceptions to the norm explained in the text. Many textbooks attempt to include this type of information and insert it in the main body of text; this tends to interrupt the thought process and cannot be pedagogically justified. By placing this information off to the side of the main text, the reader can select when to refer to it.

Highlights of this Edition—Classroom Manual

The text was updated throughout to include the latest developments. Some these new topics include dual clutch systems, various limited-slip differential designs, six-speed transmissions, constantly variable transmissions, and self-shifting manual transmissions. We also added new information in all chapters; see particularly Chapter 10, which covers basic electrical and electronic theory and the various applications for switches, speed sensors, solenoids, electromagnetic clutches, and electronic circuits.

The most notable change in this new edition is the organization of topics. The first chapter is now about the main topic of the text. This chapter introduces the purpose of the main system and how it links to the rest of the vehicle. The chapter also describes the purpose and location of the subsystems as well as the major components of the system and subsystems. The goal of this chapter is to establish a basic understanding on which students can base their learning. All systems and subsystems that will be discussed in detail later in the text are introduced and their primary purpose described. The second chapter covers the underlying basic theories of operation for the topic of the text. This is valuable to the student and the instructor because it covers the theories that other textbooks assume the reader knows. All related basic physical, chemical, and thermodynamic theories are covered in this chapter.

The order of the topics and the new chapter titles reflect the most common reviewer suggestions. In addition to the reorganization, this edition has been thoroughly updated. Current model transmissions are used as examples throughout the text. Some are discussed in detail. This includes five- and six-speed and constantly variable transmissions. This new edition also has more information on nearly all manual transmission-related topics. Finally, the art has been updated throughout the text to enhance comprehension and improve visual interest.

Highlights of this Edition—Shop Manual

Along with the Classroom Manual, the Shop Manual was updated to match current trends. Service information related to the new topics covered in the Classroom Manual is included in this manual. In addition, several new photo sequences were added. The purpose of these detailed photos is to show students what to expect when they perform the same procedure. They also can provide a student with familiarity of a system or type of equipment they may not be able to perform at their school. Although it is not the main purpose of the textbook to prepare someone to pass an ASE exam successfully, all of the information required to do so is included in the textbook.

This text was reorganized to correlate with the new organization of the classroom manual. This change created new chapters (not just new chapter titles) and allowed for more coverage of all topics.

Chapters 1 and 2 are new and cover the need-to-know transmission-related information about tools, safety, and typical services procedures. The first chapter covers safety issues. Included in this chapter are common shop hazards, safe shop practices, safety equipment, and the legislation concerning the safe handling of hazardous materials and wastes. Chapter 2 covers the basics that a transmission technician needs to know to earn a living, including basic diagnostics. Also included in this chapter are those tools and procedures that are commonly used to diagnose and service manual transmissions and drivelines. In summary, this chapter looks at what it takes to be a successful technician, typical pay plans for technicians, service information sources, ASE certification, and the laws and regulations a technician should be aware of. The main topic of the chapter is the special tools required to work on today's transmissions and drivelines.

The rest of the chapters have been thoroughly updated. Much of the updating focuses on the diagnosis and service to new systems as well as those systems on which instructors have said students need more help. Currently accepted service procedures are used as examples throughout the text. These procedures also served as the basis for new job sheets that are included in the text. Finally, the art has been updated throughout the text to enhance comprehension and improve visual interest.

Jack Erjavec

Features of the Classroom Manual include:

Cognitive Objectives

These objectives define the contents of the chapter and define what the student should have learned upon completion of the chapter.

Each topic is divided into small units to promote easier understanding and learning.

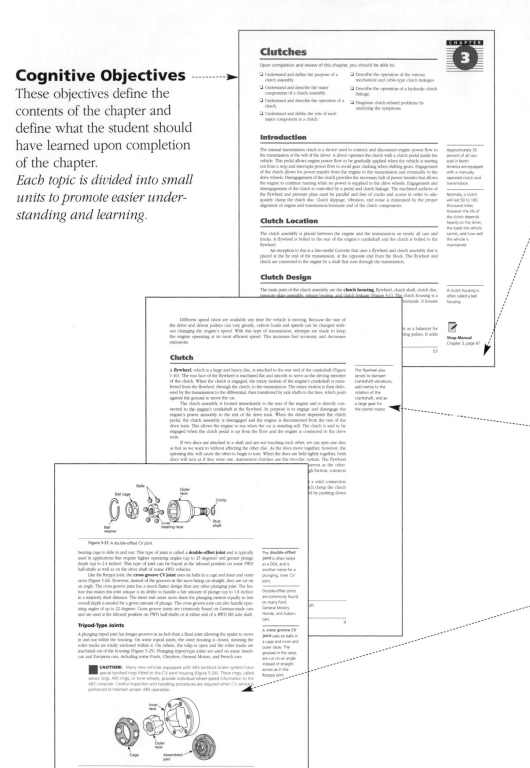

Cross-References to the Shop Manual

Reference to the appropriate page in the Shop Manual is given whenever necessary. Although the chapters of the two manuals are synchronized, material covered in other chapters of the Shop Manual may be fundamental to the topic discussed in the Classroom Manual.

Marginal Notes

These notes add "nice-to-know" information to the discussion. They may include examples or exceptions, or may give the common trade jargon for a component.

Cautions and Warnings

Throughout the text, warnings are given to alert the reader to potentially hazardous materials or unsafe conditions. Cautions are given to advise the students of things that can go wrong if instructions are not followed or if a nonacceptable part or tool is used.

A Bit of History

This feature gives the student a sense of the evolution of the automobile. This feature not only contains nice-to-know information, but also should spark some interest in the subject matter.

Author's Notes

This feature includes simple explanations, stories, or examples of complex topics. These are included to help students understand difficult concepts.

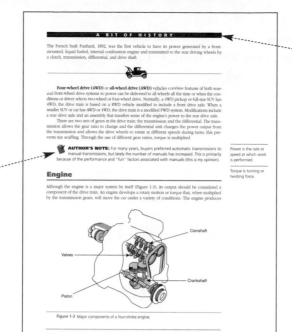

Summaries

Each chapter concludes with a summary of key points from the chapter. These are designed to help the reader review the chapter contents.

Terms to Know List

A list of new terms appears next to the Summary.

Review Questions

Short answer essay, fill-in-the-blank, and multiple-choice questions are found at the end of each chapter. These questions are designed to accurately assess the student's competence in the stated objectives at the beginning of the chapter.

Shop Manual

To stress the importance of safe work habits, the Shop Manual also dedicates one full chapter to safety. Other important features of this manual include:

Performance-Based Objectives

These objectives define the contents of the chapter and define what the student should have learned upon completion of the chapter. These objectives also correspond with the list of required tasks for ASE certification. *Each ASE task is addressed.*

Although this textbook is not designed to simply prepare someone for the certification exams, it is organized around the ASE task list. These tasks are defined generically when the procedure is commonly followed and specifically when the procedure is unique for specific vehicle models. Imported and domestic model automobiles and light trucks are included in the procedures.

Special Tools List

Whenever a Special Tool is required to complete a task, it is listed in the margin next to the procedure.

Marginal Notes

These notes add "nice-to-know" information to the discussion. They may include examples or exceptions, or may give the common trade jargon for a component.

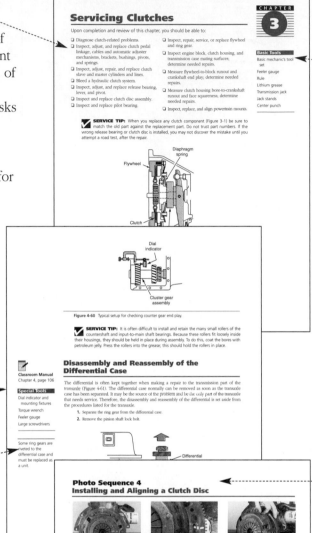

Basic Tools List

Each chapter begins with a list of the Basic Tools needed to perform the tasks included in the chapter.

Photo Sequences

Many procedures are illustrated in detailed Photo Sequences. These detailed photographs show the students what to expect when they perform particular procedures. They also can provide the student a familiarity with a system or type of equipment, which the school may not have.

Cautions and Warnings

Throughout the text, warnings are given to alert the reader to potentially hazardous materials or unsafe conditions. Cautions are given to advise the student of things that can go wrong if instructions are not followed or if a nonacceptable part or tool is used.

Service Tips

Whenever a short-cut or special procedure is appropriate, it is described in the text. These tips are generally those things commonly done by experienced technicians.

Cross-References to the Classroom Manual

Reference to the appropriate page in the Classroom Manual is given whenever necessary. Although the chapters of the two manuals are synchronized, material covered in other chapters of the Classroom Manual may be fundamental to the topic discussed in the Shop Manual.

Customer Care

This feature highlights those little things a technician can do or say to enhance customer relations.

Job Sheets

Located at the end of each chapter, the Job Sheets provide a format for students to perform procedures covered in the chapter. A reference to the ASE Task addressed by the procedure is referenced on the Job Sheet.

Often technicians use troubleshooting guides given in service manuals to aid them in diagnostics. To use them, the technicians must first describe the problem then refer to a chart or diagnostic tree to determine the most probable causes of the problem.

Clutch Slippage

Clutch **slippage** is evident when the driver has the clutch engaged and the engine's speed increases but the vehicle's road speed does not. Slipping is normally most obvious during acceleration and shifting. A road test can determine if the clutch is slipping. Normal acceleration from a stop and several gear positions should provide the conditions necessary to witness slipping.

Slippage can also be verified in the shop. Depress the clutch pedal, shift the transmission into high gear, and increase the engine's speed to approximately 2,000 rpm. Slowly release the clutch pedal until the clutch engages. The engine should stall immediately. If the engine does not stall within a few seconds, the clutch is slipping. If the clutch slips, depress the clutch pedal to end the test quickly.

CAUTION: Severe or prolonged clutch slippage may cause grooving and/or heat damage to the pressure plate. Therefore, end all testing as soon as slippage is evident.

If slippage is evident during either test, raise the vehicle and check the linkage. Check for worn or binding parts. Also check for loose or worn engine mounts. Clutch slippage can also be caused by an overadjusted clutch. In this case, the clutch is always partially released and is never fully engaged. Overadjustment is possible on a cable system, so back off the adjustment and check for slippage. On hydraulic linkage systems, make sure the return port to the master cylinder is not blocked. This can prevent the slave cylinder from returning fully. If no problem is evident, the clutch should be disassembled and repaired. Normally, clutch slippage is caused by an oil-soaked or worn disc facing, warped flywheel or pressure plate, weak pressure plate springs, or the release bearing contacting the fingers of the pressure plate. The cause of the problem should be repaired. Other causes of clutch slippage are riding the clutch pedal with the vehicle in motion and holding the vehicle on an incline by using the clutch as a brake.

Oil and grease contamination on the disc's frictional material results in a loss in the coefficient of friction. This reduces the ability of the disc to remain tightly clamped between the flywheel and the pressure plate. When the clutch slips, it generates heat, which causes it to slip even more. Late model cars are more prone to this problem because they use smaller clutch discs and discs with nonasbestos linings. Nonasbestos friction materials are prone to fail when subjected to the slightest amount of oil or grease. During a visual inspection, examine the clutch disc and the transmission's input shaft for oil residue. If oil is detected, look for leaks at the engine's rear main seal or at the transmission's front seal. Oil leaks must be corrected prior to installing a new clutch disc.

SERVICE TIP: Clutch slippage on vehicles with a dual-mass flywheel can be very difficult to diagnose. The cause of the problem may be the clutch or a bad flywheel. If the customer's complaint is a slipping clutch, but the clutch assembly shows no signs of slippage, the cause of the problem is probably the dual-mass flywheel, which should be replaced.

Clutch Chatter

Clutch **chatter** is a shaking or shuddering that occurs when the pressure plate first makes contact with clutch is fully engaged.

Start the engine and completely depress and increase the engine's speed to about 1,500 chattering as it begins to engage. Depress the to prevent damage to the clutch parts.

Figure 3-27 Checking pressure plate warpage and distortion with a straightedge.

plate, as well as the clutch disc and flywheel. If the pressure plate is found to be defective, replace it with a remanufactured or new one. Individual pressure plate parts are not available; never attempt to rebuild a pressure plate.

Distortion of the pressure plate will cause misalignment problems. Such distortion or warpage is caused by carelessness when removing or installing the pressure plate (Figure 3-27). The attaching bolts must be loosened or tightened evenly and in a staggered sequence, or the assembly may become distorted by the force of the springs.

Once the pressure plate has been removed, it should be carefully inspected for scoring, cracks, **blueing**, and hot spots, and the thrust ring or fingers should also be inspected for excessive wear. Excessive finger wear on all fingers indicates a release bearing failure. If the finger wear is uneven, the probable cause is improper tightening of the pressure plate to flywheel bolts. Also look for bent or uneven release levers, or check to see if the pressure ring is not parallel with the clutch cover. Replace the entire pressure plate assembly if any defects are found.

Blueing results from overheating.

Clutch Disc

The clutch disc (Figure 3-28) transfers engine torque from the flywheel to the transmission input shaft when it is engaged. The disc is normally replaced when its friction facing wears thin. In most modern passenger cars, the clutch disc friction material is a woven or molded compound consisting of cotton, brass, copper, rubber, and phenolic resin. Racing and heavy-duty applications may use sintered metallic or ceramic friction material. The asbestos lining material and other lining materials are normally riveted or bonded to the disc. When the lining is riveted on, the disc should be replaced before the material is worn flush with the rivets. If the rivets contact the flywheel or pressure plate, rapid wear and grooves on the surface of either will occur (Figure 3-29).

The clutch disc will wear quickly whenever it is operated in a partially engaged position. This is usually caused by inadequate pressure plate spring force or incorrect clutching and declutching. When a driver "rides the clutch," the pressure plate is unable to apply full clamping pressure on the clutch disc, which causes the disc and release bearing to wear rapidly. Other conditions that cause rapid disc wear are insufficient free-play, binding clutch linkage, and high engine rpm starts. Overloading will also cause premature wear.

Classroom Manual
Chapter 3, page 61

Special Tools
Torque wrench
An old input shaft, "dummy shaft," or a clutch alignment tool (Figure 3-30)
Flywheel locking tool

Riding the clutch describes an improper driving technique in which the driver's foot is kept partially on the clutch pedal at all times.

CUSTOMER CARE: If it appears that the cause of a clutch slippage problem is the driver, tactfully inform the customer about the driving habits that can damage the clutch. These habits include riding the clutch and holding the vehicle on an incline by using

---

Job Sheet 7

7

Name _____ Date _____

Servicing Clutches

Upon completion of this job sheet, you should be able to identify vehicles that need clutch service.

ASE Correlation

This job sheet is related to the ASE Manual Drive Train and Axles Test's content area: *Clutch Diagnosis and Repair*, Task: *Diagnose clutch noise, binding, slippage, pulsation, and chatter problems; determine needed repairs.*

Describe the vehicle being worked on:

Year _____ Make _____ VIN _____

Model _____

Procedure   Task Completed

1. Bring the vehicle into the shop area.   ☐

2. List the type of vehicle.

3. Locate and record the VIN #.

4. From the tag or sticker under the hood, determine engine size and identify the manufacturer.

5. Obtain a service manual for that year and make of vehicle.   ☐

6. Locate the section that identifies the material covering the vehicle and then find the information needed; locate and identify the following information. If you need to raise the vehicle up, use safety rules for lifting.

   Engine size _____

   Transmission tag # _____

   Manufacturer of the components: _____

7. Find the pages that cover the vehicle you are working on, then find the pages that cover clutches.   ☐

Instructor's Response _____

_____

115

Case Studies

Case Studies concentrate on the ability to properly diagnose the systems. Beginning with Chapter 3, each chapter ends with a case study in which a vehicle has a problem, and the logic used by a technician to solve the problem is explained.

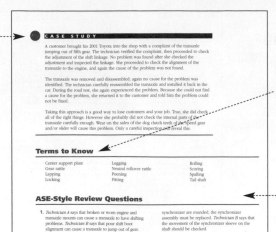

Terms to Know List

Terms in this list can be found in the Glossary at the end of the manual.

ASE-Style Review Questions

Each chapter contains ASE-style review questions that reflect the performance-based objectives listed at the beginning of the chapter. These questions can be used to review the chapter as well as to prepare for the ASE certification exam.

ASE Challenge Questions

Each technical chapter ends with five ASE challenge questions. These are not more review questions, rather they test the students' ability to apply general knowledge to the contents of the chapter.

ASE Practice Examination

A 50 question ASE practice exam, located in the appendix, is included to test students on the contents of the Shop Manual.

Reviewers

I would like to extend a special thanks to those who saw things I overlooked and for their contributions to this text:

James D. Carnahan
University of Northwestern Ohio
Lima, OH

John Eichelberger
St. Philip's College
San Antonio, TX

Russell Leonard
Ferris State University
Big Rapids, MI

Don Lumsdon
Ivy Tech State College
Terre Haute, IN

Frank D. Russo
Northern Virginia Community College
Alexandria, VA

Raymond K. Skow, Sr.
Truckee Meadows Community College
Reno, NV

David Talavera
Arapahoe Community College
Littleton, CO

Christopher VanStavoren
Pennsylvania College of Technology
Williamsport, PA

Manual Drive Trains and Axles

CHAPTER 1

Upon completion and review of this chapter, you should be able to:

❏ Identify the major components of a vehicle's drive train.

❏ State and understand the purpose of a transmission.

❏ Describe the difference between a transmission and a transaxle.

❏ Describe the construction and operation of CVTs.

❏ State and understand the purpose of a clutch assembly.

❏ Describe the differences between a typical FWD and RWD car.

❏ Describe the construction of a drive shaft.

❏ State and understand the purpose of U- and CV-joints.

❏ State and understand the purpose of a differential.

❏ Identify and describe the various gears used in modern drive trains.

❏ Identify and describe the various bearings used in modern drive trains.

Introduction

An automobile can be divided into four major systems or basic components: (1) the engine, which serves as a source of power; (2) the power train, or drive train, which transmits the engine's power to the car's wheels; (3) the chassis, which supports the engine and body and includes the brake, steering, and suspension systems; and (4) the car's body, interior, and accessories, which include the seats, heater and air conditioner, lights, windshield wipers, and other comfort and safety features.

Basically, the **drive train** has four primary purposes: to connect and disconnect the engine's power to the wheels, to select different speed ratios, to provide a way to move the car in reverse, and to control the power to the drive wheels for safe turning of the automobile. The main components of the manual drive train are the: clutch, transmission, differential, and drive axles (Figure 1-1). The exact components used in a vehicle's drive train depend on whether the vehicle is equipped with front-wheel drive, rear-wheel drive, or four-wheel drive.

Today, most cars are **front-wheel drive (FWD)**. Power flow through the drive train of FWD vehicles passes through the clutch or torque converter, the transmission, and then moves through a front differential, the driving axles, and on to the front wheels. The transmission and differential are housed in a single unit (Figure 1-2), called a **transaxle**. The gear sets in the transaxle provide the required gear ratios and direct the power flow into the differential. The differential gearing provides the final gear reduction and splits the power flow between the left and right drive axles.

The drive axles extend from the sides of the transaxle. The outer ends of the axles are fitted to the hubs of the drive wheels. **Constant velocity (CV) joints** mounted on each end of the drive axles allow for changes in length and angle without affecting the power flow to the wheels.

Some larger luxury and many performance cars are **rear-wheel drive (RWD)**. Most pickup trucks, minivans, and SUVs are also RWD vehicles. Power flow in a RWD vehicle passes through the clutch or torque converter, manual or automatic transmission, and the driveline (drive shaft assembly). Then it goes through the rear differential, the rear-driving axles, and on to the rear wheels.

About 50 percent of the questions on the ASE Manual Drive Trains and Axles Certification Test are based on transmissions. The remaining questions are related to the other drive train components.

Torque converters use fluid flow to transfer the engine's power to the transmission.

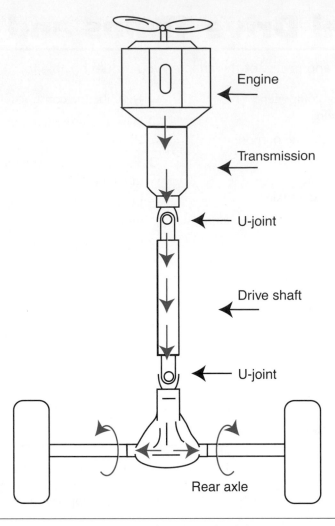

Figure 1-1 Typical drivetrain components for a rear-wheel drive car.

Figure 1-2 A transaxle.

The French built Panhard, 1892, was the first vehicle to have its power generated by a front-mounted, liquid fueled, internal combustion engine and transmitted to the rear driving wheels by a clutch, transmission, differential, and drive shaft.

Four-wheel drive (4WD) or **all-wheel drive (AWD)** vehicles combine features of both rear- and front-wheel drive systems so power can be delivered to all wheels all the time or when the conditions or driver selects two-wheel or four-wheel drive. Normally, a 4WD pickup or full-size SUV has 4WD; the drive train is based on a RWD vehicle modified to include a front drive axle. When a smaller SUV or car has AWD or 4WD, the drive train is a modified FWD system. Modifications include a rear drive axle and an assembly that transfers some of the engine's power to the rear drive axle.

There are two sets of gears in the drive train: the transmission and the differential. The transmission allows the gear ratio to change and the differential unit changes the power output from the transmission and allows the drive wheels to rotate at different speeds during turns; this prevents tire scuffing. Through the use of different gear ratios, torque is multiplied.

AUTHOR'S NOTE: For many years, buyers preferred automatic transmissions to manual transmissions, but lately the number of manuals has increased. This is primarily because of the performance and "fun" factors associated with manuals (this is my opinion).

Power is the rate or speed at which work is performed.

Torque is turning or twisting force.

Engine

Although the engine is a major system by itself (Figure 1-3), its output should be considered a component of the drive train. An engine develops a rotary motion or torque that, when multiplied by the transmission gears, will move the car under a variety of conditions. The engine produces

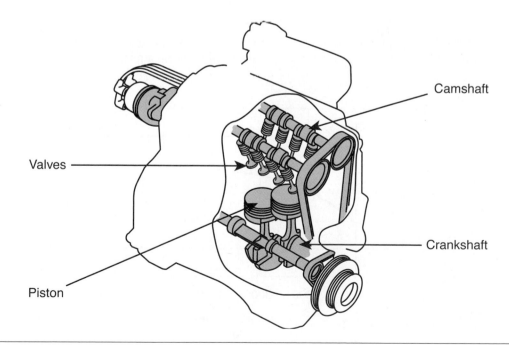

Figure 1-3 Major components of a four-stroke engine.

power by burning a mixture of fuel and air in its combustion chambers. Combustion causes a high pressure in the cylinders, which forces the pistons downward. Connecting rods transfer the downward movement of the pistons to the crankshaft, which rotates by the force on the pistons.

Most automotive engines are four-stroke cycle engines. The opening and closing of the intake and exhaust valves are timed to the movement of the piston. As a result, the engine passes through four different events or strokes during one combustion cycle. These four strokes are called the intake, compression, power, and exhaust. As long as the engine is running, this cycle of events repeats itself, resulting in the production of engine torque.

Engine Torque

The rotating or turning effort of the engine's crankshaft is called **engine torque**. Engine torque is measured in foot-pounds (ft.-lbs.) or in the metric measurement Newton-meters (N·m). Engines produce a maximum amount of torque when operating within a narrow range of engine speeds. When an engine reaches the maximum speed within that range, torque is no longer increased.

The amount of **torque** produced in relation to engine speed is called the engine's torque curve. Ideally, the engine should always be run when it is providing a maximum amount of torque (Figure 1-4). This allows the engine to provide the required amount of power while using a minimum amount of fuel.

To convert foot-pounds to Newton-meters, multiply the number of foot-pounds by 1.355.

As a car is climbing up a steep hill, its driving wheels slow down because of the increased amount of work it must do; this causes engine speed to decrease as well as reduce the engine's torque. The driver must downshift the transmission or press down harder on the gas pedal, which increases engine speed and allows the engine to produce more torque (Figure 1-5). When the car reaches the top of the hill and begins to go down, its speed and the speed of the engine rapidly increase. The driver can now upshift or let up on the gas pedal, which allows the engine's speed to decrease and places it back into its peak within the torque curve.

Downshifting is the shifting to a lower numerical gear.

Torque Multiplication

Measurements of **horsepower** indicate the amount of work being performed and the rate at which it is being done. The drive line can transmit power and multiply torque, but it cannot multiply power. When power flows through one gear to another, the torque is multiplied in proportion to the different gear sizes. Torque is multiplied, but the power remains the same, as the torque is multiplied at the expense of rotational speed. The engine's horsepower is determined by torque and engine speed. Maximum horsepower occurs when the engine is operating at a high speed and producing close to its maximum torque. Engine horsepower can be calculated by using a mathematical formula:

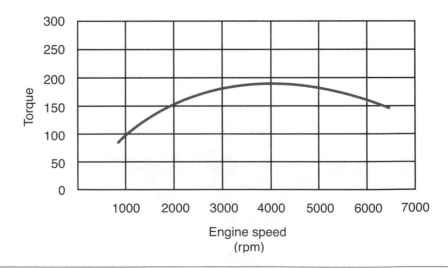

Figure 1-4 The amount of torque produced by an engine varies with the speed of the engine.

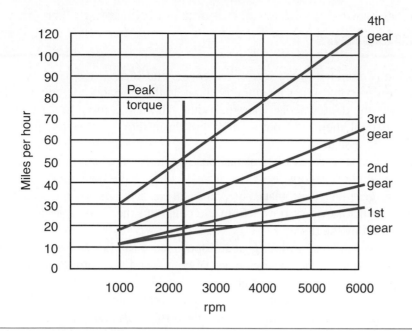

Figure 1-5 If the car represented by this chart were going up a hill in fourth gear and hill caused the car's speed to drop to 30 mph, the engine would labor heavily because it would be operating below its peak torque. To overcome the hill and allow the car to increase speed, the driver should place the car into third gear.

$$hp = T \times rpm / 5252$$

where hp = horsepower; T = the amount of torque produced by an engine, rpm = the speed of the engine when it is producing the torque, and 5252 being a mathematical constant. Figure 1-6 shows this relationship between horsepower and torque.

One horsepower is the equivalent of moving 33,000 pounds one foot in one minute.

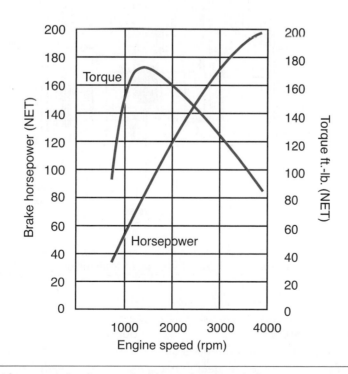

Figure 1-6 After an engine has reached its peak torque, its horsepower output increases with an increase in engine speed.

Transmissions

The transmission is mounted to the rear of the engine and is designed to allow the car to move forward and in reverse. It also has a neutral position. In this position, the engine can run without applying power to the drive wheels. Therefore, although there is input to the transmission when the vehicle is in neutral, there is no output from the transmission because the driving gears are not engaged to the output shaft.

There are two basic types of transmissions and transaxles: automatic and manual transmissions. Automatic transmissions use a combination of a torque converter and a planetary gear system to change gear ratios automatically. A manual transmission is an assembly of gears and shafts (Figure 1-7) that transmits power from the engine to the drive axle and changes in gear ratios are controlled by the driver.

By moving the shift lever, various gear and speed ratios can be selected. The gears in a transmission are selected to give the driver a choice of both speed and torque. Lower gears allow for lower vehicle speeds but more torque. Higher gears provide less torque but higher vehicle speeds. Gear ratios state the ratio of the number of teeth on the driven gear to the number of teeth on the drive gear. There often is much confusion about the terms high and low gear ratios. A gear ratio of 4:1 is lower than a ratio of 2:1. Although numerically the 4 is higher than the 2, the 4:1 gear ratio allows for lower speeds and, hence, is a low gear ratio.

Manual transmissions are commonly called Standard Shift or "Stick-Shift" transmissions.

Shift levers are typically called "shifters."

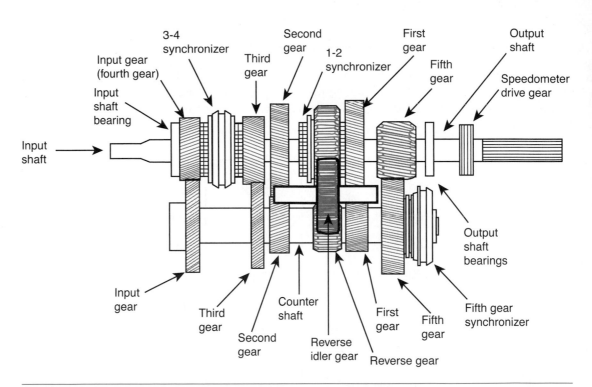

Figure 1-7 The gears and shafts of a typical five-speed transmission.

First and second gears are the low gears in a typical transmission, whereas third, fourth, and fifth are the high gears.

 AUTHOR'S NOTE: Sometimes we are our own worst enemies, especially when talking about cars. Gear ratios are a great example. *The conversations:*

"So and so put a low gear in her car and now it really flies!" "Well, so and so put in a high gear and now he claims his gas mileage is much better."

"That's great, you can put in a higher gear and go faster and get better mileage!" "Yeah, but it will take off like a slug."

"A 4.10:1 provides more torque multiplication than a 2.83:1 gear, but a four is higher than a 2 so the 4.10 gear must be the higher gear of the two!" "NO! The 4.10 gear is lower."

"A higher number gear provides more torque but lower maximum speed, that is why it is called a low gear."

High number + high torque + low speed = low gear
Low number + low torque + high speed = high gear

Today, most manual transmissions have four, five, or six forward speeds. These speeds or gears are identified as first, second, third, fourth, and fifth. Different gear ratios are necessary because an engine develops relatively little power at low engine speeds. The engine must be turning at a fairly high speed before it can deliver enough power to get the car moving. Through selection of the proper gear ratio, torque applied to the drive wheels can be multiplied. A transmission design that has been used only in pure racing is now available on a few cars. The driver changes gears with the gearshift or paddles located on the steering wheel. The **clutch** is automatically engaged and disengaged by the system, based on the numerous inputs to the control module.

Automotive manual transmissions are constant mesh, fully synchronized transmissions. "Constant mesh" means that the transmission gears are constantly in mesh, regardless of whether the car is stationary or moving. "Fully synchronized" refers to a mechanism of brass rings and synchronizers (Figure 1-8) used to bring the rotating shafts and gears of the transmission to one speed for smooth up- and downshifting.

Synchronizers serve as clutches for the brass rings and the gears.

Continuously Variable Transmission (CVT)

Another unconventional transmission design, the **continuously variable transmission (CVT)**, is found on some late-model cars and small SUVs. Basically, a CVT (Figure 1-9) is a transmission without fixed forward speeds. The gear ratio varies with engine speed and load. These transmissions are, however, fitted with a one-speed reverse gear. These automatic-like

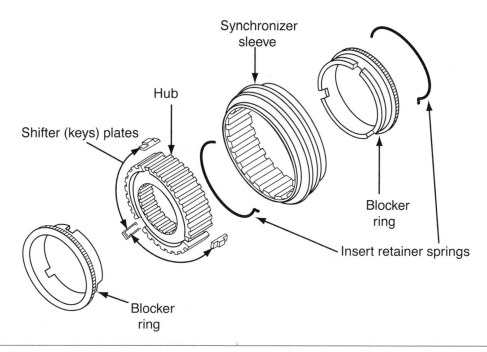

Figure 1-8 A synchronizer assembly.

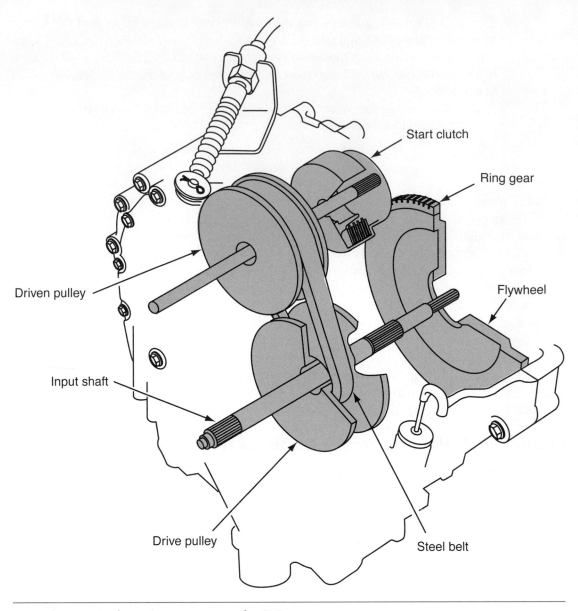

Start clutch

Ring gear

Driven pulley

Flywheel

Input shaft

Drive pulley

Steel belt

Figure 1-9 The main components of a CVT.

transaxles do not have a torque converter; rather, they use a manual transmission-type flywheel with a start clutch. Instead of relying on gear sets to provide drive ratios, a CVT uses belts and pulleys.

One pulley is the driven member and the other is the driver. Each pulley has a moveable face and a fixed face. When the moveable face moves, the effective diameter of the pulley changes. The change in effective diameter changes the effective pulley (gear) ratio. A steel belt links the driven and drive pulleys.

To achieve a low pulley ratio, high hydraulic pressure works on the moveable face of the driven pulley to make it larger. In response to this high pressure, the pressure on the drive pulley is reduced. Because the belt links the two pulleys and proper belt tension is critical, the drive pulley reduces just enough to keep the proper tension on the belt. The increase of pressure at the driven pulley is proportional to the decrease of pressure at the drive pulley. The opposite is true for high pulley ratios. Low pressure causes the driven pulley to decrease in size, whereas high pressure increases the size of the drive pulley.

Different speed ratios are available any time the vehicle is moving. Because the size of the drive and driven pulleys can vary greatly, vehicle loads and speeds can be changed without changing the engine's speed. With this type of transmission, attempts are made to keep the engine operating at its most efficient speed. This increases fuel economy and decreases emissions.

Clutch

A **flywheel**, which is a large and heavy disc, is attached to the rear end of the crankshaft (Figure 1-10). The rear face of the flywheel is machined flat and smooth to serve as the driving member of the clutch. When the clutch is engaged, the rotary motion of the engine's crankshaft is transferred from the flywheel, through the clutch, to the transmission. The rotary motion is then delivered by the transmission to the differential, then transferred by axle shafts to the tires, which push against the ground to move the car.

The clutch assembly is located immediately to the rear of the engine and is directly connected to the engine's crankshaft at the flywheel. Its purpose is to engage and disengage the engine's power smoothly to the rest of the drive train. When the driver depresses the clutch pedal, the clutch assembly is disengaged and the engine is disconnected from the rest of the drive train. This allows the engine to run when the car is standing still. The clutch is said to be engaged when the clutch pedal is up from the floor and the engine is connected to the drive train.

If two discs are attached to a shaft and are not touching each other, we can spin one disc as fast as we want to without affecting the other disc. As the discs move together, however, the spinning disc will cause the other to begin to turn. When the discs are held tightly together, both discs will turn as if they were one. Automotive clutches use this two-disc system. The flywheel serves as one of the discs and the pressure plate of the clutch assembly serves as the other. A third disc, the clutch disc, is squeezed between these two discs and, through friction, connects them (Figure 1-11).

The clutch disc is made of a high frictional material that provides for a solid connection between the two discs. The discs are forced together by strong springs, which clamp the clutch disc between the pressure plate and flywheel. The spring pressure is released by pushing down on the clutch pedal, which allows the discs to separate (Figure 1-12).

The flywheel also serves to dampen crankshaft vibrations, add inertia to the rotation of the crankshaft, and as a large gear for the starter motor.

Figure 1-10 A typical flywheel mounted to the rear of an engine's crankshaft.

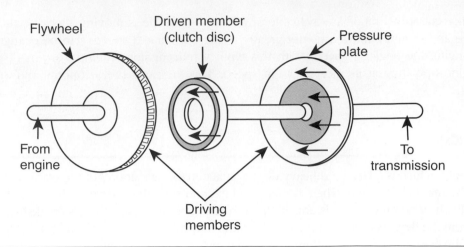

Figure 1-11 When the clutch is engaged, the driven member (clutch disc) is squeezed between the two driving members (pressure plate and flywheel). The transmission is connected to the driven member.

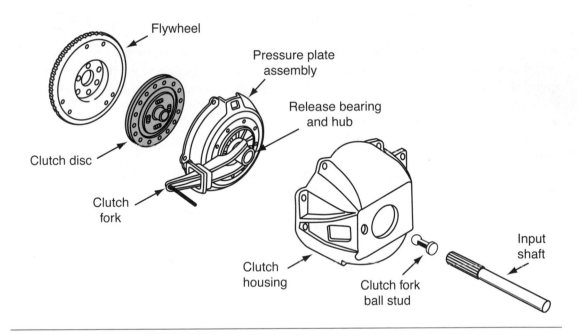

Figure 1-12 The major components of a clutch assembly.

WARNING: Some early clutch discs were made with an asbestos lining. Asbestos has been found to cause lung problems and cancer. Use approved respiratory gear and proper cleaning methods when handling and working around clutch discs.

Drive Line

The car's drive shaft and its joints are often called the **drive line**. The drive line transmits torque from the transmission to the driving wheels. RWD cars use a long drive shaft that connects the transmission to the rear axle. The engine and complete drive line of FWD cars are located between the front driving wheels. The transmission section of a transaxle is practically identical to RWD transmissions. However, a transaxle also contains the differential gear sets and the connections for the drive axles (Figure 1-13).

Drive Line for RWD Cars

A **drive shaft** is a steel or aluminum tube normally connected to at least two universal joints and a slip joint (Figure 1-14). The drive shaft transfers power from the transmission output shaft to the rear drive axle. A differential in the axle housing transmits the power to the rear wheels, which then move the car forward or backward.

Drive shafts differ in construction, lining, length, diameter, and type of slip joint. Typically, the drive shaft is connected at one end to the transmission and at the other end to the rear axle, which moves up and down with wheel and spring movement.

Drive shafts are typically made of hollow, thin-walled steel or aluminum tubing with the universal joint yokes welded at either end. **Universal joints** allow the drive shaft to change angles in response to the movements of the rear axle assembly (Figure 1-15). As the angle of the drive shaft changes, its length also must change. The slip yoke normally fitted to the front universal joint allows the shaft to remain in place as its length requirements change.

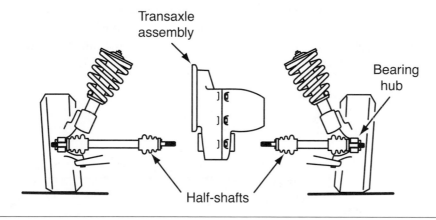

Figure 1-13 The driveline for a FWD vehicle.

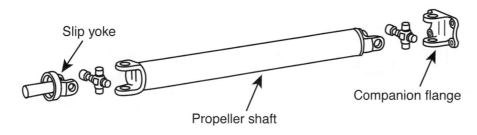

Figure 1-14 A typical drive shaft.

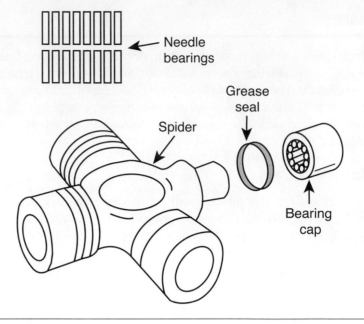

Figure 1-15 An exploded view of a U-joint.

Drive Line for 4WD and AWD Vehicles

The basic drive line for 4WD and AWD depends on whether the system uses a transfer case. If the drive train is based on a RWD, the output shaft from the transmission is connected to the transfer case. The transfer case is connected, by drive shafts, to the rear and front drive axle assemblies (Figure 1-16). If the vehicle is based on a FWD vehicle, a differential (or similar component) is attached to the output of the transaxle and a drive shaft connects that unit to the rear drive axles. Of course, there are many variations to this basic setup, all depending on the make and model of the vehicle.

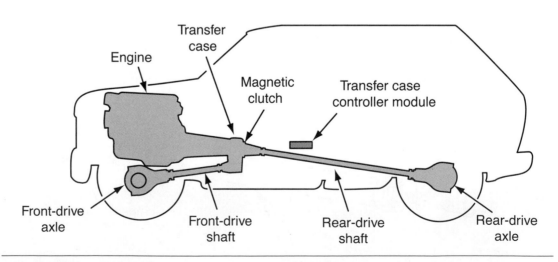

Figure 1-16 A typical 4WD system based on a RWD drive line.

Differentials

The rear axle housing encloses the complete RWD axle assembly. In addition to housing the parts, the axle housing also serves as a place to mount the vehicle's rear suspension and braking system. The rear axle assembly serves two other major functions: it changes the direction of the power flow 90 degrees and acts as the final gear reduction unit.

The rear axle consists of two sets of gears: the ring and pinion gear set and the **differential** gears. When torque leaves the transmission, it flows through the drive shaft to the ring and pinion gears, where it is further multiplied (Figure 1-17). By considering the engine's torque curve, the car's weight, and tire size, manufacturers are able to determine the best rear axle gear ratios for proper acceleration, hill-climbing ability, fuel economy, and noise level limits.

The primary purpose of the differential gear set is to allow a difference in driving wheel speed when the vehicle is rounding a corner or curve. The differential also transfers torque equally to both driving wheels when the vehicle is traveling in a straight line (Figure 1-18).

The torque on the ring gear is transmitted to the differential, where it is sent to the drive wheel that requires less turning torque. When the car is traveling in a straight line, both driving wheels travel the same distance at the same speed. However, when the car is making a turn, the outer wheel must travel farther and faster than the inner wheel. When the car is steered into a 90-degree turn to the left and the inner wheel turns on a 20-foot radius, the inner wheel travels about 31 feet. The outer wheel, which is nearly 5 feet from the inner wheel, turns on a 24⅔ foot radius and travels nearly 39 feet.

Without some means for allowing the drive wheels to rotate at different speeds, the wheels would skid when the car was turning. This would result in little control during turns as well as excessive tire wear. The differential eliminates these troubles by allowing the outer wheel to rotate faster as turns are made.

Differential Design

On FWD cars, the differential unit is normally part of the transaxle assembly (Figure 1-19). On RWD cars, it is part of the rear axle assembly. A differential unit is located in a cast iron casting, the differential case, and is attached to the center of the ring gear. Located inside the case are the differential pinion shafts and gears and the axle side gears.

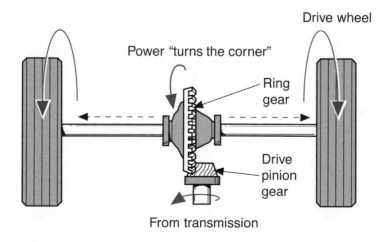

Figure 1-17 The gears of a final drive unit not only multiply torque but also transmit power "around the corner" to the drive wheels.

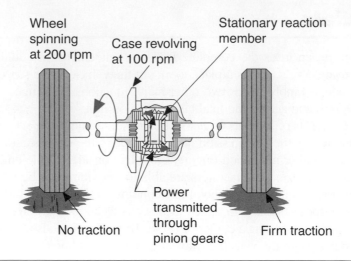

Wheel spinning at 200 rpm

Case revolving at 100 rpm

Stationary reaction member

Power transmitted through pinion gears

No traction

Firm traction

Figure 1-18 The differential allows for one drive wheel to turn at a different speed than the other when the car is turning and allows for more power to the wheel with less traction.

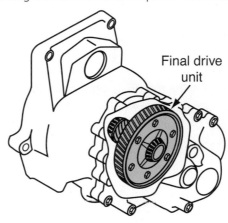

Final drive unit

Figure 1-19 The final drive assembly is part of the transaxle assembly.

Pinion gears typically are mounted or attached to a shaft and supply the input to a gear set.

The differential assembly revolves with the ring gear (Figure 1-20). Axle side gears are splined to the rear axle or front axle drive shafts.

When an automobile is moving straight ahead, both wheels are free to rotate. Engine power is applied to the drive pinion gear, which rotates the ring gear. Beveled pinion gears are carried around by the ring gear and rotate as one unit. Each axle turns at the same speed.

When the car turns a sharp corner, only one wheel rotates freely. Torque still comes in on the pinion gear and rotates the ring gear, carrying the beveled pinions around with it. However, one axle doesn't easily rotate and the beveled pinions are forced to rotate on their own axis and "walk around" their gear. The other side is forced to rotate because it is subjected to the turning force of the ring gear, which is transmitted through the pinions.

During one revolution of the ring gear, one pinion gear makes two revolutions and the other "walks around" the gear. As a result, when the drive wheels have unequal resistance applied to them, the wheel with the least resistance turns more revolutions. As one wheel turns faster, the other turns proportionally slower.

When the vehicle turns a corner or a curve, the differential pinion gears rotate around the differential pinion shaft. The differential pinion gears allow the inside axle shaft and driving wheel to slow down. On the opposite side, the pinion gears allow the outside wheels to accelerate. Both driving wheels resume equal speeds when the vehicle completes the corner or curve. This differential action improves vehicle handling and reduces driving wheel tire wear.

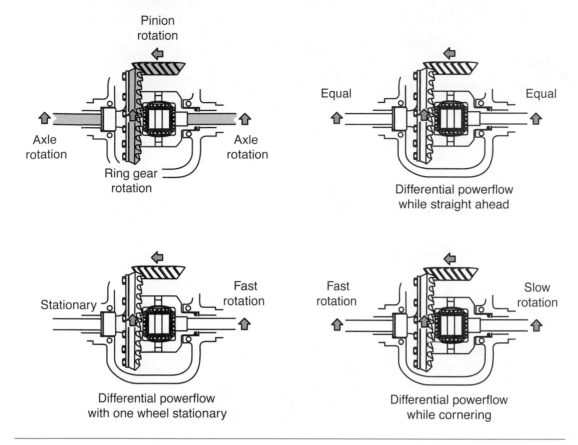

Figure 1-20 Operation of a differential assembly.

To prevent a loss of power on slippery surfaces, some differentials are fitted with clutches that allow the wheel with traction to continue to receive torque. These differentials are referred to as *limited-slip, traction-lock,* or *positraction* differentials. Rather than clutch discs, some units use cone clutches or gears to restrict the normal differential action and deliver torque to the nonslipping wheel.

Driving Axles

On RWD vehicles, the axle shafts or drive axles are located within the hollow horizontal tubes of the axle housing. The purpose of an axle shaft is to transmit the torque from the differential's side gears to the driving wheels. Axle shafts are heavy steel bars splined at the inner end to mesh with the axle side gear in the differential. The driving wheel is bolted to the wheel flange at the outer end of the axle shaft (Figure 1-21). The car's wheels rotate with the axles, which allow the car to move.

The drive axles of a FWD car extend from the sides of the transaxle to the drive wheels. CV joints are fitted to the axles to allow the axles to move with the car's suspension and steering systems.

The differential side gears are connected to inboard CV joints by splines. The drive axles extend out from each side of the differential to rotate the vehicle's wheels. The axles and joints are designed to allow the wheels to steer and to move up and down with the suspension. A short shaft extends from the differential to the inner CV joint. Connecting the inner CV joint and the outer CV joint is the axle shaft. Extending from the outer CV joint is a short shaft that fits into the hub of the wheels.

CV joints allow the angle of the axle shafts to change with no loss in rotational speed.

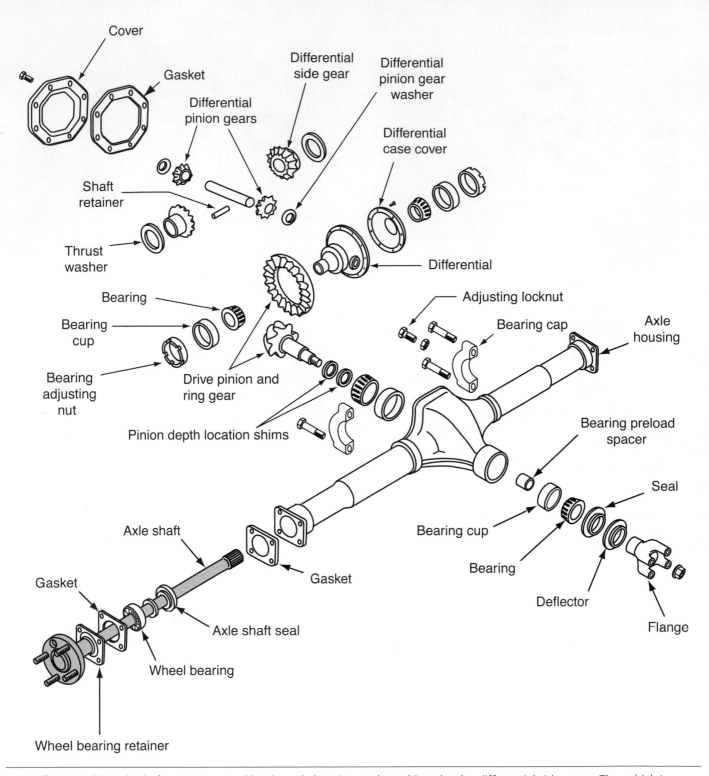

Cover

Gasket

Differential side gear

Differential pinion gear washer

Differential pinion gears

Differential case cover

Shaft retainer

Differential

Thrust washer

Bearing

Adjusting locknut

Bearing cap

Axle housing

Bearing cup

Bearing adjusting nut

Drive pinion and ring gear

Pinion depth location shims

Bearing preload spacer

Seal

Bearing cup

Bearing

Axle shaft

Deflector

Gasket

Gasket

Flange

Axle shaft seal

Wheel bearing

Wheel bearing retainer

Figure 1-21 Axle shafts are supported by the axle housing and are driven by the differential side gears. The vehicle's wheels are bolted to the axle shaft's outer flange.

Four-Wheel Drive Systems (4WD)

4WD vehicles designed for off-the-road use are normally RWD vehicles equipped with a **transfer case**, a front drive shaft, and a front differential and drive axles (Figure 1-22). Many 4WD vehicles use three drive shafts. One short drive shaft connects the output of the transmission to the transfer case. The output from the transfer case is then sent to the front and rear axles through separate drive shafts.

Some high-performance cars are equipped with all-wheel drive (AWD) to improve the handling characteristics of the car. Many of these cars are front-wheel drive models converted to four-wheel drive. Normally, FWD cars are modified by adding a transfer case, a rear drive shaft, and a rear axle with a differential. Although this is the typical modification, some cars are equipped with a center differential in place of the transfer case. This differential unit allows the rear and front wheels to turn at different speeds and with different amounts of torque.

The transfer case is usually mounted to the side or rear of the transmission. When a drive shaft is not used to connect the transmission to the transfer case, a chain, or gear drive, within the transfer case receives the engine's power from the transmission and transfers it to the drive shafts leading to the front and rear drive axles.

The transfer case itself is constructed similar to a transmission. It uses shift forks to select the operating mode, and splines, gears, shims, bearings, and other components found in transmissions. The housing is filled with lubricant that cuts friction on all moving parts. Seals hold the lubricant in the case and prevent leakage around shafts and yokes. Shims set up the proper clearance between the internal components and the case.

An electric switch or shift lever located in the passenger compartment controls the transfer case so that power is directed to the axles selected by the driver. Power typically can be directed to all four wheels, two wheels, or none of the wheels. On many vehicles, the driver can also select a low-speed range for extra torque when traveling in very adverse conditions.

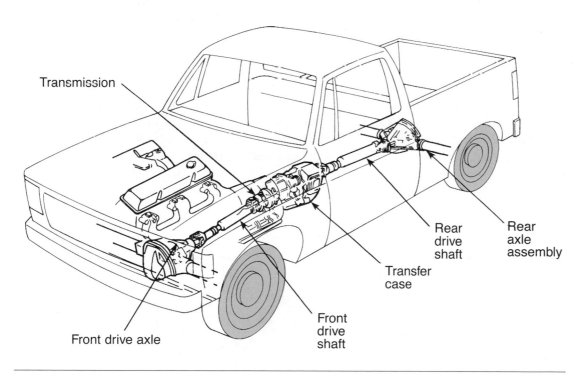

Figure 1-22 The arrangement of the drive train components in a typical 4WD vehicle.

The rear drive axle of a 4WD vehicle is identical to those used in 2WD vehicles. The front drive axle is also like a conventional rear axle, except that it is modified to allow the front wheels to steer. Further modifications are also necessary to adapt the axle to the vehicle's suspension system. The differential units housed in the axle assemblies are similar to those found in a RWD vehicle.

Types of Gears

Gears are normally used to transmit torque from one shaft to another. These shafts may operate in line, parallel to each other, or at an angle to each other. These different applications require a variety of gear designs, which vary primarily in the size and shape of the teeth.

Idler gears are gears that do not drive something. They simply are used to reduce or increase rotational speeds and to reverse the direction of rotation (Figure 1-23). If three gears were connected, the center gear would be considered an idler gear.

In order for gears to mesh, they must have teeth of the same size and design. Meshed gears have at least one pair of teeth engaged at all times. Some gear designs allow for contact between more than one pair of teeth. Gears normally are classified by the type of teeth they have and by the surface on which the teeth are cut.

Automobiles use a variety of gear types to meet the demands of speed and torque. The most basic type of gear is the **spur gear**, which has its teeth parallel to and in alignment with the center of the gear (Figure 1-24). Early transmissions used straight-cut spur gears, which were easier to machine but were noisy and difficult to shift. Today these gears are used mainly for slow speeds to avoid excessive noise and vibration. They commonly are used as reverse gears and in simple devices such as hand or powered winches.

Helical gears are like spur gears except that their teeth have been twisted at an angle from the gear center line (Figure 1-25). These gears are cut in a helix, which is a form of curve. This curve is more difficult to machine but is used because it reduces gear noise. Engagement of these gears begins at the tooth tip of one gear and rolls down the trailing edge of the teeth. This angular contact tends to cause side thrusts that are absorbed by a thrust bearing or bearings. Helical gears are widely used in transmissions today because they are quieter at high speeds and are durable.

Bevel gears are shaped like a cone with its top cut off. The teeth point inward toward the peak of the cone. These gears permit the power flow to "turn a corner." Spiral bevel gears were developed for use when higher speed and strength as well as a change in the angle of the power flow were required. Their teeth are cut obliquely on the angular faces of the gears. The most commonly used spiral beveled gear set is the ring and pinion gears used in heavy truck differentials. Bevel type gears are also used for slow-speed applications, which are not subject to high impact forces. Handwheel controls that must operate some remote device at an angle use straight bevel gears.

Figure 1-23 An idler gear between the input and output gears. The teeth of the idler gear are not used to calculate the gear ratio.

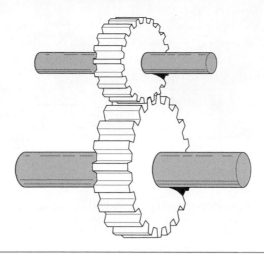

Figure 1-24 Spur gears have teeth cut straight across the gear's edge and parallel to the shaft.

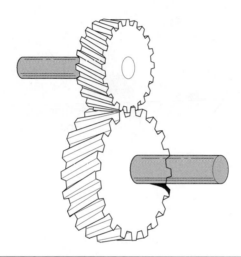

Figure 1-25 Helical gears have teeth cut at an angle to the gear's axis of rotation.

The **hypoid gear** resembles the spiral bevel gear, but the pinion drive gear is located below the center of the ring gear. Its teeth and general construction are the same as the spiral bevel gear. The most common use for hypoid gears is in modern differentials (Figure 1-26). Here, they allow for lower body styles by lowering the transmission drive shaft.

The **worm gear** is actually a screw capable of high speed reductions in a compact space. Its mating gear has teeth that are curved at the tips to permit a greater contact area. Power is supplied to the worm gear, which drives the mating gear. Worm gears usually provide right-angle power flows. The most common use for the worm gear is in applications in which the power source operates at high speed and the output is at slow speed with high torque. Many steering mechanisms use a worm gear connected to the steering shaft and wheel and a partial (sector) gear connected to the steering linkage. Small power hand tools frequently use a high-speed motor with a worm gear drive.

Rack and pinion gears convert straight-line motion into rotary motion, and vice versa. Rack and pinion gears also change the angle of power flow with some degree of speed change. The teeth on the rack are cut straight across the shaft, whereas those on the pinion are cut like a spur gear. These gear sets can provide control of arbor presses and other devices when slow speed is involved. Rack and pinion gears also are used commonly in automotive steering boxes.

Some tractors and off-road vehicles use worm gears as final drive gears.

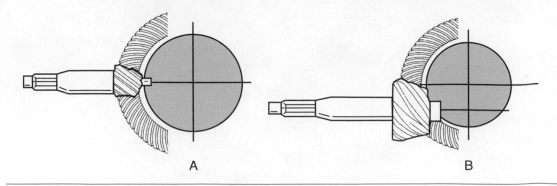

A   B

Figure 1-26 (A) Spiral bevel differential gears and (B) hypoid gears.

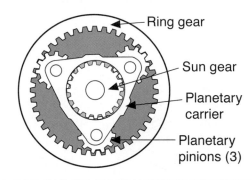

Ring gear

Sun gear

Planetary carrier

Planetary pinions (3)

Figure 1-27 A simple planetary gear set.

Internal gears have their teeth pointing inward and are commonly used in the planetary gear set used in automatic transmissions and transfer cases. These are gear sets in which an outer ring gear has internal teeth that mate with teeth on smaller planetary gears. These gears, in turn, mesh with a center or sun gear (Figure 1-27). Many changes in speed and torque are possible, depending on which parts are held stationary and which are driven. In a planetary gear set, one gear is normally the input, another is prevented from moving or held, and the third gear is the output gear. Planetary gears are widely used because each set is capable of more than one speed change. The gear load is spread over several gears, reducing stress and wear on any one gear.

Types of Bearings

Gears are either securely attached to a shaft or designed to move freely on the shaft. The ease with which the gears rotate on the shaft or the shaft rotates with the gears partially determines the amount of power needed to rotate them. If they rotate with great difficulty because of high friction, much power is lost. High friction also will cause excessive wear to the gears and shaft. To reduce the friction, bearings are fitted to the shaft or gears.

The simplest type of bearing is a cylindrical hole formed in a piece of material, into which the shaft fits freely. The hole is usually lined with a brass or bronze lining, or **bushing**, which not only reduces the friction but also allows for easy replacement when wear occurs. Bushings usually have a tight fit in the hole in which they fit.

Bushings often are referred to as plain bearings.

Ball- or roller-type bearings are used wherever friction must be minimized. With these types of bearings, rolling friction replaces the sliding friction that occurs in plain bearings. Typically, two bearings are used to support a shaft instead of a single long bushing. Bearings have four purposes: they support a load, maintain alignment of a shaft, reduce rotating friction, and control **end play**.

Most bearings are capable of withstanding only those loads that are perpendicular to the axis of the shaft. Such loads are called **radial loads** and bearings that carry them are called radial or **journal** bearings.

To prevent the shaft from moving in the axial direction, shoulders or collars may be formed on it or secured to it. If both collars are made integral with the shaft, the bearings must be split or made into halves, and the top half or cap bolted in place after the shaft has been put in place. The collars or shoulders withstand any end thrusts, and bearings designed this way are called **thrust bearings** (or Torrington bearings). Most thrust bearings are similar to roller and ball bearings except the plates that the rollers or balls ride between are designed as flat washers. These rollers or balls reduce side-to-side load.

A single row journal or radial **ball bearing** has an inner race made of a ring of case-hardened steel with a groove or track formed on its outer circumference on which a number of hardened steel balls can run (Figure 1-28). The outer race is another ring, which has a track on its inner circumference. The balls fit between the two tracks and roll around in the tracks as either race turns. The balls are kept from rubbing against each other by some form of cage. These bearings can withstand radial loads and also can withstand a considerable amount of axial thrust. Therefore, they often are used as combined journal and thrust bearings.

A bearing designed to take only radial loads has only one of its races machined with a track for the balls. Other bearings are designed to take thrust loads in only one direction. If this type of bearing is installed incorrectly, the slightest amount of thrust will cause the bearing to come apart.

End play is the end-to-end movement of a shaft or component within its housing.

A **journal** is the area on a shaft that rides on the bearing.

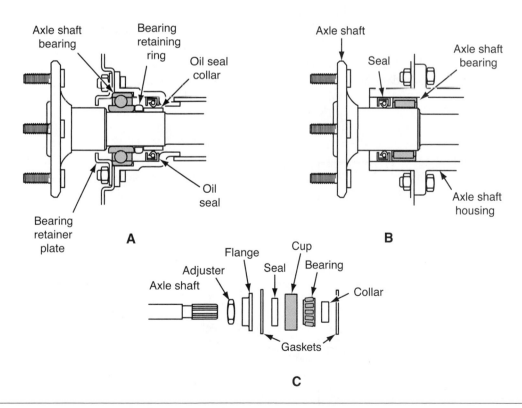

Figure 1-28 (A) Ball-type bearings supporting an axle shaft. (B) Straight roller bearing installed in an axle housing. (C) Tapered roller drive axle bearing.

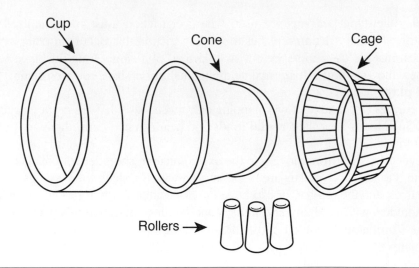

Figure 1-29 Major components of a tapered roller bearing assembly.

Another type of ball bearing uses two rows of balls. These are designed to withstand considerable amounts of radial and axial loads. Constructed like two single-row ball bearings joined together, these bearings often are used in rear axle assemblies.

Roller bearings are used wherever it is desirable to have a large bearing surface and low amounts of friction. Large bearing surfaces are needed in areas of extremely heavy loads. The rollers are usually fitted between a journal of a shaft and an outer race. As the shaft rotates, the rollers turn and rotate in the race.

Roller bearings are constructed almost the same way as ball bearings, but they have cylindrical-shaped bearings instead of spherical-shaped bearings. This gives them a greater load-bearing area. They are used in wheel bearing applications.

Tapered roller bearings are the most commonly used type of bearing in drive trains. These bearings use long tapered rollers, which are fitted into a tapered cone. This assembly rides in a tapered cup (Figure 1-29). The basic principle behind the design of a tapered roller bearing is that the apexes of the tapered surfaces converge to a common axis (Figure 1-30). The angular shape of the bearing increases the frictional surface area and limits shaft end play.

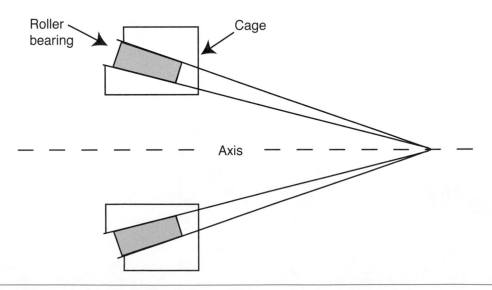

Figure 1-30 The assembly's tapered surfaces converge on a common axis.

Summary

- ❏ The drive train has four primary purposes: to connect the engine's power to the drive wheels, to select different speed ratios, to provide a way to move the vehicle in reverse, and to control the power to the drive wheels for safe turning of the vehicle.

- ❏ The main components of the drive train are the clutch, transmission, and drive axles.

- ❏ The rotating or turning effort of the engine's crankshaft is called engine torque.

- ❏ Gears are used to apply torque to other rotating parts of the drive train and to multiply torque.

- ❏ Transmissions offer various gear ratios through the meshing of various sized gears.

- ❏ Reverse gear is accomplished by adding a third gear to a two-gear set. This gear, the reverse idler gear, causes the driven gear to rotate in reverse.

- ❏ The operation of a CVT is based on a steel belt linking two variable pulleys.

- ❏ Connected to the rear of the crankshaft is the flywheel, which serves many functions, including acting as the driving member of the clutch assembly.

- ❏ The clutch assembly is comprised of another driving disc, the pressure plate, and a driven disc, the clutch disc.

- ❏ The clutch disc is mounted to the input shaft of the transmission and carries the engine's torque to the transmission when the clutch assembly is engaged.

- ❏ In FWD cars, the transmission and drive axle is located in a single assembly called a transaxle. In RWD cars, the drive axle is connected to the transmission through a drive shaft.

- ❏ The drive shaft and its joints are called the drive line of the car.

- ❏ Universal joints allow the drive shaft to change angles in response to movements of the car's suspension and rear axle assembly.

- ❏ The rear axle housing encloses the entire rear-wheel driving axle assembly.

- ❏ The primary purpose of the differential is to allow a difference in driving wheel speed when the vehicle is rounding a corner or curve. The ring and pinion in the drive axle also multiples the torque it receives from the transmission.

- ❏ On FWD cars, the differential is part of the transaxle assembly.

- ❏ The drive axles on FWD cars extend from the sides of the transaxle to the drive wheels. CV joints are fitted to the axles to allow the axles to move with the car's suspension.

- ❏ 4WD vehicles typically use a transfer case, which relays engine torque to both a front and rear driving axle.

- ❏ Bearings are used to reduce the friction caused by something rotating within something else.

Terms to Know

All-wheel drive (AWD)
Ball bearing
Bevel gear
Bushing
Clutch
Constant velocity (CV) joints
Continuously variable transmission (CVT)
Differential
Drive line
Drive shaft
Drive train
End play
Engine torque
Flywheel
Four-wheel drive (4WD)
Front-wheel drive (FWD)
Helical gear
Horsepower
Hypoid gear
Idler gears
Journal
Radial loads
Rear-wheel drive (RWD)
Roller bearing
Spur gear
Thrust bearing
Torque
Transaxle
Transfer case
Universal joints
Worm gear

Review Questions

Short Answer Essays

1. What are the primary purposes of a vehicle's drive train?
2. What is the basic advantage of a limited-slip differential?
3. What is the purpose of an idler gear?
4. Why are transmissions equipped with many different forward gear ratios?
5. What is the primary difference between a transaxle and a transmission?

6. Why are U- and CV joints used in the driveline?

7. What does a differential unit do to the torque it receives?

8. What is the purpose of the clutch assembly? How does it work?

9. What kind of gears are commonly used in today's automotive drive trains?

10. When are ball- or roller-type bearings used?

Fill-in-the-Blanks

1. The main components of the drive train are the _____,
   _____ and _____ _____.

2. The rotating or turning effort of the engine's crankshaft is called _____

3. Gears are used to apply torque to other rotating parts of the drive train and to
   _____ torque.

4. _____ _____, especially those that are used off the
   road, can deliver power to all four wheels. The driver can select two-wheel or four-wheel
   drive. Some four-wheel drive vehicles are always engaged in four-wheel drive; these
   vehicles are commonly said to have _____-_____ four-
   wheel drive or _____-_____-_____.

5. The car's drive shaft and its joints are often called the _____
   _____.

6. The simplest type of bearing is a _____.

7. Reverse gear is accomplished by adding a _____
   _____ to a two-gear set.

8. The clutch assembly is comprised of a driving disc, called the _____
   _____, and a driven disc, called the _____
   _____.

9. In FWD cars, the transmission and drive axle is located in a single assembly called a
   _____.

10. In RWD cars, the drive axle is connected to the transmission through a
    _____.

Multiple Choice

1. When discussing the purposes of a drive train, *Technician A* says that it connects the engine's power to the drive wheels. *Technician B* says that it controls the power to the drive wheels for safe turning of the vehicle. Who is correct?
   - **A.** A only
   - **B.** B only
   - **C.** Both A and B
   - **D.** Neither A nor B

2. *Technician A* says that gears are used to apply torque to other rotating parts of the drive train. *Technician B* says that gears are used to multiply torque. Who is correct?
   - **A.** A only
   - **B.** B only
   - **C.** Both A and B
   - **D.** Neither A nor B

3. The type of bearing commonly used to control axial loads is a:
   - **A.** roller bearing
   - **B.** tapered roller bearing
   - **C.** thrust bearing
   - **D.** tapered ball bearing

4. When discussing reverse gear, *Technician A* says reverse is accomplished by adding a third gear. *Technician B* says the reverse idler gear causes the driven gear to rotate in reverse. Who is correct?
   - **A.** A only
   - **B.** B only
   - **C.** Both A and B
   - **D.** Neither A nor B

5. When discussing the purpose of a flywheel, *Technician A* says that it increases engine torque. *Technician B* says that it acts as the driving member of the clutch assembly. Who is correct?
   - **A.** A only
   - **B.** B only
   - **C.** Both A and B
   - **D.** Neither A nor B

6. *Technician A* says that the clutch disc is mounted to the input shaft of the transmission. *Technician B* says that the clutch disc carries the engine's torque to the transmission when the clutch assembly is engaged. Who is correct?
   - **A.** A only
   - **B.** B only
   - **C.** Both A and B
   - **D.** Neither A nor B

7. When discussing universal joints, *Technician A* says that they eliminate vibrations caused by the power pulses of the engine. *Technician B* says that they allow the drive shaft to change angles in response to movements of the car's suspension and rear axle assembly. Who is correct?
   - **A.** A only
   - **B.** B only
   - **C.** Both A and B
   - **D.** Neither A nor B

8. When discussing the purpose of a differential, *Technician A* says that it allows for equal wheel speed when the vehicle is rounding a corner or curve. *Technician B* says that the ring and pinion in the drive axle multiples the torque it receives from the transmission. Who is correct?
   - **A.** A only
   - **B.** B only
   - **C.** Both A and B
   - **D.** Neither A nor B

9. When discussing FWD vehicles, *Technician A* says that the differential is normally part of the transaxle assembly. *Technician B* says that the drive axles extend from the sides of the transaxle to the drive wheels. Who is correct?
   - **A.** A only
   - **B.** B only
   - **C.** Both A and B
   - **D.** Neither A nor B

10. *Technician A* says that 4WD vehicles typically use a transfer case to transfer engine torque to both a front and a rear driving axle. *Technician B* says that 4WD vehicles normally have two clutches, two drive shafts, and two differentials. Who is correct?
    - **A.** A only
    - **B.** B only
    - **C.** Both A and B
    - **D.** Neither A nor B

Drivetrain Theory

Upon completion and review of this chapter, you should be able to:

❏ Describe how all matter exists.

❏ Explain what energy is and how energy is converted.

❏ Explain the forces that influence the design and operation of an automobile.

❏ Describe and apply Newton's laws of motion to an automobile.

❏ Define friction and describe how it can be minimized.

❏ Describe the various types of simple machines.

❏ Explain how a set of gears can increase torque.

❏ Define and determine the ratio between two meshed gears.

❏ Explain the difference between torque and horsepower.

❏ Explain Pascal's Law and give examples of where it is applied to an automobile.

❏ Explain how heat affects matter.

❏ Describe the origin and practical applications of electromagnetism.

Introduction

This chapter contains many of the things you have learned or will learn in other courses. It is not my intent to present this material in lieu of those other courses, but instead to emphasize those things that you will need to gain employment and be successful in an automotive career. Many of the facts presented in this chapter will be addressed again, in greater detail according to the topic covered. This chapter contains the theories that are the basis for the rest of the contents in this book. I highly recommend that you make sure you understand the contents of this chapter. Move to the end of the chapter review questions and if you have difficulty answering the questions, study the appropriate content in the chapter until you have a clear understanding and are able to answer the questions correctly.

Matter

Matter is anything that occupies space. All matter exists as a gas, liquid, or solid. Gases and liquids are considered fluids because they move or flow easily and easily respond to pressure. A gas has neither a shape nor volume of its own and tends to expand without limits. A liquid takes a shape and has volume. A solid is matter that does not flow.

Atoms and Molecules

All matter consists of countless tiny particles, called **atoms**. An atom may be defined as the smallest particle of an element in which all the chemical characteristics of the element are present. Atoms are so small they cannot be seen with an electron microscope, which magnifies millions of times. A substance with only one type of atom is referred to as an **element**. Over 100 elements are known to exist at present and, of the known elements, 92 occur naturally. The remaining elements have been manufactured in laboratories.

Small, positively charged particles called protons are located in the center, or nucleus, of each atom. In most atoms, neutrons are also located in the nucleus. Neutrons have no electrical charge, but they add weight to the atom. The positively charged protons tend to repel each other, and this repelling force could destroy the nucleus. The presence of the neutrons with the protons in the nucleus cancels the repelling action of the protons and keeps the nucleus together.

Electrons are small, very light particles with a negative electrical charge. Electrons move in orbits around the nucleus of an atom.

A proton is approximately 1,840 times heavier than an electron, and this makes the electron much easier to move. Because the electrons are orbiting around the nucleus, centrifugal force tends to move the electrons away from the nucleus. However, the attraction between the positively charged protons and the negatively charged electrons holds the electrons in their orbits. The atoms of the different elements have different numbers of protons, electrons, and neutrons. Some of the lighter elements have the same number of protons and neutrons in the nucleus, but many of the heavier elements have more neutrons than protons.

The simplest atom is the hydrogen (H) atom, which has one proton in the nucleus and one electron orbiting around the nucleus (Figure 2-1). The nucleus of a copper (CU) atom contains 29 protons and 34 neutrons, whereas 29 electrons orbit in 4 different rings around the nucleus. Because 2, 8, and 18 electrons are the maximum number of electrons on the first three electron rings next to the nucleus, the fourth ring must have one electron (Figure 2-2). The outer ring of an atom is called the valence ring, and the number of electrons on this ring determines the electrical characteristics of the element. Elements are listed on the atomic scale, or periodic chart, according to their number of protons and electrons. For example, hydrogen is number 1 on this scale, and copper is number 29.

Water is a compound that contains oxygen and hydrogen atoms. The chemical symbol for water is H_2O. This chemical symbol indicates that each molecule of water contains two atoms of hydrogen and one oxygen atom (Figure 2-3).

States of Matter

The particles of a solid are held together in a rigid structure. When a solid dissolves into a liquid, its particles break away from this structure and mix evenly in the liquid, forming a **solution**. When

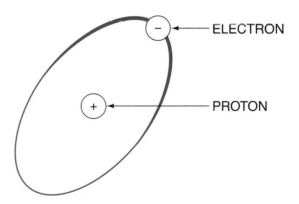

Figure 2-1 A hydrogen atom.

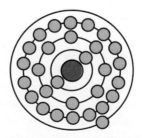

Figure 2-2 A copper atom.

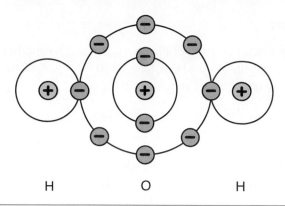

Figure 2-3 A molecule of water.

they are heated, most liquids **evaporate**. This means that the atoms or molecules of which they are made break free from the body of the liquid to become gas particles. When all of the liquid in a solution has evaporated, the solid is left behind. The particles of the solid normally arrange in a structure called a crystal.

Absorption and Adsorption. Not all solids will dissolve in a liquid; rather, the liquid will be either absorbed or adsorbed. The action of a sponge serves as the best example of absorption. When a dry sponge is put into water, the water is absorbed by the sponge. The sponge does not dissolve, the water merely penetrates into the sponge and the sponge becomes filled with water. There is no change to the atomic structure of the sponge, nor does the structure of the water change. If we take a glass and put it into water, the glass does not absorb the water. The glass, however, still gets wet as a thin layer of water adheres to the glass. This is adsorption. Materials that *absorb* fluids are **permeable** substances. **Impermeable** substances *adsorb* fluids. Some materials are impermeable to most fluids while others are impermeable to just a few.

Energy

Energy may be defined as the ability to do work. Because all matter consists of atoms and molecules that are in constant motion, all matter has energy. Energy is not matter, but it affects the behavior of matter. Everything that happens requires energy, and energy comes in many forms.

Each form of energy can change into other forms. However, the total amount of energy never changes; it can only be transferred from one form to another, not created or destroyed. This is known as the "Principle of the Conservation of Energy."

Kinetic and Potential Energy

When energy is released to do work, it is called **kinetic energy**. This type of energy also may be referred to as energy in motion. Stored energy may be called **potential energy**.

There are many components and systems that have potential energy and, at times, kinetic energy. The ignition system is a source for high electrical energy. The heart of the ignition system is the ignition coil, which has much potential energy when it has current flow through it. When it is time to fire a spark plug, current flow is stopped and energy is released and becomes kinetic energy as it creates a spark across the gap of a spark plug.

Energy Conversion

Energy conversion occurs when one form of energy is changed to another form. Because energy is not always in the desired form, it must be converted to a form we can use. Some of the most common automotive energy conversions are discussed here.

Chemical to Thermal Energy. Chemical energy in gasoline or diesel fuel is converted to thermal energy when the fuel burns in the engine cylinders.

Chemical to Electrical Energy. The chemical energy in a battery (Figure 2-4) is converted to electrical energy to power many of the accessories on an automobile.

Electrical to Mechanical Energy. In an automobile, the battery supplies electrical energy to the starting motor, and this motor converts the electrical energy to mechanical energy to crank the engine.

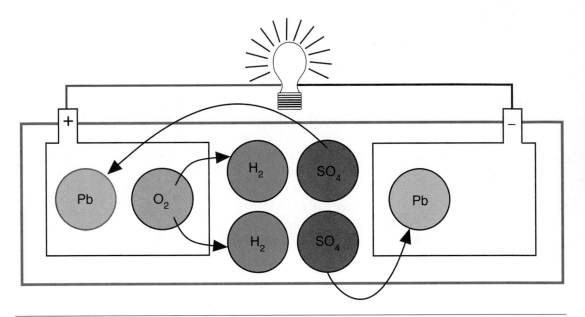

Figure 2-4 Chemical energy is converted to electrical energy in a battery.

Thermal to Mechanical Energy. The thermal energy that results from the burning of the fuel in the engine is converted to mechanical energy, which is used to move the vehicle.

Mechanical to Electrical Energy. The generator is driven by mechanical energy from the engine. The generator converts this energy to electrical energy, which powers the electrical accessories on the vehicle and recharges the battery.

Electrical to Radiant Energy. Radiant energy is light energy. In the automobile, electrical energy is converted to thermal energy, which heats up the inside of light bulbs so that they illuminate and release radiant energy.

Mass and Weight

Mass is the amount of matter in an object. **Weight** is a force and is measured in pounds or kilograms. Gravitational force gives the mass its weight. As an example, a spacecraft can weigh 500 tons (one million pounds) here on earth where it is affected by the earth's gravitational pull. In outer space, beyond the earth's gravity and atmosphere, the spacecraft is nearly weightless (Figure 2-5).

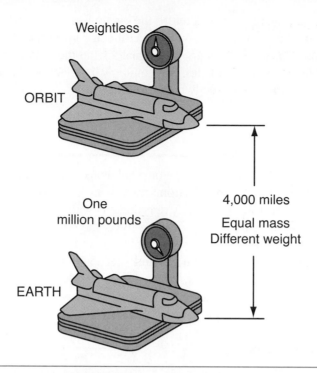

Figure 2-5 The difference in weight of a space shuttle of earth and in space.

To convert kilograms into pounds, simply multiple the weight in kilograms by 2.2046. For example, if something weighs 5 kilograms, the equation would be (5 × 2.2046 = 11.023 pounds). If you want to express the answer in pounds and ounces, you will convert the .023 pounds into ounces. Because there are 16 ounces in a pound, multiple 16 by 0.023 (16 × 0.023 = 0.368 ounces). Therefore, 5 kilograms is equal to 11 pounds and 0.368 ounces.

Size

The size of something is related to its mass. The size of an object defines how much space it occupies. Size dimensions are typically stated in terms of its length, width, and height. Length is a measurement of how long something is from one end to another. Width is a measurement of how wide something is from one side to another. Height is a measurement of the distance from something's bottom to its top. All three of these dimensions are measured in inches, feet, yards, and miles in the English system and meters in the metric system.

To convert a meter into feet, you multiply the number of meters by 3.281. If you wanted to convert the feet into inches, simply multiply your answer in feet by 12. For example, to convert 0.01 mm into inches, begin by converting 0.01 mm into meters. Because 1 mm is equal to 0.001 meters, you need to multiple 0.01 by 0.001 (0.001 × 0.01 = 0.00001). Then multiply 0.00001 meters by 3.281 (0.00001 × 3.281 = 0.00003281 feet). Now convert feet into inches by multiplying by 12 (0.00003281 × 12 = 0.00039372 inches).

An easier way to do this would be using the conversion factor that states that 1 mm is equal to 0.03937 inches. To use this conversion factor, multiply 0.01 mm by 0.03937 (0.01 × 0.03937 = 0.0003937 inches).

Sometimes distance measurements are made with a rule that has fractional rather than decimal increments. Most automotive specifications are given decimally; therefore, fractions need to be converted into decimals. It is also easier to add and subtract dimensions if they are expressed in decimal form rather than in fractions. For example, you want to find the rolling circumference of a tire. You have found the diameter of the tire to be 20⅜ inches. The distance around the tire is the circumference and it is equal to the diameter multiplied by a constant called Pi (π).

Pi is equal to approximately 3.14; therefore the circumference of the tire is equal to 20⅜ inches multiplied by 3.14. This calculation is much easier if you convert the 20⅜ inches into a whole number and a decimal. To convert the ⅜ to a decimal, divide the 3 by 8 (3 ÷ 8 = 0.375). Therefore, the diameter of the tire is 20.375 inches. Now multiply the diameter by π (20.375 × 3.14 = 63.98). The circumference of the tire is nearly 64 inches.

Force

A **force** is a push or pull, and can be large or small. Force can be applied to objects by direct contact or from a distance. Gravity and electromagnetism are examples of forces that are applied from a distance. Forces can be applied from any direction and with any intensity. For example, if a pulling force on an object is twice that of the pushing force, the object will be pulled at one half of the pulling force. When two or more forces are applied to an object, the combined force is called the resultant. The resultant is the sum of the size and direction of the forces. For example, when a mass is suspended by two lengths of wire, each wire should carry half the weight of the mass. If we move the attachment of the wires so they are at an angle to the mass, the wires now carry more force. The wires carry the force of the mass plus a force that pulls against the other wire.

Automotive Forces

When a vehicle is sitting still, gravity exerts a downward force on the vehicle. The ground exerts an equal and opposite upward force and supports the vehicle. When the engine is running and its power output transferred to the vehicle's drive wheels, the wheels exert a force against the ground in a horizontal direction. This force causes the vehicle to move, but it is opposed by the mass of the vehicle (Figure 2-6). To move the vehicle faster, the force supplied by the wheels must increase beyond the opposing forces. As the vehicle does move faster, it pushes against the air as it travels. This becomes a growing opposing force and the force at the drive wheels must overcome the force in order for the vehicle to increase speed. After the vehicle has achieved the desired speed, no additional force is required at the drive wheels.

Force—Balanced and Unbalanced. When the applied forces are balanced and there is no overall resultant force, the object is said to be in **equilibrium**. An object sitting on a solid flat surface is in equilibrium, because its weight is supported by the surface and there is no resultant force. If the surface is put on an angle, the object will tend to slide down the surface. If the surface is at a slight angle, the force will cause the object to slowly slide down the surface. If the surface is at a severe angle, the downward force will cause the object to quickly slide down the slope. In both cases, the surface is still supplying the force needed to support the object but the pull of gravity is greater and the resultant force causes the object to slide down the slope.

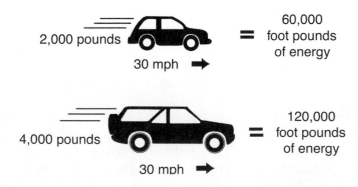

Figure 2-6 The amount of energy required to move a vehicle depends on its mass.

Turning Forces. Forces can cause rotation as well as straight-line motion. A force acting on an object that is free to rotate will have a turning effect, or turning force. This force is equal to the size of the force multiplied by the distance of the force from the turning point around which it acts.

Centrifugal/Centripetal Forces

When an object moves in a circle, its direction is continuously changing and all changes in direction require a force. The forces required to maintain circular motion are called **centripetal** and **centrifugal force**. The size of these forces depends on the size of the circle and the mass and speed of the object.

Centripetal force tends to pull the object toward the center of the circle, whereas centrifugal force tends to push the object away from the center. The centripetal force that keeps an object whirling around on the end of a string is caused by **tension** in the string. If the string breaks, there is no longer string tension and centripetal force and the object will fly off in a straight line because of the centrifugal force on it. Gravity is the centripetal force that keeps the planets orbiting around the sun. Without this centripetal force, the earth would move in a straight line through space.

Tension is the result of forces applied to a material that strain the composition or construction of that material.

Pressure

Pressure is a force applied against an object and is measured in units of force per unit of surface area (pounds per square inch or kilograms per square centimeter). Mathematically, pressure is equal to the applied force divided by the area over which the force acts. Consider two 10 pound weights sitting on a table; one occupies an area of 1 square inch and the other an area of 4 square inches. The pressure exerted by the first weight would be 10 pounds per one square inch or 10 psi. The other weight, although it weighs the same, will exert only 2.5 psi (10 pounds per 4 square inches = 10 ÷ 4 = 2.5). This illustrates an important concept; a force acting over a large area will exert less **pressure** than the same force acting over a small area.

Because pressure is a force, all principles of force apply to pressure. If more than one pressure is applied to an object, the object will respond to the resultant force. Also, all matter (liquids, gases, and solids) tends to move from an area of high pressure to a low-pressure area.

Motion

When the forces on an object do not cancel each other out, they will change the motion of the object. The object's speed, direction of motion, or both will change. The greater the mass of an object, the greater the force needed to change its motion. This resistance to change in motion is called **inertia**. Inertia is the tendency of an object at rest to remain at rest, or the tendency of an object in motion to stay in motion. The inertia of an object at rest is called static inertia, whereas dynamic inertia refers to the inertia of an object in motion. Inertia exists in liquids, solids, and gases. When you push and move a parked vehicle, you overcome the static inertia of the vehicle. If you catch a ball in motion, you overcome the dynamic inertia of the ball.

When a force overcomes static inertia and moves an object, the object gains momentum. **Momentum** is the product of an object's weight times its speed. Momentum is a type of mechanical energy. An object loses momentum if another force overcomes the dynamic inertia of the moving object.

Rates

Speed is the distance an object travels in a set amount of time. It is calculated by dividing distance covered by time taken. We refer to the speed of a vehicle in terms of miles per hour (mph) or kilometers per hour (km/h). **Velocity** is the speed of an object in a particular direction. **Acceleration**, which only occurs when a force is applied, is the rate of increase in speed. Acceleration is calculated by dividing the change in speed by the time it took for that change. **Deceleration** is the reverse of acceleration, as it is the rate of a decrease in speed.

Newton's Laws of Motion

How forces change the motion of objects was first explained by Sir Isaac Newton. These explanations are known as Newton's Laws. Newton's first law of motion is called the Law of Inertia. It states that an object at rest tends to remain at rest and an object in motion tends to remain in motion, unless some force acts on it. When a car is parked on a level street, it remains stationary unless it is driven or pushed.

Newton's second law states that when a force acts on an object the motion of the object will change. This change in motion is equal to the size of the force divided by the mass of the object on which it acts. Trucks have a greater mass than cars. Because a large mass requires a larger force to produce a given acceleration, a truck needs a larger engine than a car.

Newton's third law says that for every action there is an equal and opposite reaction. A practical application of this law occurs when the wheel on a vehicle strikes a bump in the road surface. This action drives the wheel and suspension upward with a certain force, and a specific amount of energy is stored in the spring. After this action occurs, the spring forces the wheel and suspension downward with a force equal to the initial upward force caused by the bump.

Friction

Friction is a force that slows or prevents motion of two moving objects or surfaces that touch. Friction may occur in solids, liquids, and gases. It is the joining or bonding of the atoms at each of the surfaces that causes the friction. When you attempt to pull an object across a surface, the object will not move until these bonds have been overcome. Smooth surfaces produce little friction; therefore, only a small amount of force is needed to break the bonds between the atoms. Rougher surfaces produce a larger friction force because stronger bonds are made between the two surfaces (Figure 2-7). To move an object over a rough surface, such as sandpaper, a great amount of force is required. Friction causes heat to build on the surfaces and can cause wear.

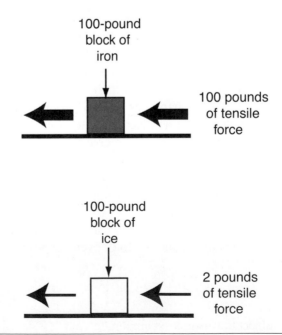

Figure 2-7 Sliding ice across a surface produces less friction than sliding a rougher material, such as iron, across a surface.

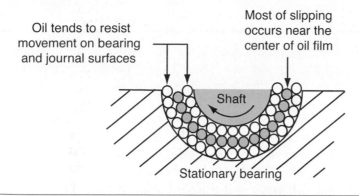

Oil tends to resist movement on bearing and journal surfaces

Most of slipping occurs near the center of oil film

Shaft

Stationary bearing

Figure 2-8 Oil separates the rotating shaft from the stationary bearing.

Lubrication. Friction can be reduced in two main ways: by lubrication or by the use of rollers. The presence of oil or another fluid between two surfaces keeps the surfaces apart. Because fluids (liquids and gases) flow, they allow movement between surfaces. The fluid keeps the surfaces apart, allowing them to move smoothly past one another (Figure 2-8).

Rollers. Rollers placed between two surfaces keep the surfaces apart. An object placed on rollers will move smoothly if pushed or pulled. Rollers actually use friction to grip the surfaces and produce rotation. Instead of sliding against one another, the surfaces produce turning forces, which cause each roller to spin. This leaves very little friction to oppose motion. Bearings are a type of roller (Figure 2-9) used to reduce the friction between moving parts such as a wheel and its axle. As the wheel turns on the axle, the balls in the bearing roll around inside the bearing, drastically reducing the friction between the wheel and axle.

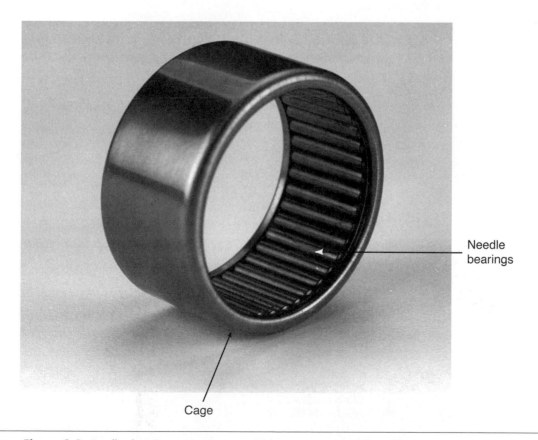

Needle bearings

Cage

Figure 2-9 A roller bearing.

Work

When a force moves a certain mass a specific distance, **work** is done. When work is accomplished, the mass may be lifted or slid on a surface against a resistance or opposing force (Figure 2-10). Work is equal to the applied force multiplied by the distance the object moved (force × distance = work) and is measured in foot-pounds (Figure 2-11), watts, or Newton-meters (N · m). For example if a force moves a 3,000 pound car 50 feet, 150,000 foot-pounds of work was done.

During work, a force acts on an object to start, stop, or change the direction of the object. It is possible to apply a force to an object and not move the object. For example, you may push with all your strength on a car stuck in a ditch, and not move the car. Under this condition, no work is done. Work is only accomplished when an object is started, stopped, or redirected by a force.

Simple Machines

A machine is any device that can be used to transmit a force and, in doing so, change the amount of force or its direction. The force applied to a machine is called the effort, whereas the force it overcomes is called the **load**. The effort is often smaller than the load, because a small effort can overcome a heavy load if the effort is moved a larger distance. The machine is then said to give a mechanical advantage. Although the effort will be smaller when using a machine, the amount of work done, or energy used, will be equal to or greater than that without the machine.

Inclined Plane. The force required to drag an object up a slope (Figure 2-12) is less than that required to lift it vertically. However, the overall distance moved by the object is greater when pulled up the slope than if it were lifted vertically. A screw is like an inclined plane wrapped around a shaft. The force that turns the screw is converted to a larger one, which moves a shorter distance and drives the screw in.

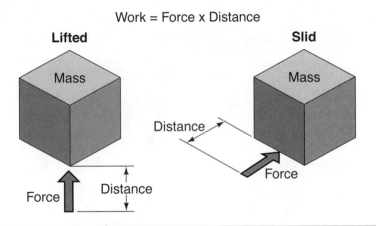

Figure 2-10 When work is performed, a mass is moved a certain distance.

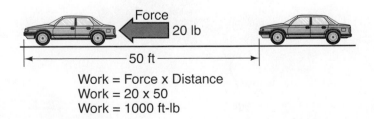

Figure 2-11 1,000 foot-pounds of work.

Figure 2-12 It takes less energy to pull a mass up an inclined plane than lifting the mass vertically would require.

Pulleys. A **pulley** is a wheel with a grooved rim in which a rope, belt, or chain runs to raise something by pulling on the other end of the rope, belt, or chain. A simple pulley changes the direction of a force but not its size. Also, the distance the force moves does not change. By using several pulleys connected together as a block and tackle, the size of the force can be changed, too, so that a heavy load can be lifted using a small force. With a double pulley, the applied force required to move an object can be reduced by one half, but the distance the force must be moved is doubled. A quadruple pulley can reduce the force by four times, but the distance will be increased by four times. Pulleys of different sizes can change the amount of required applied force, as well as the speed or distance the pulley needs to travel to accomplish work (Figure 2-13).

Levers. A **lever** is a device made up of a bar turning about a fixed pivot point, called the fulcrum, that uses a force applied at one point to move a mass on the other end of the bar. Types of levers are divided into classes. In a class one lever, the fulcrum is between the effort and the load (Figure 2-14). The load is larger than the effort, but it moves through a smaller distance. A pair of pliers is an example of a class one lever. In a class two lever, the load is between the fulcrum and effort. Here again, the load is greater than the effort and moves through a smaller distance (Figure 2-15). In a class three lever, the effort is between the fulcrum and the load. In this case, the load is less than the effort, but it moves through a greater distance.

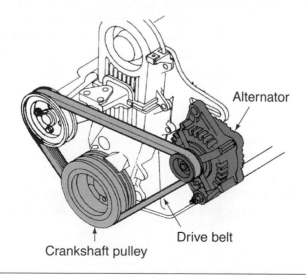

Alternator

Drive belt

Crankshaft pulley

Figure 2-13 Pulleys of different sizes can change the amount of required applied force and the speed or distance the pulley needs to travel to accomplish work.

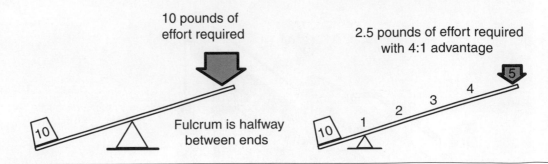

Figure 2-14 A mechanical advantage can be gained with a class one lever.

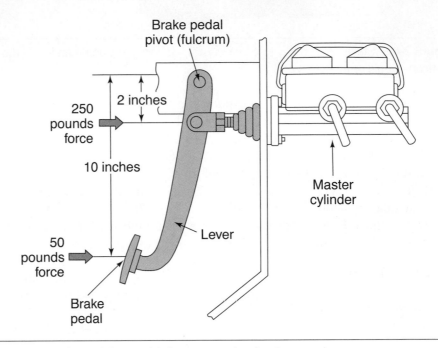

Figure 2-15 A brake pedal assembly is an example of a class two lever.

Gears. A **gear** is a toothed wheel that becomes a machine when it is meshed with another gear. The action of one gear is that of a rotating lever and moves the other gear meshed with it. Based on the size of the gears in mesh, the amount of force applied from one gear to the other can be changed. Keep in mind that this does not change the amount of work performed by the gears, as although the force changes so does the distance of travel (Figure 2-16). The relationship of force and distance is inverse. **Gear ratios** express the mathematical relationship (diameter and number of teeth) of one gear to another.

Wheels and Axles. The most obvious application of a wheel and axle is a vehicle's tires and wheels. These units revolve around an axle and limit the amount of area of a vehicle that contacts the road. Wheels function as rollers to reduce the amount of friction between a vehicle and the road. Basically, the larger the wheel, the less force is required to turn it. However, the wheel must move farther as it gets larger. An example of this is a steering wheel. A steering wheel that is twice the size of another will require one half the force to turn it but also will require twice the distance to accomplish the same work.

Torque

Torque is a force that tends to rotate or turn things and is measured by the force applied and the distance traveled. The technically correct unit of measurement for torque is pounds per

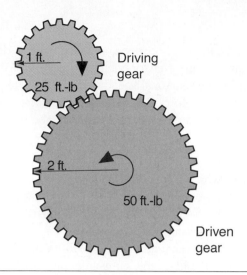

Figure 2-16 When a small gear drives a larger gear, the larger gear turns with more force but travels less; therefore, the amount of work stays the same.

foot (lb.-ft.). However, it is rather common to see torque stated in terms of foot-pounds (ft.-lb.). In the metric or SI system, torque is stated in N·m or kilogram-meters (kg-m).

An engine creates torque and uses it to rotate the crankshaft. The combustion of gasoline and air in the cylinder creates pressure against the top of a piston. That pressure creates a force on the piston and pushes it down. The force is transmitted from the piston to the connecting rod, and from the connecting rod to the crankshaft. The engine's crankshaft rotates with a torque that is transmitted through the drive train to turn the drive wheels of the vehicle.

Torque is force times leverage, the distance from a pivot point to an applied force. Torque is generated any time a wrench is turned with force. If the wrench is a foot long, and you put 20 pounds of force on it, 20 pounds per foot are being generated. To generate the same amount of torque when exerting only 10 pounds of force, the wrench needs to be 2 feet long (Figure 2-17). To have torque, it is not necessary to have movement. When you pull a wrench to tighten a bolt, you supply torque to the bolt. If you pull on a wrench to check the torque on a bolt, and the bolt torque is sufficient, torque is applied to the bolt but no movement occurs. If the bolt turns during torque application, work is done. When a bolt does not rotate during torque application, no work is accomplished.

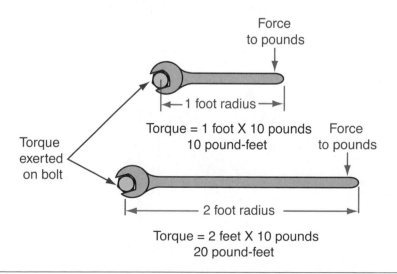

Figure 2-17 The amount of torque applied to a wrench is changed by the length of the wrench.

Power

Power is a measurement for the rate, or speed, at which work is done. The metric unit for power is the watt. A watt is equal to one Newton-meter per second. You can multiply the amount of torque in Newton-meters by the rotational speed to determine the power in watts. Power is a unit of speed combined with a unit of force. For example if you were pushing something with a force of 1 N·m and it moved at a speed of 1 meter per second, the power output would be 1 watt.

Horsepower

Horsepower is the rate at which torque is produced. James Watt is credited with being the first person to calculate horsepower and power. He measured the amount of work that a horse could do in a specific time. He found that a horse could move 330 pounds 100 feet in one minute (Figure 2-18). Therefore, he determined that one horse could do 33,000 foot-pounds of work in one minute. Thus, one horsepower is equal to 33,000 foot-pounds per minute, or 550 foot-pounds per second. Two horsepower could do this same amount of work in one half minute. If you push a 3,000-pound (1,360-kilogram) car for 11 feet (3.3 meters) in one quarter minute, you produce four horsepower.

An engine that produces 300 pounds-feet torque at 4,000 rpm produces 228 horsepower at 4,000 rpm. This is based on the formula that horsepower is equal to torque multiplied by engine speed and that quantity divided by 5252 ([torque × engine speed] ÷ 5252 = horsepower). The constant, 5252, is used to convert the rpm for torque and horsepower into revolutions per second.

Basic Gear Theory

The primary components of the drive train are gears. Gears apply torque to other rotating parts of the drivetrain and are used to multiply torque. As gears with different numbers of teeth mesh, each rotates at a different speed and torque. Torque is calculated by multiplying the force by the distance from the center of the shaft to the point where the force is exerted (Figure 2-19).

For example, if you tighten a bolt with a wrench that is one foot long and apply a force of 10 pounds to the wrench, you are applying 10 ft.-lbs. of torque to the bolt. Likewise, if you apply a force of 20 pounds to the wrench, you are applying 20 ft.-lbs. of torque. You also could apply 20 ft.-lbs. of torque by applying only 10 pounds of force if the wrench were 2 feet long (Figure 2-20).

The distance from the center of a circle to its outside edge is called the radius. On a gear, the radius is the distance from the center of the gear to the point on its teeth at which force is applied.

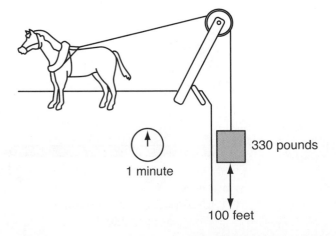

Figure 2-18 This is how James Watt defined one horsepower.

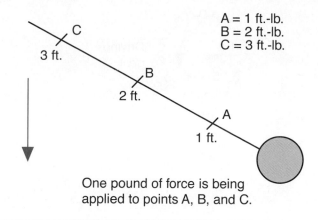

A = 1 ft.-lb.
B = 2 ft.-lb.
C = 3 ft.-lb.

3 ft.

C

B

2 ft.

A

1 ft.

One pound of force is being
applied to points A, B, and C.

Figure 2-19 Torque is calculated by multiplying the force (1 pound) by the distance from the center of the shaft to the point (Points A, B, and C) where force is exerted.

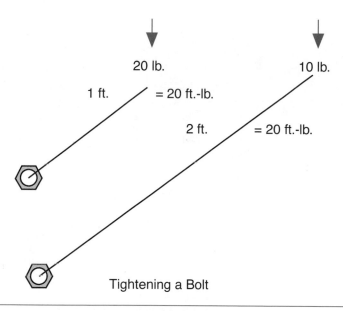

20 lb.

10 lb.

1 ft.   = 20 ft.-lb.

2 ft.   = 20 ft.-lb.

Tightening a Bolt

Figure 2-20 The torque applied to both bolts is 20 ft.-lbs.

If a tooth on the driving gear is pushing against a tooth on the driven gear with a force of 25 pounds and the force is applied at a distance of 1 foot, which is the radius of the driving gear, a torque of 25 ft.-lbs. is applied to the driven gear. The 25 pounds of force from the teeth of the smaller (driving) gear is applied to the teeth of the larger (driven) gear. If that same force were applied at a distance of 2 feet from the center, the torque on the shaft at the center of the driven gear would be 50 ft.-lbs. The same force is acting at twice the distance from the shaft center (Figure 2-21).

The amount of torque that can be applied from a power source is proportional to the distance from the center at which it is applied. If a fulcrum or pivot point is placed closer to object being moved, more torque is available to move the object, but the lever must move farther than if the fulcrum was farther away from the object (Figure 2-22). The same principle is used for gears in **mesh**: a small gear will drive a large gear more slowly but with greater torque.

A drivetrain consisting of a driving gear with 24 teeth and a radius of 1 inch and a driven gear with 48 teeth and a radius of 2 inches will have a torque multiplication factor of 2 and a speed reduction of ½. Thus, it doubles the amount of torque applied to it at half the speed (Figure 2-23). The radii between the teeth of a gear act as levers; therefore, a gear that is twice the size of another has twice the lever arm length of the other.

The **meshing** of gears describes the fit of one tooth of one gear between two teeth of the other gear.

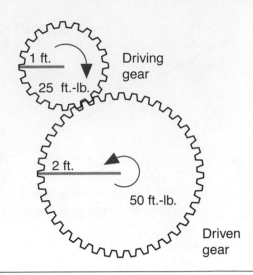

Figure 2-21 The driven gear will turn at half the speed but twice the torque because it is two times larger than the driving gear.

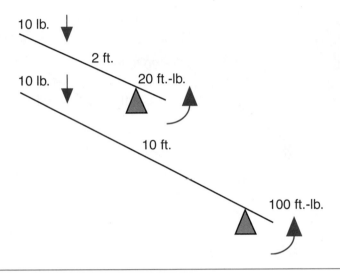

Figure 2-22 The principle of lever action is that a long lever is able to perform more work with less force than a short lever.

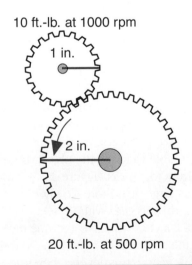

Figure 2-23 The one-inch gear will turn the two-inch gear at half its speed but twice the torque.

Gear ratios express the mathematical relationship of one gear to another. Gear ratios can be varied by changing the diameter and number of teeth of the gears in mesh. A gear ratio also expresses the amount of torque multiplication between two gears. The ratio is obtained by dividing the diameter or number of teeth of the driven gear by the diameter or teeth of the drive gear. If the smaller driving gear had 10 teeth and the larger gear had 40 teeth, the ratio would be 4:1 (Figure 2-24). The gear ratio tells you how many times the driving gear has to turn to rotate the driven gear once. With a 4:1 ratio, the smaller gear must turn four times to rotate the larger gear once.

Gear ratios are normally expressed in terms of some number to one (1) and use a colon (:) to show the numerical comparison, for example, 3.5:1, 1:1, 0.85:1.

Transmission Gears

Transmissions contain several combinations of large and small gears. In low or first gear, a small gear on the input shaft drives a large gear on another shaft. This reduces the speed of the larger gear but increases its turning force or torque. Connected to the second shaft is a small gear, which drives a larger gear, which is connected, to the drive shaft, which in turn is connected to the driving axle. This reduces the speed and increases the torque still more, giving a higher gear ratio for starting movement or pulling heavy loads. First gear is primarily used to initiate movement. It has the lowest gear ratio of any gear in a transmission. It also allows for the most torque multiplication (Figure 2-25).

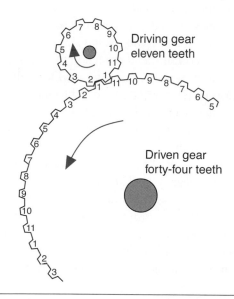

Figure 2-24 The driving gear must rotate four times to rotate the driven gear once. The ratio of the gear set is 4:1.

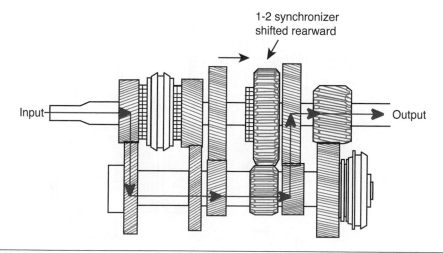

Figure 2-25 Power flow through first gear.

Power moves through the transmission via four gears (two sets of two gears). Speed and torque was altered in steps. To explain how this works, let us put numbers to each of the gears. The small gear on the input shaft has 20 teeth. The gear it meshes with has 40. This provides a gear ratio of 2:1. The output of this gear set moves along the shaft of the 40-tooth gear and rotates other gears. The gear involved with first gear has 15 teeth. This gear rotates with the same speed and with the same torque as the 40-tooth gear. However, the 15-tooth gear is meshed with a larger gear with 35 teeth. The gear ratio of the 15-tooth and the 35-tooth gearset is 2.33:1. However, the ratio of the entire gear set (both sets of two gears) is 4.67:1.

To calculate this gear ratio, divide the driven (output) gear of the first set by the drive (input) gear of the first set. Do the same for the second set of gears, then multiply the answer from the first by the second. The result is equal to the gear ratio of the entire gear set. The mathematical formula is as follows:

$$\frac{\text{driven (A)}}{\text{drive (A)}} \times \frac{\text{driven (B)}}{\text{drive (B)}} = \frac{40}{20} \times \frac{35}{15} = 4.67\text{:}1$$

Second gear uses the same first pair of gears as low does. However, the second pair of gears is disconnected and the power flows through another gear set. This set consists of a larger drive gear and smaller driven gear, A larger gear on the second shaft drives a smaller gear connected to the drive shaft, which results in less overall speed reduction than in first gear. Second gear is used when the need for torque multiplication is less than the need for vehicle speed and acceleration. Because the car is already in motion, less torque is needed to move the car (Figure 2-26).

Third gear allows for a further decrease in engine speed and torque multiplication, while increasing vehicle speed and encouraging fuel economy (Figure 2-27).

Fourth gear typically provides a **direct drive** (1:1) ratio, so that the amount of torque that enters the transmission is also the amount of torque that passes through and out of the transmission output shaft (Figure 2-28). This gear is used at cruising speeds and promotes fuel economy. When the car is in fourth gear, it lacks the performance characteristics of the second and third gears. To pass slower moving vehicles, the transmission often must be downshifted to third to take advantage of third gear's slight torque multiplication, resulting with improved acceleration.

Many of today's transmissions have a fifth gear, called overdrive gear. **Overdrive** gears have ratios of less than 1:1. These ratios are achieved by using a driving gear meshed with a smaller

Direct drive is characterized by the transmission's output shaft rotating at the same speed as its input shaft.

Overdrive causes the output shaft of the transmission to rotate faster than the input shaft.

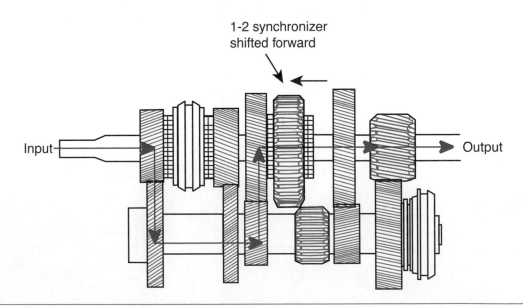

1-2 synchronizer shifted forward

Input

Output

Figure 2-26 Power flow through second gear.

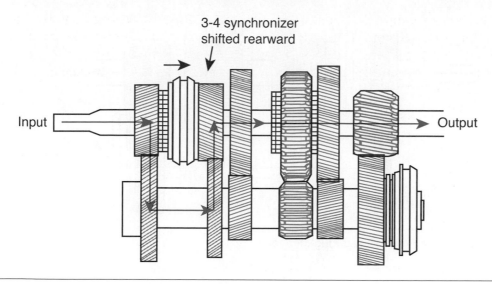

Figure 2-27 Power flow through third gear.

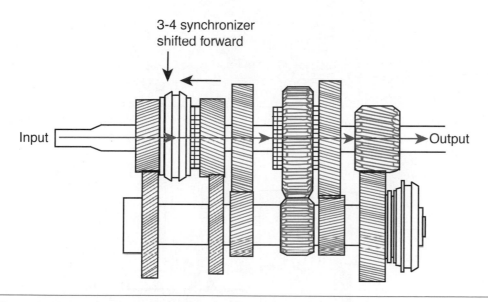

Figure 2-28 Power flow through fourth gear.

driven gear (Figure 2-29). Output speed is increased and torque is reduced. The purpose of over-drive is to promote fuel economy and reduce operating noise when maintaining highway cruising speed.

Through the use of a reverse idler gear (Figure 2-30), the direction of the incoming torque is reversed and the transmission output shaft rotates in the opposite direction of the forward gears (Figure 2-31). Normally, reverse uses a ratio similar to first gear with the addition of the reverse idler gear. Therefore, only low speeds and high torque can be obtained in reverse.

Gear Spacing. Another aspect of transmission gear ratios is their spacing. Spacing is the "distance" between gear ratios of the various gears. For example, in a four-speed transmission, first gear may have a gear ratio of 3.63:1, a second gear a 2.37:1 ratio, third gear a 1.41:1 ratio, and fourth gear a 1:1 ratio, this is a wide ratio transmission. In a close ratio transmission for the same car, the ratios could be 2.57:1 in first, 1.72:1 in second, 1.26:1 in third, and 1:1 in fourth. Fourth gear is the same in both transmissions, but the close ratio gearbox moves the three lower gears closer to fourth. This makes the car more difficult to start from a dead stop but allows for faster

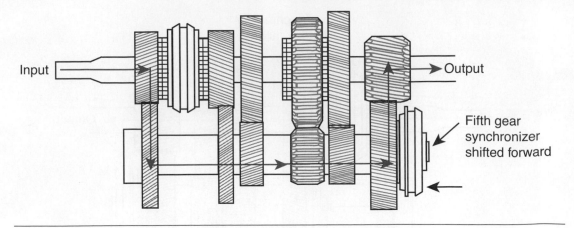

Figure 2-29 Power flow through fifth gear.

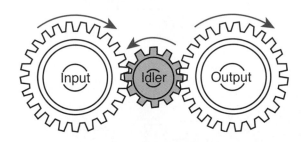

Figure 2-30 An idler gear is used to transfer motion without changing rotational direction to obtain reverse gear.

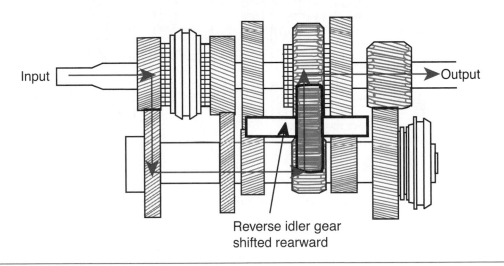

Reverse idler gear
shifted rearward

Figure 2-31 Power flow through reverse gear.

acceleration once the car is rolling. Quicker acceleration is possible because there is less loss of engine speed between gear changes.

Reverse Gear Ratios. Calculating reverse gear ratios is done in the same way as the forward gears, but the math is a little more complicated because reverse gear ratios involve two driving gears and two driven gears:

- the input gear is driver #1
- the idler gear is driven #1

- the idler gear is also driver #2
- the output gear is driven #2

For example: the input gear has 20 teeth, the idler gear has 28, and the output gear has 48 teeth. Because a single idler gear is used, the teeth of it are not used in the calculation of gear ratio. The idler gear merely transfers motion from one gear to another. The calculations for determining reverse gear ratio with a single idler gear is as follows:

$$\text{Reverse gear ratio} = \frac{\text{driven \#2}}{\text{driver \#1}}$$

$$= \frac{48}{20}$$

$$= 2.40$$

If the gear set uses two idler gears (one with 28 teeth and the other with 40 teeth), the gear ratio involves three driving gears and three driven gears:

- the input gear is driver #1
- the #1 idler gear is driven #1
- the #1 idler gear is also driver #2
- the #2 idler gear is driven #2
- the #2 idler gear is also driver #3
- the output gear is driven #3

The ratio of this gear set would be calculated as follows:

$$\text{Reverse gear ratio} = \frac{\text{driven \#1} \times \text{driven \#2} \times \text{driven \#3}}{\text{driver \#1} \times \text{driver \#2} \times \text{driver \#3}}$$

$$= \frac{28 \times 40 \times 48}{20 \times 28 \times 40}$$

$$= \frac{53,760}{22,400}$$

$$= 2.40$$

Final Drive Gears

The transmission's gear ratios are further increased by the gear ratio of the ring and pinion gears in the drive axle assembly. Typical axle ratios are between 2.5 and 4.5:1. The final (overall) drive gear ratio is calculated by multiplying the transmission gear ratio by the final drive ratio. If a transmission is in first gear with a ratio of 3.63:1 and has a final drive ratio of 3.52:1, the overall gear ratio is 12.87:1 (Figure 2-32). If fourth gear has a ratio of 1:1, using the same final drive ratio, the overall gear ratio is 3.52:1.

Final drive ratio is also called the overall gear ratio.

Heat

Heat is a form of energy and is used in many ways. The main sources of heat are the sun, the earth, chemical reactions, electricity, friction, and nuclear energy. Heat is the result of the kinetic energy that is present in all matter; therefore, everything has heat. Cold objects have low kinetic energy because their atoms and molecules are moving very slowly, whereas hot objects have more kinetic energy because their atoms and molecules are moving fast.

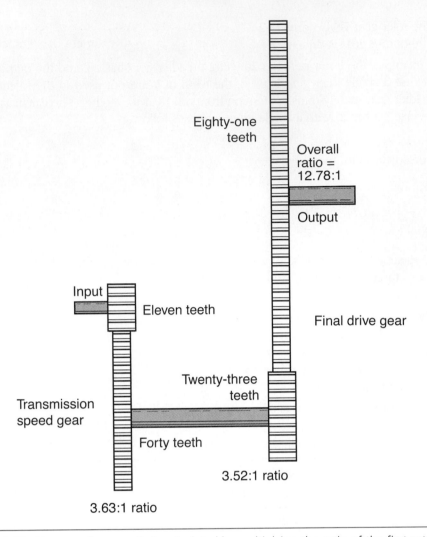

Eighty-one teeth

Overall ratio = 12.78:1

Output

Input

Eleven teeth

Final drive gear

Twenty-three teeth

Transmission speed gear

Forty teeth

3.52:1 ratio

3.63:1 ratio

Figure 2-32 The overall gear ratio is calculated by multiplying the ratio of the first set of gears by the ratio of the second (3.63 × 3.52 = 12.78).

Temperature is an indication of an object's intensity (not volume) of kinetic energy. Temperature is measured with a thermometer, which has either a Fahrenheit (F) or Celsius (Centigrade) (C) scale. At absolute zero (−273°C, also referred to as 0°Kelvin), particles of matter do not vibrate but, at all other temperatures, particles have motion. The temperature of an object is also a statement of how well the object will transfer heat or kinetic energy to or from another object. Heat and temperature are not the same thing; heat is the volume (commonly measured in British Thermal Units—BTUs) of kinetic energy something has. Temperature is an indication of the intensity of kinetic energy something has. Energy from something hot will always move to an object that is colder, until both are at the same temperature. The greater the difference in temperature between the two objects, the faster the heat will flow from one to the other.

Heat is measured in BTUs and calories. One BTU is the amount of heat required to heat one pound of water by 1°F. One calorie is equal to the amount of heat needed to raise the temperature of one gram of water 1°C.

The Effects of Temperature Change

Any time the temperature of an object has changed, a transfer of heat has occurred. A transfer of heat also may cause the object to change size or its state of matter. The amount of heat required to raise the temperature of one gram of mass 1°C is called the specific heat capacity. Every substance has its own specific heat capacity and this factor is assigned to material based on its difference from water, which has a specific heat capacity of 1. For example, the temperature of one gram of water will increase by 10°C if 10 calories of heat were transferred to it. But if 10 calories of heat were added to one gram of copper, the temperature would increase by 111°C. This is because copper has a specific heat capacity of only 0.09, as compared to the 1.0 specific heat capacity of water.

As heat moves in and out of a mass, the movement of atoms and molecules in that mass increases or slows down. With an increase in motion, the size of the mass tends to get bigger or expand. This is commonly called thermal **expansion**. Thermal **contraction** takes place when a mass has heat removed from it and the atoms and molecules slow down. All gases and most liquids and solids expand when heated, with gases expanding the most. Water is an exception; it expands when enough heat is removed to turn it into a solid—ice. Solids, because they are not fluid, expand and contract at a much lower rate. It is important to realize that all materials do not expand and contract at the same rate. For example, an aluminum component will expand at a faster rate than the same component made of iron. This explains why aluminum cylinder heads have unique service requirements and procedures when compared to iron cylinder heads.

Thermal expansion takes place every time fuel and air are burned in an engine's cylinders. The sudden temperature increase inside the cylinder causes a rapid expansion of the gases, which pushes the piston downward and causes engine rotation.

Typically, when heat is added to a mass, the temperature of the mass increases. This does not always happen, however. In some cases, the additional heat causes no increase in temperature but causes the mass to change its state (solid to liquid or liquid to gas). For example, if we take an ice cube and heat it to 32°F (0°C) and continue to apply heat to it, it will begin to melt (Figure 2-33). As heat is added to the ice cube, the temperature of the ice cube will not increase until it becomes a liquid. The heat added to the ice cube that did not raise its temperature but caused it to melt is called **latent heat** or the heat of fusion. Each gram of ice at 0°C requires 80 calories of heat to melt it to water at 0°C. As more heat is added to the 0°C water, the water's temperature will once again increase. This continues until the temperature of the water reaches 212°F (100°C). This is the boiling temperature of water. At this point, any additional heat applied to the water is latent heat causing the water to change its state to that of a gas. This added heat is called the heat of evaporation.

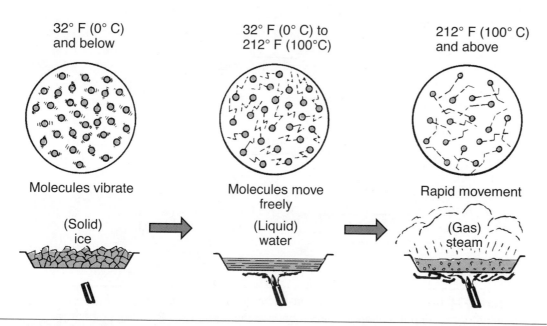

Figure 2-33 Water can exist in three different states of matter.

To change the water gas or steam back to liquid water, the same amount of heat required to change the liquid to a gas must be removed from the gas. At that point, the steam begins to condense into a liquid. As additional heat is removed, the temperature will drop until enough heat is removed to bring its temperature back down to freezing (melting in reverse) point. At that time, latent heat must be removed from the liquid before the water turns to ice again.

Electricity and Electromagnetism

All electrical effects are caused by electric charges. There are two types of electric charge: positive and negative. These charges exert electrostatic forces on each other because of the strong attraction of electrons to protons. An electric field is the area on which these forces have an effect. In atoms, protons carry positive charge, whereas electrons carry negative charge. Atoms are normally neutral; they have an equal number of protons and electrons, but an atom can gain or lose electrons, for example, by being rubbed. It then becomes a charged atom, or ion. Electricity has many similarities with magnetism. For example, the lines of the electric fields between charges take the same form as the lines of magnetic force, so magnetic fields can be said to be an equivalent to electric fields. Charges of the same type repel, whereas charges of a different type attract (Figure 2-34).

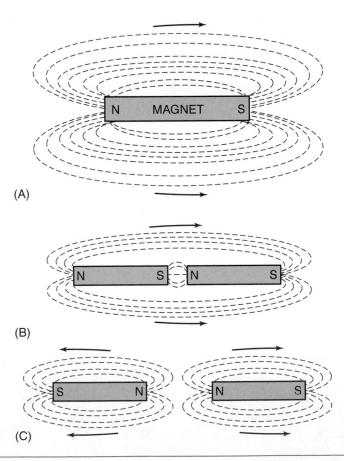

Figure 2-34 (A) In a magnet, lines of force emerge from the north pole and travel to the south pole before passing through the magnet back to the north pole. (B) Unlike poles attract, whereas (C) similar poles repel each other.

Electricity

An electric circuit is simply the path along which an electric current flows. Electrons carry negative charge and can be moved around a circuit by electrostatic forces. A circuit usually consists of a conductive material, such as a metal, where the electrons are held very loosely to their atoms, thus making movement possible. The strength of the electrostatic force is the voltage. The resulting movement of the electric charge is called an electric current. The higher the voltage, the greater the current will be. But the current also depends on the thickness, length, temperature, and nature of the materials that conducts it. The resistance of a material is the extent to which it opposes the flow of electric current. Good conductors have low resistance, which means that a small amount of voltage will produce a large current. In batteries, chemical reactions in the metal electrode causes the freeing of electrons, which results in their movement to another electrode and the formation of a current.

Magnets

Some materials are natural magnets; however, most magnets are produced. The materials typically used to make a permanent magnet are called ferromagnetic materials. These materials consist of mostly iron compounds that are heated. The heat causes the atoms to shift direction. Once they all point in the same direction, the metal becomes a magnet. This sets up two distinct poles, called the north and south poles. The poles are at the ends of the magnet and there is an attraction between the north pole and the south pole. This attraction or force set up by a magnet can be observed but the type of force is not known.

The lines of a magnetic field form closed lines of force, from the north to the south. If another iron or steel object enters into the magnetic field, it is pulled into the magnet. If another magnet is introduced into the magnetic field, it will either move into the field or push away from it. This is the result of the natural attraction of a magnet from north to south. If the north pole of one magnet is introduced to the north pole of another, the two poles will oppose each other and will push away. If the south pole of a magnet is introduced to the north pole of another, the two magnets will join together because the opposite poles are attracted to each other.

The strength of the magnetic force is uniform all around the outside of the magnet. The force is strongest at the surface of the magnet and weakens with distance. If you double the distance from a magnet, the force is reduced by ⅛.

The strength of a magnetic field is typically measured with devices known as magnetometers and in units of Gauss (G).

Electromagnetism

Any electrical current will produce magnetism that affects other objects in the same way as permanent magnets. The arrangement of force lines around a current-carrying conductor, its magnetic field, is circular. The magnetic effect of electrical current is increased by making the current carrying wire into a coil (Figure 2-35).

When a coil of wire is wrapped around an iron bar, it is called an **electromagnet**. The magnetic field produced by the coil magnetizes the iron bar, strengthening the overall effect. A field like that of a bar magnet is formed by the magnetic fields of the wires in the coil. The strength of the magnetism produced depends on the number of coils and the size of the current flowing in the wires. Electromagnetic coils and permanent magnets are arranged inside an electric motor so that the forces of electromagnetism create rotation of the armature.

Producing Electrical Energy

There are many ways to generate electricity. The most common is to use coils of wire and magnets in a generator. Whenever a wire and magnet are moved relative to each other, a voltage is produced (Figure 2-36). In a generator, the wire is wound into a coil. The more turns in the coil

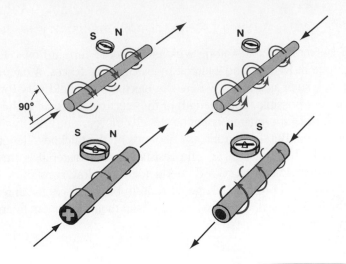

Figure 2-35 When current is passed through a conductor such as a wire, magnetic lines of force are generated around the wire at right angles to the direction of the current flow.

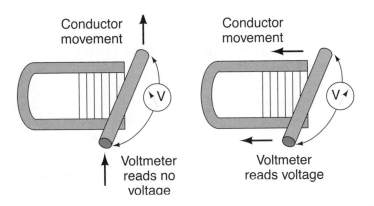

Conductor movement

Conductor movement

V

V

Voltmeter reads no voltage

Voltmeter reads voltage

Figure 2-36 Moving a conductor across magnetic lines of force induces a voltage in the conductor.

and the faster the coil moves, the greater the voltage. The coils or magnets spin around at high speed, typically turned by steam pressure. The steam is usually generated by burning coal or oil, a process that creates pollution. Renewable sources of electricity, such as hydroelectric power, wind power, solar energy, and geothermal power, produce only heat as a pollutant. In automobiles, the generator is spun by a belt driven by the engine's crankshaft. In a generator, the kinetic energy of a spinning object is converted into electrical energy.

A solar cell converts the energy of sunlight directly into electrical energy, using layers of semiconductors. Electricity is produced by causing electrons to leave the atoms in the semiconductor material. Each electron leaves behind a hole or gap. Other electrons move into the hole, leaving holes in their atoms. This process continues all the way around a circuit. The moving chain of electrons is an electrical current.

Summary

❏ Matter is anything that occupies space, and it exists as a gas, liquid, or solid.

❏ When a solid dissolves into a liquid, a solution is formed. Not all solids will dissolve in a liquid; rather, the liquid will be either absorbed or adsorbed.

❏ Materials that *absorb* fluids are permeable substances. Impermeable substances *adsorb* fluids.

❏ Energy is the ability to do work and all matter has energy.

❏ The total amount of energy never changes; it can only be transferred from one form to another, not created or destroyed.

❏ When energy is released to do work, it is called kinetic energy. Stored energy may be called potential energy.

❏ Mass is the amount of matter in an object. Weight is a force and is measured in pounds or kilograms. Gravitational force gives the mass its weight.

❏ A force is a push or pull, and can be large or small and can be applied to objects by direct contact or from a distance.

❏ When an object moves in a circle, its direction is continuously changing and all changes in direction require a force. The forces required to maintain circular motion are called centripetal and centrifugal force.

❏ Pressure is a force applied against an opposing object and is measured in units of force per unit of surface area (pounds per square inch or kilograms per square centimeter).

❏ The greater the mass of an object, the greater the force needed to change its motion. Inertia is the tendency of an object at rest to remain at rest, or the tendency of an object in motion to stay in motion.

❏ When a force overcomes static inertia and moves an object, the object gains momentum. Momentum is the product of an object's weight times its speed.

❏ Speed is the distance an object travels in a set amount of time. Velocity is the speed of an object in a particular direction. Acceleration, which only occurs when a force is applied, is the rate of increase in speed. Deceleration is the reverse of acceleration, as it is the rate of decrease in speed.

❏ Newton's laws of motion state that: an object at rest tends to remain at rest and an object in motion tends to remain in motion, unless some force acts on it; when a force acts on an object, the motion of the object will change; and for every action there is an equal and opposite reaction.

❏ Friction is a force that slows or prevents motion of two moving objects that touch.

❏ Friction can be reduced in two main ways: by lubrication, or by the use of rollers.

❏ When a force moves a certain mass a specific distance, work is done.

❏ A machine is any device that can be used to transmit a force and, in doing so, change the amount of force or its direction. Examples of simple machines are inclined planes, pulleys, levers, gears, and wheels and axles.

❏ Torque is a force that tends to rotate or turn things and is measured by the force applied and the distance traveled.

❏ Gear ratios express the mathematical relationship of one gear to another.

❏ Gears are used to apply torque to other rotating parts of the drivetrain and to multiply torque.

❏ Power is a measurement of the rate at which work is done. It is measured in watts.

❏ Horsepower is the rate at which torque is produced.

❏ Heat is a form of energy caused by the movement of atoms and molecules and is measured in British Thermal Units (BTUs) and calories.

Terms to Know

Acceleration

Atoms

Centrifugal force

Centripetal force

Contraction

Deceleration

Direct drive

Electromagnet

Element

Equilibrium

Evaporate

Expansion

Force

Friction

Gear

Gear ratios

Heat

Horsepower

Impermeable

Inertia

Kinetic energy

Latent heat

Lever

Load

Mass

Matter

Mesh

Momentum

Overdrive

Permeable

Potential energy

Power

Pressure

Pulley

Solution

Speed

Tension

Velocity

Weight

Work

❑ Temperature is an indication of an object's kinetic energy and is measured with a thermometer, which has either a Fahrenheit (F) or Celsius (Centigrade) (C) scale.

❑ As heat moves in and out of a mass, the size of the mass tends to change.

❑ Any electrical current will produce magnetism. When a coil of wire is wrapped around an iron bar, it is called an electromagnet.

❑ The most common way to produce electricity is to use coils of wire and magnets in a generator.

Review Questions

Short Answer Essays

1. Describe Newton's first law of motion and give an application of this law in automotive theory.

2. Explain Newton's second law of motion and give an example of how this law is used in automotive theory.

3. Describe four different forms of energy.

4. Describe four different types of energy conversion.

5. What is the difference between speed and velocity?

6. Why does torque increase when a smaller gear drives a larger gear?

7. How are gear ratios calculated?

8. Why does the size of something change with a change in heat?

9. What is torque?

10. What is a solution and how is it formed?

Fill-in-the-Blanks

1. The nucleus of an atom contains _____ and _____.

2. Work is calculated by multiplying _____ × _____.

3. Energy may be defined as the ability to do _____.

4. When one object is moved over another object, the resistance to motion is called _____.

5. Weight is the measurement of the earth's _____ _____ on an object.

6. Torque is a force that does work with a _____ action.

7. Torque is calculated by multiplying the applied force by the _____ from the center of the _____ to the point where the force is exerted.

8. Torque is measured in _____-_____ and _____-_____.

9. Gear ratios are determined by dividing the number of teeth on the _____ gear by the number of teeth on the _____ gear.

10. One horsepower is equal to _____ foot-pounds per minute, or _____ foot-pounds per second.

Multiple Choice

1. Which of the following statements about friction is not true?
   A. Friction can be reduced by lubrication.
   B. Bearings are a type of roller used to increase the friction between moving parts such as a wheel and its axle.
   C. The presence of oil or another fluid between two surfaces keeps the surfaces apart and thereby reduces friction.
   D. Friction can be reduced by the use of rollers.

2. When discussing different types of energy, *Technician A* says that when energy is released to do work, it is called potential energy. *Technician B* says that stored energy is referred to as kinetic energy. Who is correct?
   A. A only   C. Both A and B
   B. B only   D. Neither A nor B

3. When discussing friction in matter, *Technician A* says that friction is a force that slows or prevents motion of two moving objects or surfaces that touch. *Technician B* says that friction occurs in liquids, solids, and gases. Who is correct?
   A. A only   C. Both A and B
   B. B only   D. Neither A nor B

4. When discussing mass and weight, *Technician A* says that mass is the measurement of an object's inertia. *Technician B* says that mass and weight may be measured in cubic inches. Who is correct?
   A. A only   C. Both A and B
   B. B only   D. Neither A nor B

5. When applying the principles of work and force,
   A. work is accomplished when force is applied to an object that does not move.
   B. in the metric system the measurement for work is cubic centimeters.
   C. no work is accomplished when an object is stopped by mechanical force.
   D. if a 50-pound object is moved 10 feet, 500 ft.-lbs. of work are produced.

6. All these statements about energy and energy conversion are true, EXCEPT:
   A. thermal energy may be defined as light energy.
   B. chemical to thermal energy conversion occurs when gasoline burns.
   C. mechanical energy is defined as the ability to do work.
   D. mechanical to electrical energy conversion occurs when the engine drives the generator.

7. When discussing gear ratios, *Technician A* says that they express the mathematical relationship, according to the number of teeth, of one gear to another. *Technician B* says that they express the size difference of two gears by stating the ratio of the smaller gear to the larger gear. Who is correct?
   A. A only   C. Both A and B
   B. B only   D. Neither A nor B

8. Which of the following statements about mass and weight is true?
   A. Mass is the amount of matter in an object.
   B. Weight is a measurement of mass expressed in pounds or kilograms.
   C. Gravitational force gives a weight its mass.
   D. Something that weighs much on earth will weigh nothing once it is lifted from the earth's surface.

9. A screw is a simple machine that operates as a(n):
   A. gear.   C. inclined plane.
   B. pulley.   D. lever.

10. Which of the following statements is not true?
    A. Materials that absorb fluids are permeable substances.
    B. When a solid dissolves into a liquid, its particles break away from this structure and mix evenly in the liquid, forming a solution.
    C. When most liquids are heated, they evaporate.
    D. Permeable substances adsorb fluids.

Clutches

Upon completion and review of this chapter, you should be able to:

❏ Understand and define the purpose of a clutch assembly.

❏ Understand and describe the major components of a clutch assembly.

❏ Understand and describe the operation of a clutch.

❏ Understand and define the role of each major component in a clutch.

❏ Describe the operation of the various mechanical and cable-type clutch linkages.

❏ Describe the operation of a hydraulic clutch linkage.

❏ Diagnose clutch-related problems by analyzing the symptoms.

Introduction

The manual transmission clutch is a device used to connect and disconnect engine power flow to the transmission at the will of the driver. A driver operates the clutch with a clutch pedal inside the vehicle. This pedal allows engine power flow to be gradually applied when the vehicle is starting out from a stop and interrupts power flow to avoid gear clashing when shifting gears. Engagement of the clutch allows for power transfer from the engine to the transmission and eventually to the drive wheels. Disengagement of the clutch provides the necessary halt of power transfer that allows the engine to continue running while no power is supplied to the drive wheels. Engagement and disengagement of the clutch is controlled by a pedal and clutch linkage. The machined surfaces of the flywheel and pressure plate must be parallel and free of cracks and scores in order to adequately clamp the clutch disc. Clutch slippage, vibration, and noise is minimized by the proper alignment of engine and transmission/transaxle and of the clutch components.

Approximately 35 percent of all cars sold in North America are equipped with a manually operated clutch and transmission.

Clutch Location

The clutch assembly is placed between the engine and the transmission on nearly all cars and trucks. A flywheel is bolted to the rear of the engine's crankshaft and the clutch is bolted to the flywheel.

An exception to this is a late-model Corvette that uses a flywheel and clutch assembly that is placed at the far end of the transmission, at the opposite end from the block. The flywheel and clutch are connected to the engine by a shaft that runs through the transmission.

Normally, a clutch will last 50 to 100 thousand miles. However the life of the clutch depends heavily on the driver, the loads the vehicle carries, and how well the vehicle is maintained.

Clutch Design

The main parts of the clutch assembly are the **clutch housing**, flywheel, clutch shaft, clutch disc, pressure plate assembly, release bearing, and clutch linkage (Figure 3-1). The clutch housing is a large bell-shaped metal casting that connects the engine and the transmission/transaxle. It houses the clutch assembly and supports the transmission/transaxle.

A clutch housing is often called a bell housing.

Flywheel

A flywheel is bolted to the engine's crankshaft to serve many purposes. It acts as a balancer for the engine and it smoothens out, or dampens, engine vibrations caused by firing pulses. It adds

Shop Manual
Chapter 3, page 97

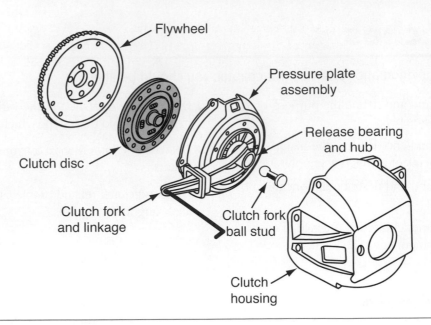

Figure 3-1 Basic clutch components.

A flywheel is a large diameter heavy disc, usually made of nodular cast iron with a high graphite content.

A heat sink is a piece of material that absorbs heat to prevent the heat from settling on another component.

On some flywheels, a special plate is bolted onto the flywheel to provide a frictional surface for the clutch.

A pressure plate may be referred to as a clutch cover.

inertia to the rotating crankshaft. It provides a machined surface from which the clutch can contact and pick up engine torque and transfer it to the transmission. The flywheel also acts as a friction surface and heat sink for one side of the clutch disc.

The flywheel is normally made of nodular cast iron, which has a high graphite content to lubricate the engagement of the clutch. The rear surface of the flywheel is a friction surface, machined very flat to assure smooth clutch engagement.

The flywheel is bolted to the engine's crankshaft and the clutch assembly's pressure plate is bolted to the flywheel. In the center of the flywheel or crankshaft is the bore for the pilot bearing or bushing. The teeth around the circumference of the flywheel form a ring gear for the engine starting motor to contact. The ring gear is not actually a part of the flywheel, rather it is pressed around the outside of the flywheel (Figure 3-2).

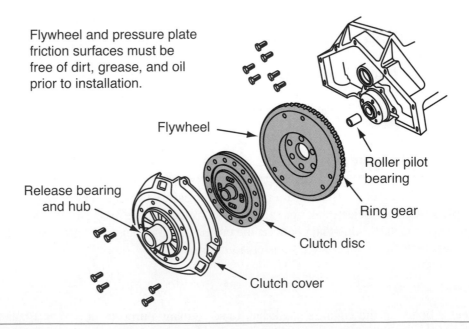

Figure 3-2 Location and mounting of typical flywheel.

A few cars and light trucks use a dual-mass flywheel. These flywheels are normally used to reduce engine vibrations transmitted through the transmission, provide for smoother shifting, and reduce gear noise. Although a conventional (single-mass) flywheel along with the engine's vibration damper smoothens out most of the vibrations of the crankshaft, some remain and move through the transmission. The vibrations result from oscillations set up by the firing of the cylinders and the surges of power on the crankshaft.

Dual-mass flywheels can reduce the oscillations of the crankshaft before they move through the transmission (Figure 3-3). The flywheel consists of two rotating plates, connected by a spring and damper system (Figure 3-4). The forward-most portion of the flywheel is bolted to the end of the crankshaft and smoothens out the crankshaft's oscillations. The pressure plate of the clutch is

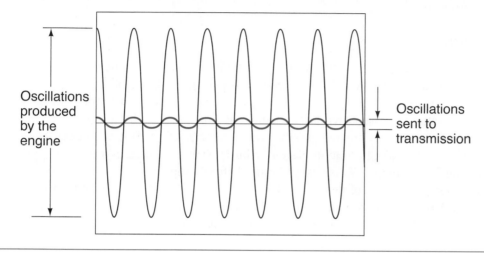

Figure 3-3 A dual-mass flywheel dampens crankshaft vibrations before they enter the transmission.

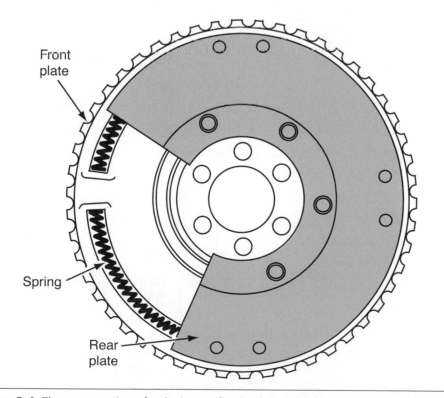

Figure 3-4 The construction of a dual-mass flywheel.

bolted to the rearward portion of the flywheel. Engine torque moves from the front plate through the damper and spring assembly to the rear plate before it enters the transmission.

Some have a torque-limiting feature that prevents damage to the transmission during peak torque loads. The rotation of the two flywheel plates can differ by as much as 360 degrees. This allows the forward plate to absorb torque spikes and not pass them along through the transmission.

Vehicles with automatic transmissions do not have a flywheel. Instead, they use a **flexplate** and the weight of a torque converter to dampen the engine vibrations and provide inertia for the crankshaft. Flexplates are lightweight, stamped steel discs and are used as the attaching point for the torque converter to the engine's crankshaft. They have no clutch friction surface and will not interchange with manual transmission flywheels. They do have a ring gear for the engine's starter motor. However, the ring gear is normally part of the plate and is not replaceable except on some Chrysler products.

Pilot Bearing

The **clutch shaft** projects from the front of the transmission. Most clutch shafts have a smaller shaft or pilot that projects from its outer end. This pilot rides in the pilot bearing in the engine crankshaft flange or flywheel. The pilot bearing or bushing serves as a support for the outer end of the input shaft and it maintains proper alignment of the shaft with the crankshaft (Figure 3-5). A pilot bushing is normally pressed into the bore of the crankshaft. Some flywheels are fitted with a ball or needle-type pilot bearing in place of a bushing. Most transaxles are not equipped with a pilot bearing or bushing because their input shaft is supported by two bearings inside the transaxle. The splined area of the shaft allows the clutch disc to move laterally a small amount along the **splines**, while preventing the disc from rocking on the shaft. When the clutch is engaged, the clutch disc drives the transaxle's input shaft through these splines (Figure 3-6).

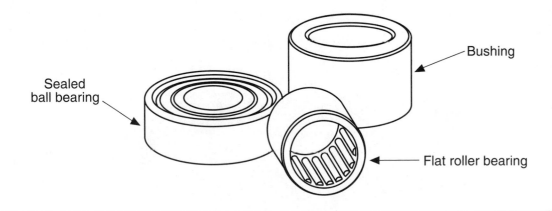

Figure 3-5 Examples of pilot bushings and bearings.

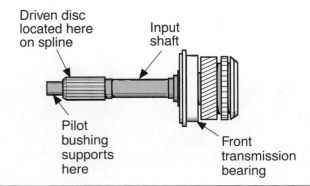

Figure 3-6 Typical clutch shaft shaft.

60

Clutch Disc

The **clutch disc** is a steel plate covered with frictional material that fits between the flywheel and the pressure plate. In the center of the disc is the hub, which is splined to fit over the splines of the input shaft. When the clutch is engaged, the disc is firmly squeezed between the flywheel and pressure plate and power from the engine is transmitted by the disc's hub to the transmission's input shaft. The width of the hub prevents the disc from rocking on the shaft while it moves between the flywheel and the pressure plate.

A clutch disc (Figure 3-7) has frictional material riveted or bonded to both sides. Frictional facings are either woven or molded. Molded facings are preferred because they can withstand high pressure plate loading forces without being damaged. Woven facings are used when additional cushioning during clutch engagement is desired. Grooves are cut across the face of the friction facings to allow for smooth clutch action, increased cooling, and a place for the facing dust to go as the clutch disc wears. Like brake lining material, the frictional facing wears as the clutch is engaged. **Asbestos** wire-woven material was the most common facing for clutch discs. Due to recent awareness of the health hazards resulting from asbestos, new lining materials have been developed and are widely used on newer vehicles. The most commonly used are paper-based and ceramic materials that are strengthened by the addition of cotton and brass particles and wire. These increase the torsional strength of the facings and prolong the life of the clutch disc.

The facings are attached to wave springs that cause the contact pressure on the facings to rise gradually as the springs flatten out when the clutch is engaged. These springs eliminate chatter when the clutch is engaged and also help to move the disc away from the flywheel when the clutch is disengaged. The wave springs and facings are attached to the steel disc.

There are two types of clutch discs: rigid and flexible. A **rigid clutch disc** is a solid circular disc fastened directly to a center splined **hub**. The **flexible clutch disc** is easily recognized by the torsional dampener springs that circle the center hub. The dampener is a shock-absorbing feature built into a flexible clutch disc. A flexible disc absorbs power impulses from the engine that would otherwise be transmitted directly to the gears in the transmission. A flexible

Shop Manual
Chapter 3, page 95

Asbestos is a mineral fiber composed of a silicate of calcium and magnesium that occurs in long threadlike fibers. It has a high resistance to heat.

A **hub** is the center part of a wheel or disc.

Figure 3-7 Typical clutch disc.

A clutch disc is also called a friction disc.

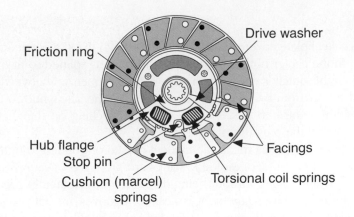

Figure 3-8 Major components of a clutch disc.

clutch disc has torsion springs and friction discs between the plate and hub of the clutch. When the clutch is engaged, the springs cushion the sudden loading by flexing and allowing some twist between the hub and plate. When the "surge" is past, the springs release and the disc transmits power normally. The number and tension of these springs is determined by the amount of engine torque and the weight of the vehicle. Stop pins limit this torsional movement to approximately $\frac{3}{8}$ inch (Figure 3-8).

Some high performance clutch assemblies use multiple clutch discs. An intermediate plate is used in these assemblies to separate the clutch discs. When the clutch is engaged, the first clutch disc is held between the clutch pressure plate and intermediate plate, and the second clutch disc is held between the intermediate plate and the flywheel. When disengaged, the intermediate plate, flywheel, and pressure plate assembly rotate as a unit, while the clutch discs, which are not in contact with the plates, rotate freely within the assembly and do not transmit power to the transmission.

Pressure Plate Assembly

The pressure plate (Figure 3-9) squeezes the clutch disc onto the flywheel when the clutch is engaged and moves away from the disc when the clutch pedal is depressed. These actions allow the clutch disc to transmit or not transmit the engine's torque to the transmission. A pressure plate is basically a large spring-loaded clamp that is bolted to and rotates with the flywheel. A pressure plate assembly includes a sheet metal cover, heavy release springs, a metal pressure ring that provides a friction surface for the clutch disc, a thrust ring or fingers for the release bearing, and release levers. The release levers release the holding force of the springs when the clutch is disengaged. The spring used in most pressure plates is a single **Belleville spring** or **diaphragm-type spring**, however a few use multiple coil springs. Some pressure plates are of the semicentrifugal design and use centrifugal weights that increase the clamping force on the thrust springs as engine speed increases.

Diaphragm spring pressure plate (Figure 3-10) assemblies use a cone-shaped diaphragm spring between the pressure plate and the cover to clamp the pressure plate against the clutch disc. This spring is normally secured to the cover by **rivets**. When pressure is exerted on the center of the spring, the outer diameter of the spring tends to straighten out. As soon as the pressure is released, the spring resumes its normal cone shape. The center portion of the spring is slit into numerous fingers that act as release levers. When the clutch is disengaged, these fingers are depressed by the release bearing. The diaphragm spring pivots over the fulcrum ring and its outer rim moves away from the flywheel. The retracting springs pull the pressure plate away from the clutch disc, thereby disengaging the clutch.

Figure 3-9 Typical pressure plate.

Some manufacturers refer to a pressure plate as a clutch cover.

Some diaphragm-type springs are fitted with a thrust pad at the ends of the release fingers.

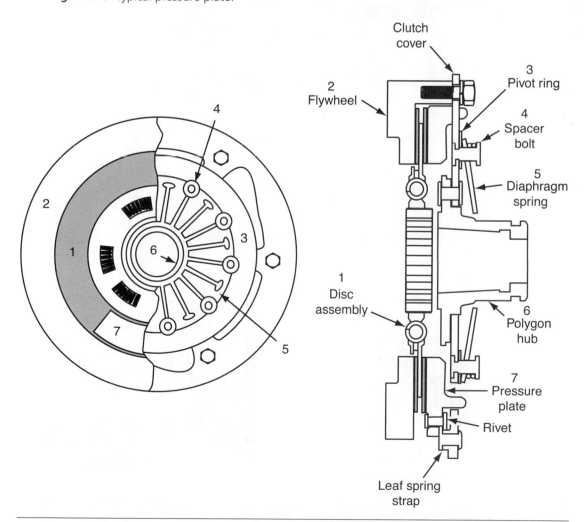

Clutch cover

2
Flywheel

3
Pivot ring

4
Spacer bolt

5
Diaphragm spring

4

2

3

6

1

5

7

1
Disc assembly

6
Polygon hub

7
Pressure plate

Rivet

Leaf spring strap

A coil spring clutch assembly uses coil springs to hold the pressure plate against the friction disc.

Figure 3-10 A typical diaphragm spring clutch.

When the clutch is engaged, the release bearing is moved away from the diaphragm spring's release fingers. As the spring pivots over the **fulcrum** ring, its outer rim forces the pressure plate tightly against the clutch disc. At this point the clutch disc is clamped between the flywheel and pressure plate. At the outer rim, the spring is moved outward and the pressure plate is forced against the clutch disc. Diaphragm-type pressure plates have another distinguishing characteristic. As the clutch disc wears, the pressure plate load increases. For all other types of pressure plates, the load decreases. Diaphragm-type pressure plates are preferred because they are compact, lightweight, require less pedal effort, and have few moving parts to wear.

Coil spring pressure plate assemblies (Figure 3-11) use helical springs that are evenly spaced around the inside of the pressure plate cover. These springs exert pressure to hold the pressure plate tightly against the flywheel. During clutch disengagement, release levers release the holding force of the springs and the clutch disc no longer rotates with the pressure plate and flywheel. Normally these pressure plates are equipped with three release levers and each lever has two pivot points. One pivot point attaches the lever to a pedestal cast into the pressure plate and the other attaches the lever to a release yoke that is bolted to the cover. The levers pivot on the pedestals and release lever yokes to move the pressure plate through its engagement and disengagement operations (Figure 3-12).

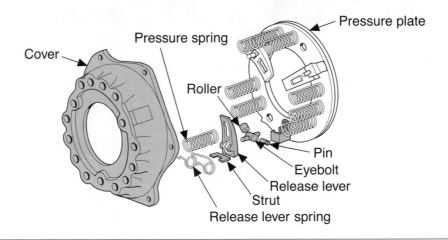

Figure 3-11 Coil spring pressure plate assembly.

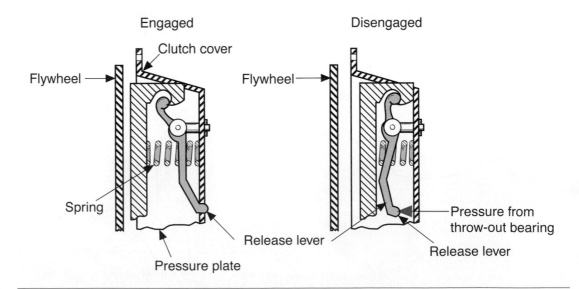

Figure 3-12 Action of a coil spring pressure plate's release levers.

64

To disengage the clutch, the release bearing pushes the inner ends of the release levers toward the flywheel. The release levers act as a fulcrum for the levers and the outer ends of the release levers move to pull the pressure plate away from the clutch disc. This action compresses the coil springs and disengages the clutch. When the clutch is engaged, the release bearing moves and allows the springs to exert pressure to hold the pressure plate against the clutch disc. This forces the disc against the flywheel and the engine's power is transmitted to the transmission through the clutch disc.

A **semicentrifugal pressure plate** is a design variation of a coil-spring pressure plate. These assemblies alter the holding force on the clutch disc according to engine speed. As engine speed increases, so does the holding force. A weighted end on the release levers (Figure 3-13) act to increase centrifugal force as the rotational speed of the pressure plate increases (Figure 3-14). The centrifugal force adds to the spring pressure to produce greater holding force against the clutch disc. This design allows for the use of coil springs with less tension, as the centrifugal force compensates for the decrease in spring tension. Therefore, less pedal effort is required to operate the clutch without a loss of clamping pressure on the disc.

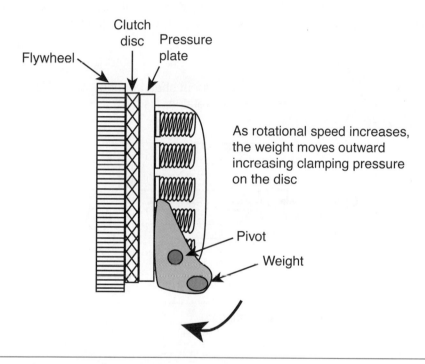

Figure 3-13 Semicentrifugal pressure plate assembly.

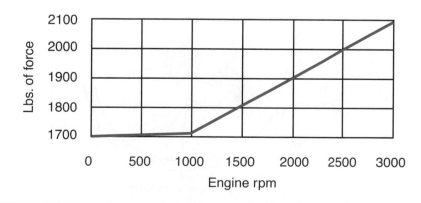

Figure 3-14 Graph showing how a semicentrifugal pressure plate changes plate loading according to engine speed.

The individual parts of a pressure plate assembly are contained in the cover. Some covers are vented to allow heat to escape and air to enter. Other covers are designed to provide a fan action to force air circulation around the clutch assembly. The effectiveness of the clutch is affected by heat, therefore, by allowing the assembly to cool, it is able to work better.

Release Bearing

Shop Manual
Chapter 3, page 105

The **clutch release bearing** (Figure 3-15) is a ball-type bearing located in the bell housing and operated by the clutch linkage (Figure 3-16). Release bearings are usually sealed and prelubricated to provide smooth and quiet operation as they move against the pressure plate to disengage the clutch. A clutch release bearing is often referred to as a throw-out bearing. When the clutch pedal is depressed to disengage the clutch, the release bearing moves toward the flywheel, depressing the pressure plate's release fingers or thrust pad and moving the pressure plate fingers or levers against pressure plate spring force. This action moves the pressure plate away from the clutch disc, thus interrupting power flow. A few vehicles have release bearings that pull on the springs of the pressure plate rather than push them.

The hollow shaft on the front of the front bearing retainer is often referred to as the *quill shaft*.

Release bearings are mounted on an iron casting called a hub, which slides on a hollow shaft at the front of the transmission housing. This hollow shaft is part of the transmission's front bearing retainer.

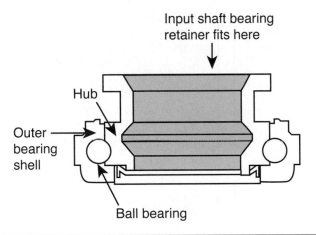

Figure 3-15 A typical clutch release bearing.

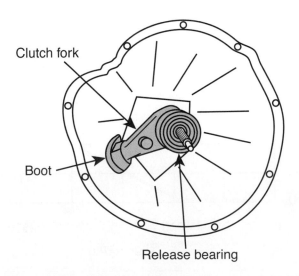

Figure 3-16 Location of clutch fork and release bearing.

The release bearing is mounted on a sleeve that is designed to slide back and forth on the transmission's bearing retainer (Figure 3-17). The sleeve is grooved or has raised flat surfaces and retaining springs that hold the inner ends of the release fork in place on the release bearing assembly. The fork and connecting linkage convert the movement of the clutch pedal to the back and forth movement of the clutch release bearing.

To disengage a clutch, the release bearing is moved toward the flywheel by the clutch fork. As the bearing contacts the release levers or fingers, it begins to rotate with the pressure plate assembly. As the release bearing continues to move forward, the pressure on the release levers or fingers causes the force of the pressure plate's spring to move away from the clutch disc.

To engage the clutch, the clutch pedal is released and the release bearing moves away from the pressure plate. This action allows the pressure plate's springs to force against the clutch disc, engaging the clutch to the flywheel. Once the clutch is fully engaged, the release bearing is normally stationary and does not rotate with the pressure plate.

Some release bearings must be adjusted so they do not touch the release fingers, beveled springs, or thrust pad when the pedal is released. However, most release bearings are designed to ride lightly against the pressure plate assembly; these are called constant running release bearings (Figure 3-18) and are used on transmissions equipped with self-adjusting cable or hydraulically operated clutch linkages.

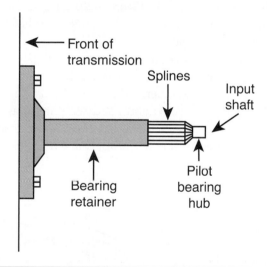

Figure 3-17 The release bearing slides on the hollow shaft nose of the transmission's front bearing retainer.

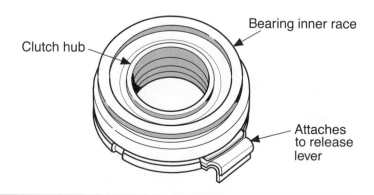

Figure 3-18 Typical constant running release bearing.

The clutch linkage connects the driver-operated clutch pedal to a bell housing-mounted release fork that acts directly on the release bearing. The release bearing slides back and forth on the transmission's front bearing retainer in response to clutch pedal movement.

Shop Manual
Chapter 3, page 78

Clutch pedal free-play is the amount the pedal can move without applying pressure on the pressure plate.

Clutch Linkages

Clutches are normally operated by either mechanical or hydraulic linkages. Two types of mechanical linkages are used: the cable type and the shaft and lever type. The shaft and lever clutch linkage has many parts and pivot points and transfers the movement of the clutch pedal to the release bearing via shafts, levers, and bell cranks. In older vehicles, the pivot points were equipped with grease fittings. Current systems pivot on low-friction plastic grommets and bushings. As the pivot points wear, the extra play in the linkage makes precise clutch pedal free-play adjustments difficult.

A typical shaft and lever clutch control assembly (Figure 3-19) includes a release lever and rod, an equalizer or cross shaft, a pedal to equalizer rod, an assist or overcenter spring, and the pedal assembly. Depressing the pedal moves the equalizer that in turn, moves the release rod. When the pedal is released, the assist spring returns the linkage to its normal position and removes the pressure on the release rod. This action causes the release bearing to move away from the pressure plate.

A cable-type clutch linkage is simple and lightweight. Normally the cable connects the pivot of the clutch pedal directly to the release fork (Figure 3-20). This simple set-up is compact, flexible, and eliminates the wearing pivot points of a shaft and lever linkage. However, cables will gradually stretch and can break due to electrolysis.

Typically, one end of the cable is connected to the pedal assembly. At the assembly is a spring that keeps the pedal in the up position. The cable is held under tension by the spring and a cable stop located on the fire wall. The other end of the cable is connected to the outer end of the clutch release fork. This end is threaded and fitted with an adjusting nut and locknut that allows for pedal free-play adjustments. When the clutch pedal is depressed, the cable pulls on the clutch fork, which causes the release bearing to move against the pressure plate.

A control cable is an assembly with a flexible outer housing anchored at the upper and lower ends. Moving back and forth inside the housing is a braided stainless steel wire cable that transfers pedal movement to the release lever.

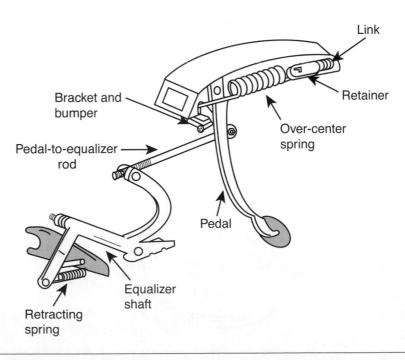

Figure 3-19 Typical shaft and lever-type clutch control.

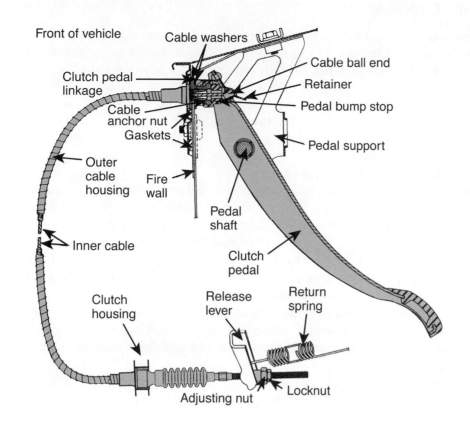

Figure 3-20 Typical clutch cable system.

On many late model vehicles, the cable is self-adjusting. At the pedal pivot, the cable is wrapped around and attached to a toothed wheel (Figure 3-21). Slight contact of the release bearing on the pressure plate is maintained by a ratcheting, spring-loaded pawl that is engaged to the toothed wheel. When the clutch pedal is released, the pawl takes any slack out of the cable by engaging the next tooth of the wheel. Self-adjusting clutches use a constant running release bearing and have no built-in free-play.

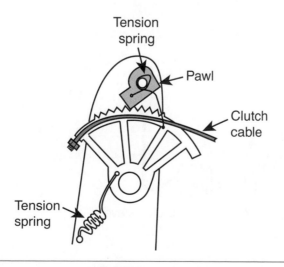

Figure 3-21 Typical automatic clutch cable adjusting mechanism.

Hydraulic Linkage

Another method used to move the clutch fork is with a hydraulic system. The hydraulic clutch linkage is often used when the design of the vehicle makes it difficult to use levers and cables. A hydraulic system also multiplies force, which reduces the force required to depress the clutch pedal.

A hydraulic clutch linkage, like a brake system, consists of a master cylinder, hydraulic tubing, and a slave cylinder (Figure 3-22). The **master cylinder** is attached to and activated by the clutch pedal through the use of an actuator rod. The **slave cylinder** is connected to the master cylinder by flexible pressure hose or metal tubing. The slave cylinder is positioned so that it can work directly on the clutch release yoke lever.

Depressing the clutch pedal pushes the actuator rod into the bore of the master cylinder. This action forces a plunger or piston up the bore of the master cylinder (Figure 3-23). During the initial $\frac{1}{32}$ inch of pedal travel, the valve seal at the end of the master cylinder bore closes the port to the fluid reservoir. As the pedal is further depressed, the movement of the plunger forces fluid from the master cylinder through a line to the slave cylinder (Figure 3-24). This fluid is under pressure and causes the piston of the slave cylinder to move. Moving the slave cylinder's piston causes its pushrod to move against the release fork and bearing, thus disengaging the clutch. When the clutch pedal is released, the springs of the pressure plate push the slave cylinder's pushrod back, forcing the hydraulic fluid back into the master cylinder. The final movement of the clutch pedal and the master cylinder's piston opens the valve seal and fluid flows from the reservoir into the master cylinder.

A clutch master cylinder performs the following functions: moves fluid through the hydraulic line to the slave cylinder, compensates for temperature change and minimal fluid loss to maintain the correct fluid volume through the use of a bleed port and compensation port, and compensates for a worn clutch disc and pressure plate by displacing fluid through the reservoir bleed port, thereby eliminating the need for periodic adjustment.

The slave cylinder disengages the clutch by extending the slave cylinder pushrod and makes sure the clutch release bearing is contacting the pressure plate through the use of the slave cylinder preload spring.

Hydraulic clutch controls are normally used on large units that require high pedal pressure to disengage. Hydraulic clutch linkages are also used in vehicles where mechanical linkage would be difficult to route, due to the compactness of the vehicle's design. By using hydraulics, a small force can be multiplied and used effectively to disengage the clutch.

The **master cylinder** is a liquid-filled cylinder in the hydraulic brake system or clutch in which hydraulic pressure is developed when the driver depresses a foot pedal.

The **slave cylinder** is located at a lower part of the clutch housing that receives fluid pressure from the master cylinder, allowing it to engage or disengage the clutch.

The clutch fork is a Y-shaped member into which the release bearing is attached.

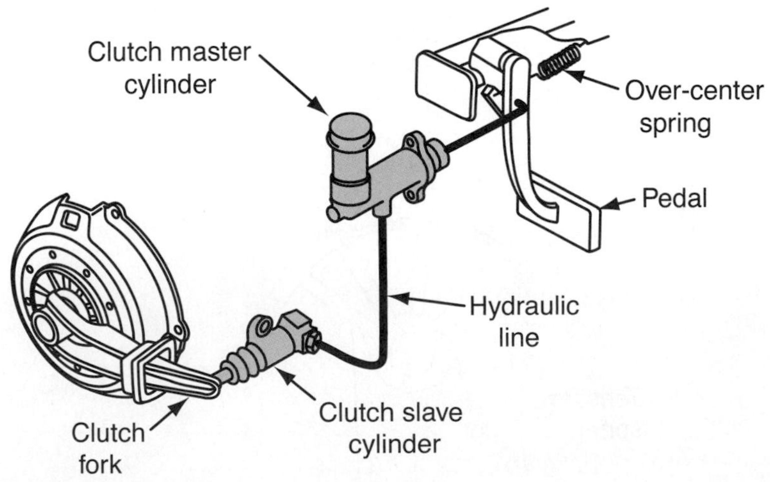

Figure 3-22 Typical hydraulic clutch linkage.

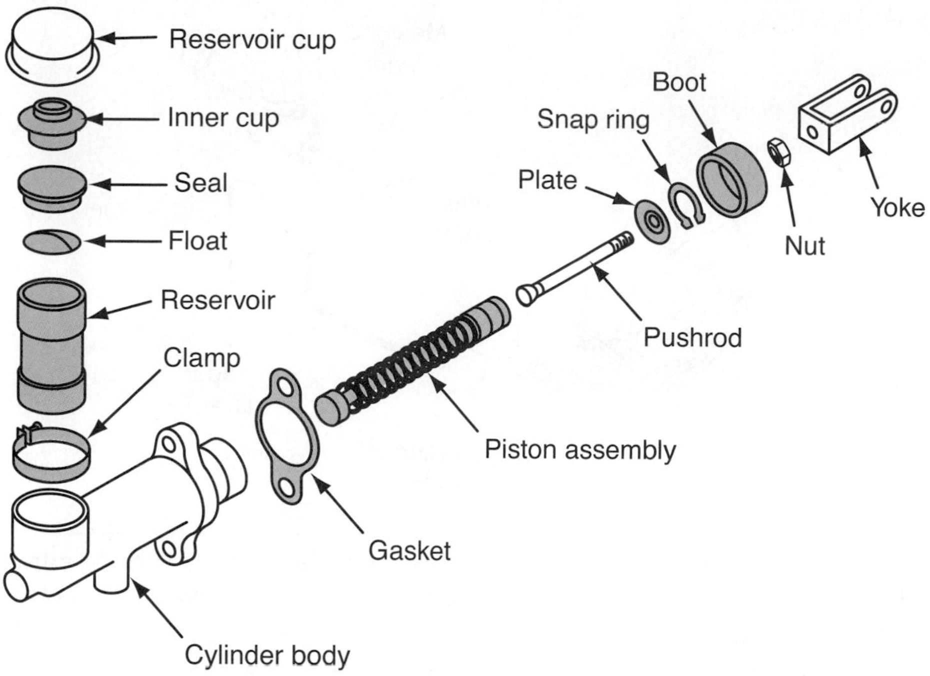

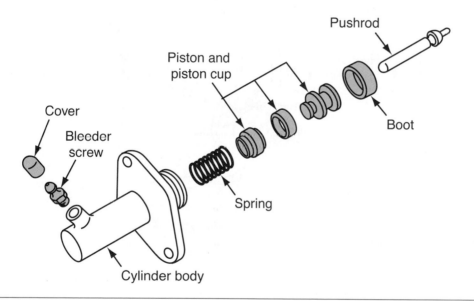

Reservoir cup

Inner cup

Seal

Float

Reservoir

Clamp

Cylinder body

Boot

Snap ring

Plate

Yoke

Nut

Pushrod

Piston assembly

Gasket

Hydraulic clutch linkage systems are often called hydraulic clutches.

Figure 3-23 Parts of a hydraulic clutch master cylinder.

Pushrod

Piston and piston cup

Cover

Bleeder screw

Boot

Spring

Cylinder body

The slave cylinder is appropriately named, as it only works in response to the master cylinder.

Figure 3-24 Parts of a hydraulic clutch slave cylinder.

Some hydraulic clutch systems have a clutch damper (Figure 3-25). The damper absorbs torsional vibrations during engagement and disengagement of the clutch.

A concentric slave cylinder is found on some cars and light trucks. These units are actually both the slave cylinder and the clutch release bearing (Figure 3-26). By having the slave cylinder directly behind the release bearing, the movement of the release bearing is linear, sliding on the bearing retainer.

A concentric slave cylinder is a doughnut-shaped unit that mounts to the front of the transmission and the transmission's input shaft passes through it. The slave cylinder is either bolted to the transmission's front bearing cover or is secured by a pressed pin.

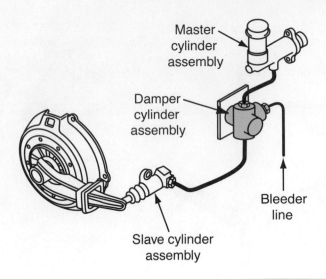

Master cylinder assembly

Damper cylinder assembly

Bleeder line

Slave cylinder assembly

Figure 3-25 A hydraulic clutch circuit with a damper cylinder assembly. Note the extra hydraulic line from the damper. This line is not connected to any other component and is used to bleed the system.

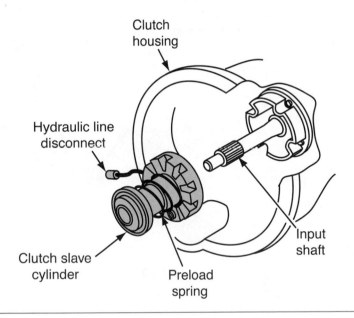

Clutch housing

Hydraulic line disconnect

Clutch slave cylinder

Preload spring

Input shaft

Figure 3-26 A concentric slave cylinder.

AUTHOR'S NOTE: I have seen many students search and search for the clutch slave cylinder on Ford Rangers and other vehicles. These are concentric slaves and are located inside the bell housing. You can only see them when you separate the transmission from the engine. Always refer to the service manual when servicing a clutch assembly.

Clutch Operation

The clutch disc is sandwiched between the flywheel and pressure plate assembly to become the driven member of the clutch assembly. The flywheel drives the front side while the pressure plate drives the rear side of the clutch disc. In the center of the clutch disc is a splined hub that meshes with the splines on the clutch shaft. The pressure plate is held in contact with the rear friction facing of the clutch disc by spring tension. When the clutch is disengaged, the pressure plate is released by a release bearing that utilizes a lever action to pull the pressure plate away from the clutch disc. The release yoke, with the release bearing clipped to it, is mounted on a pivot located inside the clutch housing and is operated by the clutch linkage and pedal assembly.

To disengage the clutch, the driver presses the clutch pedal. The linkage forces the release bearing and release yoke forward to move the pressure plate away from the disc (Figure 3-27). Because the disc is no longer in contact with the flywheel and pressure plate, the clutch is said to be disengaged. When the clutch pedal is depressed, the flywheel, clutch disc, and pressure plate are disengaged, thus power flow is interrupted. As the clutch pedal is released, the pressure plate moves closer to the clutch disc, clamping the disc between the pressure plate and the flywheel. Therefore, if the transmission is in gear, the drive wheels will turn when the clutch disc turns.

To engage the clutch, the driver eases the clutch pedal up from the floor. The control linkage moves the release bearing and lever rearward, permitting the pressure plate spring tension to force the pressure plate and the driven disc against the flywheel. Engine torque again acts on the disc's friction facings and splined hub to drive the transmission input shaft.

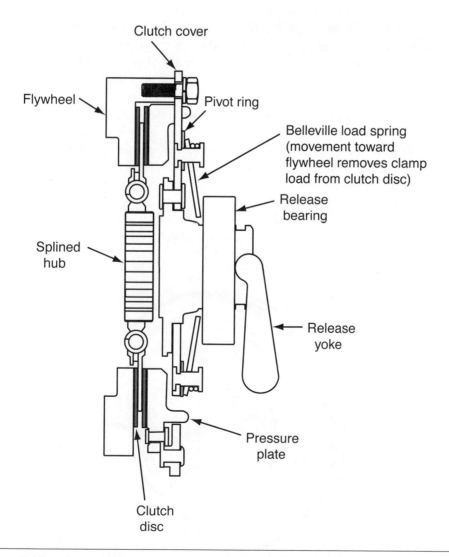

Figure 3-27 Purpose of the major components of a clutch assembly.

Cone clutches are still used in some limited-slip differentials and have been used in overdrive units.

In a **cone clutch**, the driving and driven parts are conically shaped to connect and disconnect power flow. The cones, one fitting inside the other, create friction between them, forcing them to rotate together.

Cone clutches were used almost exclusively on early automobiles. As technology changed, expanding shoe, band-type clutches, and wet or dry disc and plate-type clutches became more prominent. By 1950, nearly all automobiles were equipped with a dry disc clutch system.

Dual Clutch Transmissions

One of the latest trends in manual transmission design is that of a self-shifting manual unit. Some of these units use a dual clutch arrangement, which commonly are found in VW and Audi vehicles. Duals clutches are used in these units to improve the shifting quality and driveability in self-shifting manual transmissions. These transmissions have two separate gear set shafts, each with its own corresponding clutch, to engage and disengage the gears. For example, in a five-speed manual transmission, one shaft takes care of the one-, two-, and five-speed gears, and the second shaft handles the two-, four-, and reverse-speed gears. By clutching the shaft for the outgoing gear when simultaneously, engaging the shaft for the next desired gear, the unit greatly reduces the time required for the gearshift but, more important, drastically improves shift quality by nearly eliminating shift shock and harsh response to throttle inputs.

The dual clutches can be dry or wet (similar to those used in automatic transmissions) design. Dry dual clutches are used primarily in vehicles with low engine torque. Dual wet clutch systems (Figure 3-28) are used on more powerful engines and typically include a torsional damper, hydraulic control unit (HCU) for actuating and cooling the dual clutch, and transmission actuators with an electronic transmission and system control unit (TCU).

The two discs in the dual clutch setup can be engaged independently of each other through hydraulic pressure of pistons (Figure 3-29). When the clutch is engaged, torque is transmitted to the allocated input shaft (Figure 3-30) via the respective clutch assembly.

Band-type clutches are used in automatic transmissions.

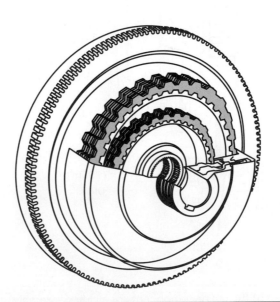

Figure 3-28 A duel wet clutch assembly.

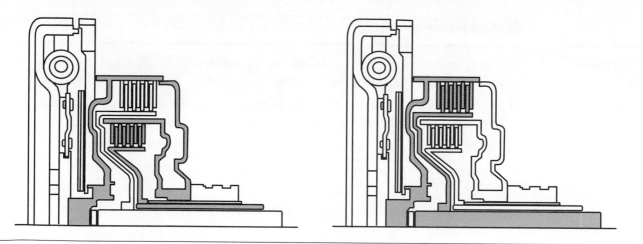

Figure 3-29 In the dual clutch arrangement, the two disc sets can be engaged independently of each other through hydraulic pressure of pistons.

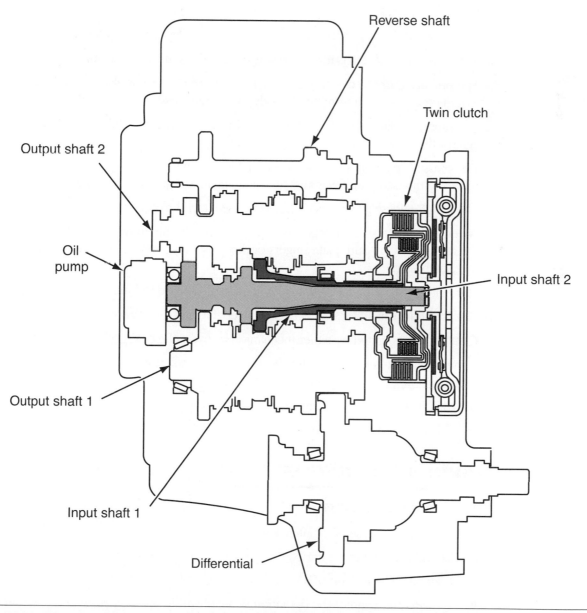

Reverse shaft

Twin clutch

Output shaft 2

Oil pump

Input shaft 2

Output shaft 1

Input shaft 1

Differential

Figure 3-30 When the clutch is engaged, torque is transmitted to the allocated input shaft via the respective clutch assembly.

Summary

❏ The main parts of the clutch assembly are the clutch housing, flywheel, input shaft, disc, pressure plate assembly, release bearing, and linkage.

❏ The flywheel acts as a balancer and smoothens out, or dampens, engine vibrations caused by firing pulses and adds inertia to the rotating crankshaft of the engine.

❏ The flywheel also provides a machined surface for the clutch friction disc.

❏ Vehicles with automatic transmission are equipped with a drive plate or flexplate rather than a heavy flywheel.

❏ The clutch disc is splined to the input shaft, which allows the disc to move without rocking on the shaft.

❏ The clutch disc is a steel plate with friction material bonded to both sides that fits between the flywheel and the pressure plate.

❏ Some older friction discs contain asbestos fibers. Always follow safety precautions when handling asbestos.

❏ A rigid or solid clutch disc is a solid disc fastened directly to a center splined hub.

❏ A flexible clutch disc has torsional dampener springs in its center hub.

❏ The primary purpose of a flexible disc is to absorb power impulses from the engine that would otherwise be transmitted directly to the transmission.

❏ The pressure plate is a large spring-loaded plate that engages the clutch by pressing the disc against the flywheel surface.

❏ The pressure plate moves away from the flywheel when the clutch pedal is depressed, releasing the clamping force and stopping engine torque from reaching the transmission.

❏ The clutch release bearing is operated by the clutch linkage.

❏ When the clutch pedal is depressed, the bearing typically moves toward the flywheel, depressing the pressure plate fingers or thrust pad and moving the pressure plate away from the clutch disc.

❏ The clutch linkage connects the clutch pedal to a release fork that acts on the release bearing.

❏ The clutch is usually located between the engine and the transmission.

❏ Clutches are operated by either mechanical or hydraulic linkages.

❏ A mechanical clutch linkage transfers the clutch pedal movement to the release bearing via shafts, levers, and bell cranks, or by a cable.

❏ A hydraulic clutch linkage consists of a master cylinder, hydraulic tubing, and a slave cylinder.

Review Questions

Short Answer Essays

1. Define the purpose of a clutch assembly.

2. List and describe the major components of a clutch assembly.

3. Describe the operation of a clutch.

4. Compare and contrast the operation of a coil spring pressure plate and a diaphragm spring pressure plate.

5. Define the role of each major component in a clutch assembly.

6. Describe the operation of a mechanical lever-type clutch linkage.

7. Describe the operation of a cable-type clutch linkage.

8. Describe the operation of a hydraulic clutch linkage.

9. Explain why some vehicles are equipped with a semicentrifugal pressure plate.

10. Describe the construction of a flexible clutch disc and state why it is different than a rigid disc.

Fill-in-the-Blanks

1. A clutch housing is also called a _____ - _____ .

2. The teeth around the outside of the flywheel form a ring gear for the

   _____ - _____ .

3. The clutch disc slides over the splines of the _____ shaft.

4. There are two types of dry clutch discs: _____ and _____ .

5. The pressure plate engages and disengages the _____

   _____ .

6. A flexible clutch disc has _____ springs in its center hub.

7. The pressure plate moves away from the flywheel when the clutch pedal is

   _____ .

8. When the clutch is disengaged, the pressure plate is released by a _____

   _____ .

9. A hydraulic clutch linkage consists of a _____ _____ ,

   _____ _____ , and a _____

   _____ .

10. Tension is maintained in a cable-type clutch linkage by the use of a

    _____ and a toothed _____ .

Multiple Choice

1. When discussing where a clutch disc is normally mounted, *Technician A* says that it is bolted to the flywheel. *Technician B* says that it is splined to the transmission input shaft. Who is correct?
   - **A.** A only
   - **B.** B only
   - **C.** Both A and B
   - **D.** Neither A nor B

2. When discussing the purpose of the torsional springs in a clutch disc, *Technician A* says that they cushion the sudden load on the disc when it is quickly engaged. *Technician B* says that they allow the alignment of the clutch disc and hub to shift slightly. Who is correct?
   - **A.** A only
   - **B.** B only
   - **C.** Both A and B
   - **D.** Neither A nor B

3. When discussing the purpose of the flywheel, *Technician A* says that it serves as a heat sink for the clutch disc. *Technician B* says that it serves as a vibration dampener for the engine. Who is correct?

A. A only   **C.** Both A and B
B. B only   **D.** Neither A nor B

4. When discussing the operation of the clutch assembly, *Technician A* says that when the pedal is depressed, the release bearing is pushed into the center of the pressure plate, which releases the plate's pressure on the clutch disc. *Technician B* says that normally the clutch disc is pressed against the flywheel by the pressure plate.
Who is correct?

     A. A only   **C.** Both A and B
     B. B only   **D.** Neither A nor B

5. When discussing the major components of a hydraulic clutch linkage, *Technician A* says that the master cylinder for the car's brakes is also used for the clutch. *Technician B* says that the slave cylinder is directly connected to the clutch pedal and increases hydraulic pressure as the pedal is depressed.
Who is correct?

     A. A only   **C.** Both A and B
     B. B only   **D.** Neither A nor B

6. *Technician A* says that constant running release bearings are used with hydraulically controlled clutches. *Technician B* says that release bearings move the pressure plate to disengage the clutch.
Who is correct?

     A. A only   **C.** Both A and B
     B. B only   **D.** Neither A nor B

7. When discussing cable-type clutch linkages, *Technician A* says that they are not commonly used

because they are expensive and complicated. *Technician B* says that many late model cars use self-adjusting cables for their clutch linkage.
Who is correct?

     A. A only   **C.** Both A and B
     B. B only   **D.** Neither A nor B

8. When discussing the different types of clutches, *Technician A* says that flexible clutch discs are most commonly used. *Technician B* says that flexible clutch discs apply the power forces from the engine directly to the transmission.
Who is correct?

     A. A only   **C.** Both A and B
     B. B only   **D.** Neither A nor B

9. When discussing the purpose of a pilot bearing or bushing, *Technician A* says that they are only necessary when the transmission's input shaft is long. *Technician B* says that they serve as a low-friction pivot point for the clutch linkage.
Who is correct?

     A. A only   **C.** Both A and B
     B. B only   **D.** Neither A nor B

10. When discussing the different types of pressure plates, *Technician A* says that coil spring types are commonly used because they have strong springs. *Technician B* says that Belleville-type pressure plates are not commonly used because they require excessive space in the bell housing.
Who is correct?

     A. A only   **C.** Both A and B
     B. B only   **D.** Neither A nor B

Manual Transmissions/ Transaxles

Upon completion and review of this chapter, you should be able to:

❑ Understand and discuss the purpose and operation of typical manual transmissions.

❑ Understand and describe the purpose, design, and operation of a synchronizer assembly.

❑ Understand and discuss the flow of power through a manual transmission.

❑ Compare and contrast the design and operation of a transmission and a transaxle.

❑ Understand and discuss the flow of power through a manual transaxle.

❑ Understand and describe how gears are shifted in a manual transmission/ transaxle.

❑ Identify the accessories controlled by a manual transmission/transaxle.

❑ Discuss the importance of gear oil in a manual transmission/transaxle.

❑ Understand and describe the basic operation of a continuously variable transmission.

Introduction

Transmissions and transaxles serve basically the same purpose and operate by the same basic principles. Although the assembly of a transaxle is different than a transmission, the fundamentals and basic components are the same. In fact, a transaxle is basically a transmission with other driveline components housed within the assembly. All basics covered in this chapter will refer to both a transaxle and a transmission unless otherwise noted.

A transmission (Figure 4-1) is a system of gears that transfers the engine's power to the drive wheels of the car. The transmission receives torque from the engine through its input shaft when

The purpose of a transmission is to use various sized gears to keep the engine in its best operating range in a variety of vehicle speeds and loads.

A gear transfers power and motion from one shaft to another shaft.

Transmission gears are normally helical-type gears; however, some reverse speed gears are spur-type gears.

Some manufacturers call the output shaft the main shaft.

Figure 4-1 A late model transaxle.

the clutch is engaged. The torque is then transferred through a set of gears that either multiply it, transfer it directly, or reduce it. The resultant torque turns the transmission's output shaft, which is indirectly connected to the drive wheels. All transmissions have two primary purposes: They select different speed ratios for a variety of conditions and provide a way to reverse the movement of the vehicle.

A manual transaxle is a single unit composed of a manual transmission, final drive, and differential. Most front-wheel-drive (FWD) cars are equipped with a transaxle. Transaxles are also found on some front engined and rear-wheel-drive (RWD) and four-wheel-drive (4WD) cars and on rear-engined and rear-wheel-drive cars (Figure 4-2).

With a transaxle, the engine is normally mounted transversely across the front of front-wheel-drive (FWD) cars (Figure 4-3). The compactness of a transaxle allows designers to offer increased passenger room, reduce vehicle weight, and reduce vibration and alignment problems normally caused by the long drive shafts of rear-wheel-drive cars.

A transaxle is a final drive unit and transmission housed in a single unit.

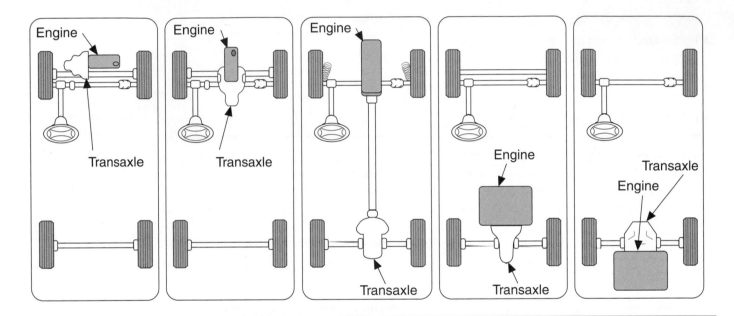

Figure 4-2 Different transaxle locations.

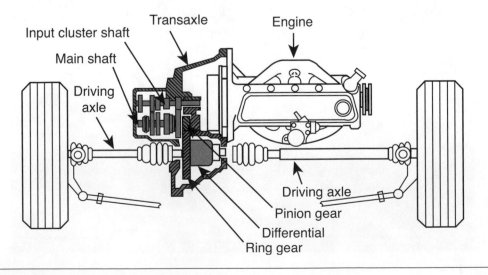

Figure 4-3 Location of typical FWD components.

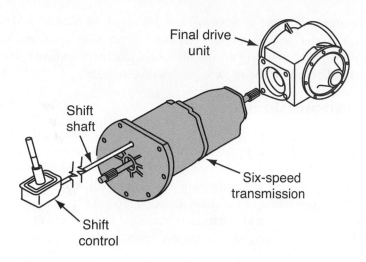

Figure 4-4 Later model Corvettes and a few other cars have the transmission mounted directly to the rear drive axle.

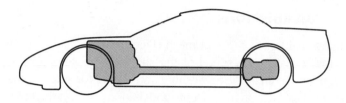

Figure 4-5 With the weight of the transmission a the rear, the weight on the rear wheels about equals the weight on the front wheels.

Remote-Mounted Transmissions

Late-model Corvettes, as well as other cars, have a driveline setup that is somewhat unique. These vehicles are front-engined, RWD vehicles. The six-speed transmission is mounted directly to the rear axle assembly (Figure 4-4) instead of the engine. The clutch assembly is still at the rear of the engine and a drive shaft (torque tube) connects the output from the clutch to the transmission (Figure 4-5). Shifting is often accomplished through a shift shaft located at the front of the transmission. This shaft extends from the transmission to the shift control inside the car. On some later model vehicles, shifting is done through electronically controlled solenoids. This setup reduces the bulk of the shift rod and offers precise shifting.

The primary advantage to moving the transmission to the rear of the car is to shift the weight of the unit to the rear. This gives the car a near balance of weight on the front and rear wheels. This weight balance provides for improved handling and braking.

Types of Manual Transmissions/Transaxles

Three major types of manual transmissions have been used by the automotive industry: the sliding gear, collar shift, and synchromesh designs.

Sliding Gear Transmissions

A **sliding gear** transmission uses two or more shafts mounted in parallel, with sliding spur gears arranged to mesh with each other to provide a change in speed or direction. The driver moves

In a **sliding gear** transmission, gears are moved on their shafts to change gear ratios.

the shifter, which in turn moves the appropriate gear into mesh with its mate. If either of the gears are rotating, the changing of gears is very difficult and will normally grind into mesh. This type of transmission is currently used only on farm and industrial machines. However, some transmissions use a sliding gear mechanism for the engagement of reverse gear.

Collar Shift Transmissions

When a transmission is called constant mesh, that means the gears are always mated. However, they do not always transfer power.

A **collar shift** transmission has parallel shafts fitted with gears in **constant mesh**. The change of gear ratios is accomplished by locking the free-running gears to their shafts using sliding collars. The sliding collars connect and lock the appropriate gear to its shaft. One side of the gear has short splines into which the internal splines of the sliding collar mesh. Because the collar is splined to the shaft, the connection of the collar to the gear connects the gear to the rotating shaft. Although the splines on the gears and collars have rounded ends for easier shifting, gear clashing will occur when the gears are changed or engaged, unless speeds are matched by the driver by double-clutching.

 AUTHOR'S NOTE: If you don't know what double-clutching is, ask your instructor to explain.

Synchromesh Transmissions

A **synchromesh** transmission is a constant mesh (Figure 4-6), collar shift transmission equipped with synchronizers that equalize the speed of the shaft and gear before they are engaged. The action of the synchronizer eliminates gear clashing and allows for smooth changing of gears. Synchromesh transmissions are used on all current models of cars and are commonly found in other machines wherever shifting while moving is required.

The **main shaft** is also referred to as the output shaft on most transmissions.

The counter shaft is also called the **cluster gear**.

Engine torque is applied to the transmission's input shaft when the clutch is engaged. The input shaft enters the transmission case, where it is supported by a bearing and fitted with a gear. The gear on the input shaft is called the input gear or **clutch gear**. The output shaft (**main shaft**) is inserted into, but rotates independently of the input shaft. The main shaft is supported by the input shaft bearing and a bearing at the rear of the transmission case. The various speed gears rotate on the main shaft. Located below or to the side of the input and main shaft assembly is a counter shaft that is fitted with several sized gears. All of these gears, except one, are in constant mesh with the gears on the main shaft. The remaining gear is in constant mesh with the input gear.

Gear changes occur when a gear is selected by the driver and is locked or connected to the main shaft. This is accomplished by the movement of a collar that connects the gear to the shaft

Free-running gears are gears that rotate independently on their shafts.

Figure 4-6 The gears in a transmission transmit the rotating power from the engine.

Figure 4-7 The shift forks and rails are assembled in the cover of this transmission with a direct shift linkage.

(Figure 4-7). Smooth and quiet shifting can only be possible when the gears and shaft are rotating at the same speed. This is the primary function of the synchronizers.

Forward Speeds

All transmissions are equipped with a varied number of forward speed gears and one reverse speed. Transmissions can be classified by the number of forward gears (speeds) it has. For many years, the three-speed manual transmission was the most commonly used and four-speed transmissions were found only in heavy-duty vehicles and high-performance cars. The concern for improved fuel mileage led to smaller engines with four-speed transmissions. The additional gear allowed the smaller engines to perform better by matching the engine's torque curve with vehicle speeds.

Five-speed transmissions and transaxles are now the most commonly used units. Some earlier units were actually four speeds with an add-on fifth or overdrive gear. Many late model transmissions incorporate fifth gear in their main assemblies. This is also true of late model six-speed transmissions and transaxles. The fifth and sixth gears are part of the main gear assembly. Most often, the two high gears in a six-speed provide overdrive gear ratios.

Synchronizers

A synchronizer's primary purpose is to bring components that are rotating at different speeds to one synchronized speed. It also serves to lock these parts together. The forward gears of all current automotive transmissions are synchronized. Some older transmissions and truck transmissions were not equipped with synchronizers on first or reverse gears. These gears could only be easily engaged when the vehicle was stopped. Reverse gear on some late-model transmissions and transaxles is also synchronized. A single synchronizer is placed between two different speed gears, therefore transmissions have two or three synchronizer assemblies.

In the past, reverse gear was not normally synchronized. Today most late model transmissions have a synchronized reverse gear, or at least a blocking ring in the design to stop the shaft from turning during reverse engagement. Also, in keeping with the very high cost of new vehicles the manufacturers are demanding helical gearing to eliminate noise concerns from the customer.

Shop Manual
Chapter 4, page 147

AUTHOR'S NOTE: Sometimes the best way to really understand how a synchronizer works is to hold a complete synchro assembly in your hands and move each part around and back and forth. Notice what else moves when the parts are in their various positions, then think about the rest of the transmission parts and their movements.

Synchronizer Designs

There are four types of synchronizers used in synchromesh transmissions: block, disc and plate, plain, and pin. The most commonly used type on current transmissions is the block type. All synchronizers use friction to synchronize the speed of the gear and shaft before the connection is made.

Block synchronizers consist of a hub (called a **clutch hub**), sleeve, blocking ring, and inserts or spring-and-ball detent devices (Figure 4-8). The synchronizer sleeve surrounds the synchronizer assembly and meshes with the external splines of the hub. The hub is internally splined to the transmission's main shaft. The outside of the sleeve is grooved to accept the shifting fork. Three slots are equally spaced around the outside of the hub and are fitted with the synchronizer's inserts or spring-and-ball detent assemblies.

These inserts are able to freely slide back and forth in the slots. The inserts are designed with a ridge on their outer surface and insert springs hold this ridge in contact with an internal groove in the synchronizer sleeve. When the transmission is in the "neutral" position, the inserts keep the sleeve lightly locked into position on the hub. If the synchronizer assembly uses spring-and-ball detents, the balls are held in this groove by their spring. The sleeve is machined to allow it to slide smoothly on the hub.

Bronze or brass blocking rings are positioned at the front and rear of each synchronizer assembly. Blocking rings are made of brass or bronze because of the high heat produced by friction. The use of these metals minimizes the wear on the hardened steel gear's cone. A common trend in the automotive industry today is to reduce frictional losses as much as possible. The materials used to manufacture blocking rings has not been overlooked. Powdered metal and organic frictional materials are currently being used and/or being tested by a few manufacturers. Blocking rings have notches to accept the insert keys that cause them to rotate at the same speed as the hub (Figure 4-9). Around the outside of the blocking ring is a set of beveled **dog teeth.** These teeth are used for alignment during the shift sequence. The inside of the blocking ring is shaped like a cone, the surface of which has many sharp grooves. The inner surfaces of the blocking rings match the conical shape of the shoulders of the driven gear. These cone-shaped surfaces serve as the frictional surfaces for the synchronizer. The shoulder of the gear also has a ring of beveled dog teeth designed to align with the dog teeth on the blocking ring (Figure 4-10).

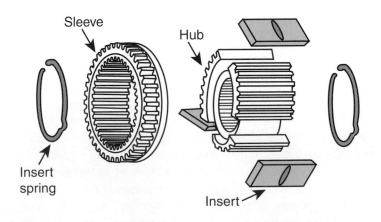

Sleeve · Hub · Insert spring · Insert

Figure 4-8 Typical block synchronizer assembly.

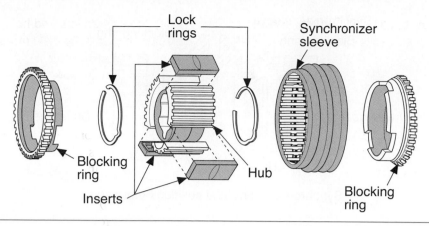

Figure 4-9 The notches in the blocking ring correspond with the spacing of the inserts.

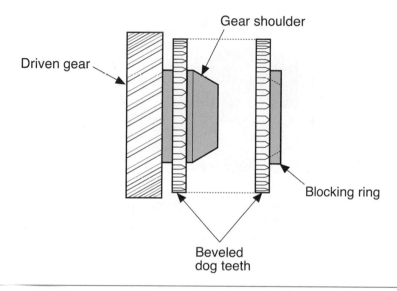

Figure 4-10 Beveled dog teeth on the speed gear and the blocking ring.

Synchronizer Operation

When the transmission is in neutral (Figure 4-11), the synchronizers are in their neutral position and are not rotating with the main shaft. The main shaft's gears are meshed with the counter gears and are rotating with the counter shaft. However, they turn freely, at various speeds, on the main shaft and do not cause the shaft to rotate because they are not connected to it.

When a gear is selected, the shifting fork forces the sleeve toward the selected gear. As the sleeve moves, so do the inserts because they are locked in the sleeve's internal groove. The movement of the inserts pushes the blocking ring into contact with the shoulder of the driven gear. When this contact is made, the grooves on the blocking ring's cone cut through the film of lubrication on the gear's shoulder. If the film of lubrication is not cut by the grooves, synchronization cannot take place. Destroying the film allows for metal-to-metal contact and begins the speed synchronization of the two parts. (Figure 4-12). The resultant friction between the two brings the gear's cone to the blocking ring cone's speed.

As the components reach the same speed, the synchronizer sleeve can now slide over the external dog teeth on the blocking ring and then over the dog teeth on the speed gear's shoulder (Figure 4-13). This completes the engagement of the synchronizer and the gear is now locked to

The term *synchro* is a commonly used slang word for synchronizer.

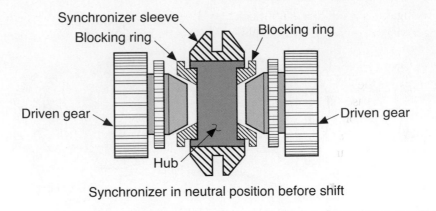

Synchronizer in neutral position before shift

Figure 4-11 A block synchronizer in its neutral position.

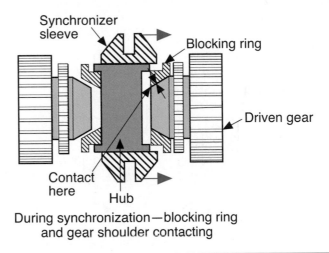

During synchronization—blocking ring
and gear shoulder contacting

Figure 4-12 Initial contact of the blocking ring to the gear's cone.

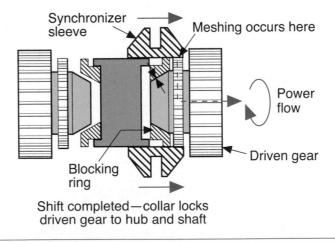

Shift completed—collar locks
driven gear to hub and shaft

Figure 4-13 Shift collar movement locks the gear to the synchronizer hub.

the main shaft. The blocking ring will only allow the sleeve to mesh with the gear when its teeth are lined up with the locking teeth on the gear.

Power now flows from input gear to the counter gear, back up to the speed gear locked by the synchronizer. Power then flows from the gear through the locking teeth to the sleeve, then to the hub, and finally to the main shaft.

To disengage a gear, the shifter is moved to the neutral position, which causes the synchronizer sleeve to move away from the previous gear, thereby disconnecting it from the shaft.

In summary, synchronization occurs in three stages. In the first stage, the sleeve is moved toward the gear by the shift lever and engages the hub assembly. In the second stage, the movement of the sleeve causes the inserts to press the blocking ring onto the cone of the gear. In the third stage, the synchronizer ring completes its friction fit over the gear cone and the gear is brought up to the same speed as the synchronizer assembly. The sleeve slides onto the gear's teeth and locks the gear and its synchronizer assembly to the main shaft.

Advanced Synchronizer Designs

Synchronizers act as clutches to speed up or slow down the gear sets that are being shifted to, and greater friction area results in easier shifting for the driver. This is the reason manufacturers use multiple cone-type synchronizers in their transmissions. Many transmissions are fitted with single-cone, double-cone, or triple-cone synchronizers on their forward gears. For example, first and second gears may have triple-cone synchronization, third and fourth may have double-cone, and fifth and sixth have single-cone.

Double-cone synchronizers (Figure 4-14) have friction material on both sides of the synchronizer rings. The extra friction surfaces spread the workload of accelerating and decelerating driven gears to the desired speed over a larger area. This results in decreased shift effort and greater synchronizer durability. Triple-cone synchronizers provide a third surface in the synchronizer assembly lined with friction material.

With these multiple-cone synchronizers, the size of the transmission can be reduced. The multiple-cone synchronizers offer a high synchronizer capacity in a smaller package. To obtain the

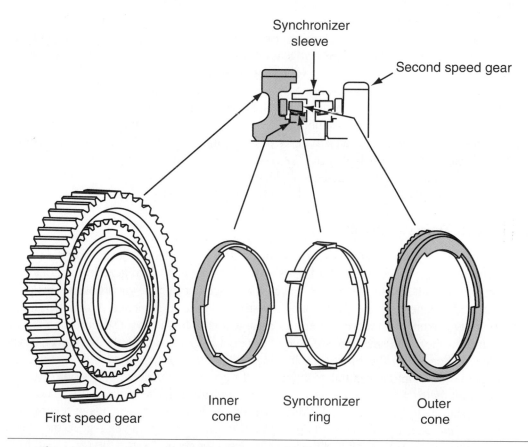

Figure 4-14 A dual cone synchronizer assembly.

same results in shifting, a single-cone synchronizer would need to have a larger diameter, which would increase the overall size and weight of the transmission.

Stepped clutch teeth. Stepped clutch teeth on the synchronizer sleeves engage the gears sequentially in the transmissions of the new Corvettes. This makes gear engagement feel lighter and more positive.

Lengthened Sleeves. Synchronizer sleeves can be lengthened, which will reduce the distance the synchronizers must travel when engaging a gear. This reduces the shift throw distance and provides for quicker gear changes.

Materials. A variety of materials are used as the friction surface on today's synchronizers. Most use sintered bronze, whereas other use carbon and molybdenum. The latter two materials are extremely durable friction surfaces that remain stable even under extreme heat. Some synchronizers are made of a paper friction material, which offers more durability and clash resistance than brass. It is common to find different materials used for the different synchronizers in a transmission. For example, in the commonly used five-speed truck transmission, the Aisin AR-5, there are sintered bronze double-cone blocker rings on the synchronizers for first and second gears, whereas third and fourth gears use carbon fiber blocker rings, and fifth and reverse gears use molybdenum on their synchronizers.

Transmission Designs

All automotive transmissions/transaxles are equipped with a varied number of forward speed gears and one reverse speed. Transmissions can be divided into groupings according to the number of forward gears (speeds) it has. For many years, the three-speed manual transmission was the most commonly used; four-speed transmissions were found only in heavy-duty or utility vehicles and high-performance cars. The growing concern for improved fuel mileage led to smaller engines with four-speed transmissions. The additional gear allowed the smaller engines to perform better by matching the engine's torque curve with vehicle speeds.

Five-speed transmissions and transaxles are now the most commonly used units. Some of the earlier units were actually four-speeds with an add-on fifth or overdrive gear. Most late-model units incorporate fifth gear in their main assemblies. This is also true of late-model six-speed transmissions and transaxles. The fifth and sixth gears of these units are part of the main gear assembly. Most often each of the two high gears in a six-speed provides an overdrive gear. Again, overdrive reduces engine speed at a given vehicle speed. This typically improves fuel economy and lowers engine noise.

Three-Speed Transmissions

In a typical three-speed transmission, one synchronizer assembly shifts first gear and reverse and another synchronizer shifts second and third gears. Actually, there is no third gear. Instead of a separate gear, the input shaft and the main shaft are coupled together by the synchronizer. This gives a 1:1 gear ratio. The benefit of directly coupling the shafts is greater efficiency. A direct coupling transmits about 99 percent of its received torque, and a gear coupling is about 96 percent efficient.

When a three-speed transmission is shifted out of neutral into first gear, the first/reverse synchronizer sleeve slides onto the first gear's locking teeth. When it is shifted into second, the first/reverse sleeve slides into its neutral position, and the second/third sleeve slides onto the second gear's locking teeth. When third is selected, the second/third sleeve moves off second gear, slides through neutral, and engages the locking teeth on the input shaft. To engage reverse, the second/third synchronizer is pulled into its neutral position. To engage reverse gear, the clutch must be disengaged to stop the turning of the input and counter gears. To stop the main shaft from turning, the vehicle must be brought to a complete stop. Then the first/reverse sleeve can be

The locking teeth on a gear are also called the gear's dog or clutch teeth.

pushed onto reverse gear's locking teeth. All shafts must be stopped because there is no synchronizing action on the reverse gear.

Four-Speed Transmissions

Fourth gear (commonly called **"high" gear**) usually is a 1:1 ratio like the third gear of a three-speed transmission. The additional gear is used to provide an extra speed ratio in the first and second gear range of the three-speed gearbox (Figure 4-15). Synchronizer assemblies are placed between first and second gears and between third and fourth gears. A third hub and sleeve assembly is added to shift the reverse gear. Some designs use a sliding reverse gear that is moved by the shift fork until it meshes with the reverse idler gear.

Five-Speed Transmissions

A five-speed transmission is usually a four-speed plus an overdrive gear. The fifth gear is added to the rear of the main shaft, near the reverse gear (Figure 4-16). The hub-and-sleeve shifting assembly used only for reverse in the four-speed transmission becomes the fifth/reverse synchronizer. The 1:1 ratio of fourth gear is usually retained, and fifth gear is made into an overdrive. Typical fifth-gear overdrive ratios range from 0.70 to 0.90:1. Such ratios greatly reduce engine rpm at freeway speeds and increase fuel mileage and engine life.

Shop Manual
Chapter 4, page 145

Six-Speed Transmissions

Six-speed transmissions are normally found only in performance cars. These units are typically based on a five-speed transmission and have an extra gearset to provide for a sixth gear. Typically fourth gear is a direct drive and fifth and sixth gears are overdrives. The use of two overdrive gears allows the manufacturers to use lower final drive gears for acceleration. The overdrive gears reduce the overall gear ratio so that engine speed is kept low at highway speeds. This allows for relatively good fuel economy. To illustrate this point, look at these gear ratios of a commonly used six-speed matched with a final drive ratio of 3.42:1.

Gear	Ratio	Overall ratio
First	2.66:1	9.10:1
Second	1.78:1	6.10:1
Third	1.30:1	4.45:1
Fourth	1.00:1	3.42:1
Fifth	0.74:1	2.53:1
Sixth	0.50:1	1.71:1

The 1.71:1 final drive ratio in sixth gear allows this engine to run at just 1400 rpm at 60 mph. This low engine speed conserves fuel and reduces engine noise at highway speeds.

The ZF S6–40 six-speed manual transmission (Figure 4-17) is used in 1988–96 Corvettes. Its design is similar to other six-speeds used in other cars. All forward gears and reverse gear are fully synchronized. Reverse gear is part of the main shaft assembly. Reverse is provided by a sliding reverse idler gear located between the countershaft and the main shaft.

1997 and newer Corvettes use a T56, which is a six-speed transmission mounted at the rear axle. This arrangement moves the weight of the transmission to the rear of the car and helps to improve vehicle weight distribution.

Overdrive Units

Overdrive gears can be an integral part of a transmission, as in a four- or five-speed transmission, or they can be add-on units. Overdrive units were popular before the widespread use of four- and

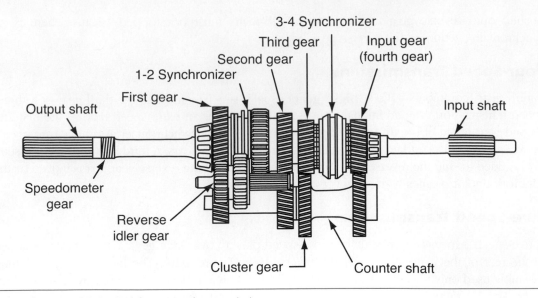

Figure 4-15 Typical four-speed transmission.

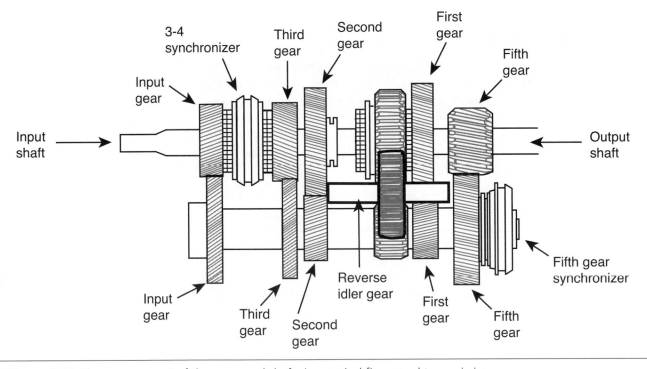

Figure 4-16 The arrangement of the gears and shafts in a typical five-speed transmission.

five-speed transmissions. They gave three-speed manual transmissions a freeway cruising gear for higher top speeds and better fuel mileage.

A BIT OF HISTORY

The first American car to use a four-speed gear box was the Locomobile in 1903. In 1934, Chrysler introduced the first transmission with overdrive.

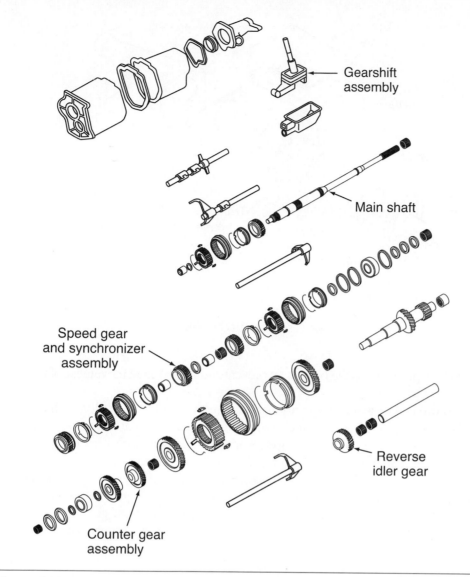

Figure 4-17 An exploded view of a ZF S6-40 six-speed transmission.

Basic Operation of Manual Transmissions

All manual transmissions function in much the same way and have similar parts. (Although transaxles operate in basically the same way as transmissions, there are enough differences to warrant a separate discussion. The operation of transaxles is explained after this section on transmissions.) Before you can completely understand how a transmission works, you must be familiar with the names, purposes, and descriptions of the major components of all transmissions.

The transmission case normally is a cast aluminum case that houses most of the gears in the transmission. The housing is shaped to accommodate the gears within the transmission. It also has machined surfaces for a cover, rear extension housing, and mounting to the clutch's bell housing (Figure 4-18). On transmissions, the clutch bell housing is a separate stamped or cast metal part, which houses the clutch and connects the transmission to the back of the engine. Most bell housings have a pivot point mounted to it where the clutch throw-out bearing lever arm is mounted. On most transaxles, the bell housing is part of the transmission housing.

The front bearing retainer is a cast-iron piece bolted to the front of the transmission case. The retainer serves several functions: It houses an oil seal that prevents oil from leaking out the input shaft and into the clutch disc area, it holds the input shaft bearing rigid, and because

Shop Manual
Chapter 4, page 125

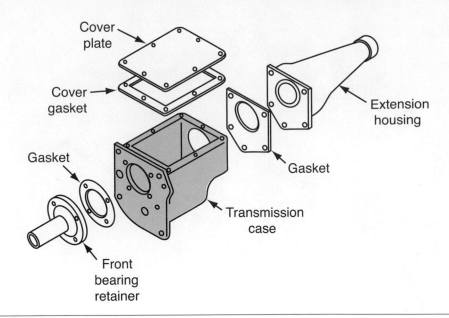

Figure 4-18 Typical transmission case with attachments.

The counter gear assembly is often referred to as the cluster gear assembly.

the input shaft is pressed into the bearing, it prevents undesired axial movement of the input shaft (Figure 4-19).

The **counter gear assembly** is normally located in the lower portion of the transmission case and is constantly in mesh with the input gear. The countershaft has several different sized gears on it that rotate as one solid assembly. Normally the counter gear assembly rotates on the countershaft using several rows of roller or needle bearings (Figure 4-20).

The speed gears are located on the main shaft (Figure 4-21). These gears are not fastened to the main shaft and rotate freely on the main shaft journals. The gears are constantly in mesh with the counter gears and are turned by the counter gears. A worm gear is machined into or pressed onto the rear of the main shaft to drive a speedometer pinion gear. The outer end of the main shaft has splines for the slip-joint yoke of the drive shaft.

A reverse gear is not meshed with the counter gear like the forward gears; rather, the **reverse idler gear** is meshed with the counter gear (Figure 4-22). Normally, reverse gear is engaged by sliding it into mesh with the reverse idler gear. The addition of this third gear causes the reverse gear to rotate in the opposite direction as the forward gears.

Most **shift forks** have two fingers that ride in the groove on the outside of the sleeve. The forks are bored to fit over shift rails. Tapered pins are commonly used to fasten the shift forks to

The counter gear assembly is a one-piece machined unit, containing first, second, third, and fourth counter gears. The fifth speed counter gear is often a separate assembly that is splined to the countershaft.

Reverse idler gear in a transmission is an additional gear that must be meshed to obtain reverse gear; it is a gear used only in reverse that does not transmit power when the transmission is in any other position.

Shift forks move the synchronizer sleeves to engage and disengage gears.

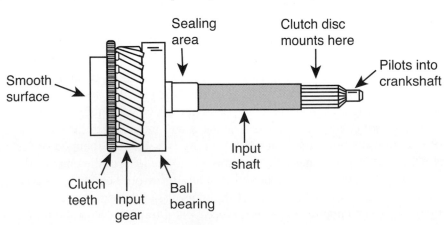

Figure 4-19 Typical input shaft assembly.

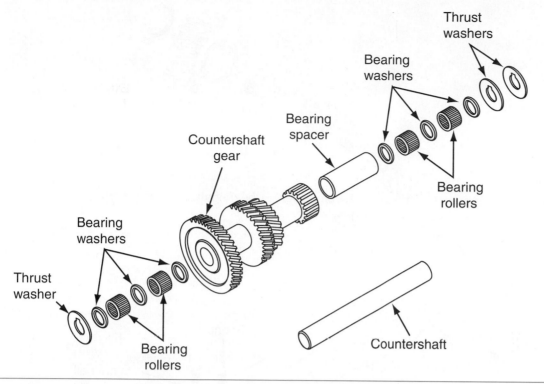

Figure 4-20 Typical countershaft assembly.

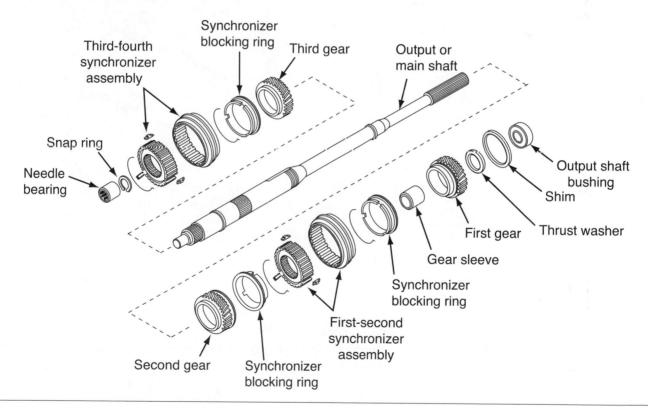

Figure 4-21 Typical main shaft assembly.

the rails (Figure 4-23). Each of the shift rails have shift lugs that the shift lever fits into. These lugs are also fastened to the rails by tapered pins. The shift rails slide back and forth in bores of the transmission case. Each shift rail has three notches, in which a spring-loaded ball or bullet rides

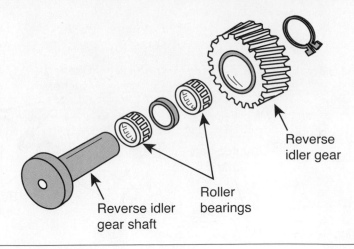

Figure 4-22 Reverse idler gear assembly.

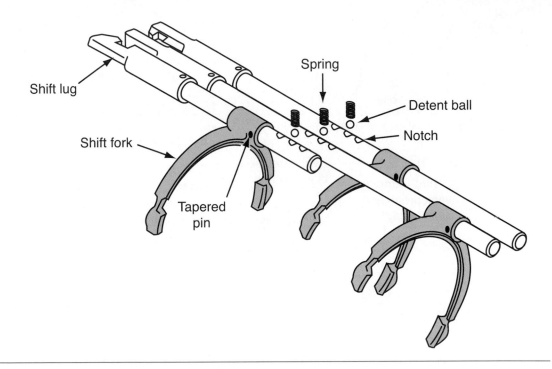

Figure 4-23 Typical shift fork and rail arrangement.

to give a detent feel to the shift lever and locate the proper position of the shift fork during gear changes. The shift rails also have notches cut in their sides for interlock plates or pins to fit in (Figure 4-24). Interlock plates prevent the engagement of two gears at the same time. The lower portion of the shifter assembly fits into the shift rail lugs and moves the shift rails for gear selection.

The arrangement of the shafts in a transmission require bearings to support them. The ends of the shafts are fitted with large roller or ball bearings pressed onto the shaft. Some transmissions with long shafts use an intermediate bearing that is fitted into the intermediate bearing housing to give added strength to the shaft (Figure 4-25). Small roller or needle bearings are often used on the countershaft, reverse idler gear shaft, and at the connection of the output shaft to the input shaft.

Operation of a Five-Speed Transmission

In a typical five-speed transmission, all five forward helical gear assemblies are in constant mesh. They are activated by the first/second speed synchronizer, the third/fourth speed synchronizer,

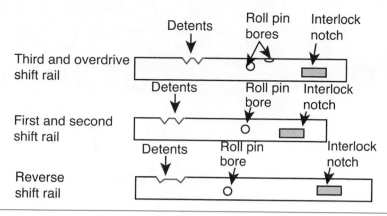

Figure 4-24 Notches in typical shift rails.

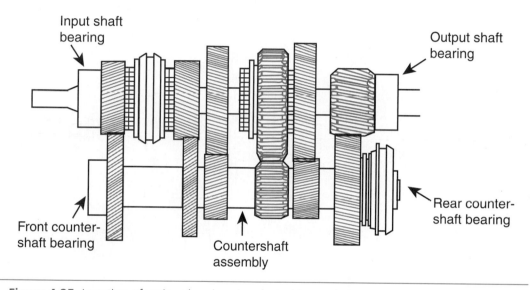

Figure 4-25 Location of various bearings in a typical transmission.

and the fifth speed synchronizer. Each synchronizer is activated by its own shift fork. All three shifter forks slide along the transmission's single shifter rail. Only the reverse idler gear slides along another rail to engage the reverse spur gear. The transmission's floor-mounted shift lever is spring loaded, so the driver must push down or pull up on the lever in order to engage reverse gear.

With the gears in constant mesh, all the gears will rotate when the input shaft is supplying input power. However, the gears will not transfer power to the main shaft until one of the synchronizers engages with a gear. If the gears are not engaged by a synchronizer, they are free-wheeling on the main shaft. The individual output shaft gears are mechanically locked to the output shaft only when the synchronizers are activated. At other times they rotate independently of the output shaft.

Power Flow in Neutral

In the neutral position (Figure 4-26), the input shaft drives the countershaft gears but no power is transferred out of the transmission. Because they are in mesh, all of the gears on the main shaft rotate, but power is not transferred to the output shaft because the synchronizers are not engaged with any of the gears.

Power Flow in First Gear

In the first gear (Figure 4-27), power enters the transmission through the input shaft and rotates the counter shaft gear. The first/second synchronizer sleeve is engaged with the dog teeth on the

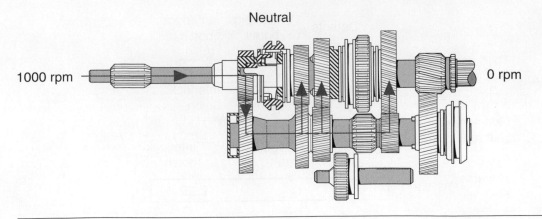

Neutral

1000 rpm →   0 rpm

Figure 4-26 Power flow through a five-speed transmission when it is in neutral.

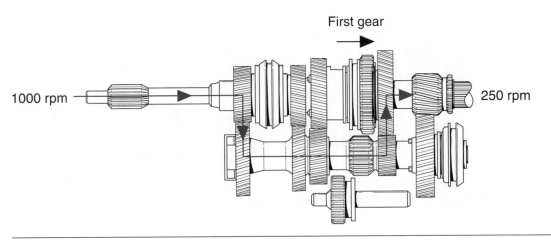

First gear

1000 rpm →   250 rpm

Figure 4-27 Power flow in first gear.

first-speed gear, locking the gear to the main shaft. The power coming in the input shaft transfers through the counter gear and up into the first gear. The gear rotates the synchronizer sleeve, which rotates the hub and main shaft for output power. All other gears mounted on the main shaft rotate freely.

Power Flow in Second Gear

In the second-gear position (Figure 4-28), the input shaft again drives the counter shaft gear. The first/second synchronizer sleeve is moved to engage with the dog teeth of the second gear, locking it to the output shaft. Power comes in through the input shaft, down to the counter gear and up to the second gear. The dog teeth of the second gear rotate the synchronizer sleeve, which rotates the hub and main shaft for output power.

Power Flow in Third Gear

The third-gear position causes the countershaft gear, which is driven by the input shaft, to be mechanically locked to the third gear on the output shaft (Figure 4-29). The third/fourth synchronizer sleeve is moved to engage with the dog teeth of the third gear. The power comes from the input shaft to the counter gear and then to the third gear. The dog teeth on the third gear rotate the synchronizer sleeve, which rotates the hub and main shaft for output power.

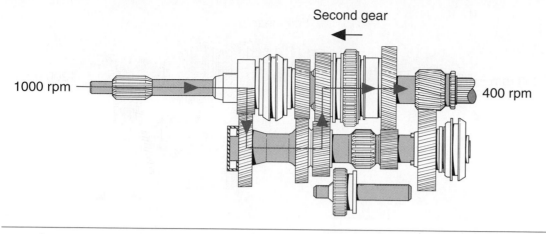

Second gear

1000 rpm

400 rpm

Figure 4-28 Power flow in second gear.

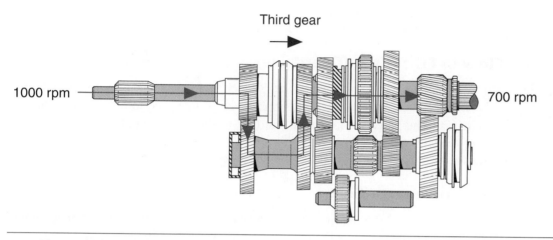

Third gear

1000 rpm

700 rpm

Figure 4-29 Power flow in third gear.

Power Flow in Fourth Gear

The fourth-gear position mechanically locks the output shaft to the input shaft (Figure 4-30). The third/fourth synchronizer sleeve is moved to engage with the dog teeth of the input gear. The power flows in from the input shaft, through the synchronizer sleeve and hub, and then through

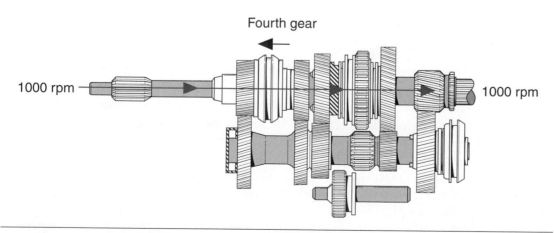

Fourth gear

1000 rpm

1000 rpm

Figure 4-30 Power flow in fourth gear.

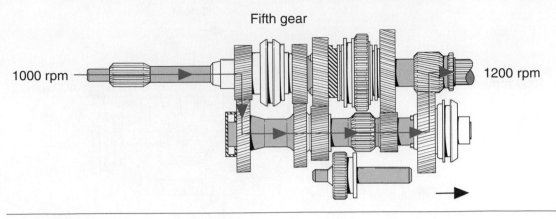

Fifth gear

1000 rpm   1200 rpm

Figure 4-31 Power flow in fifth gear.

the main shaft for output power. This directly links the two shafts, and the output shaft rotates at the same speed as the input shaft to provide for direct drive.

Power Flow in Fifth Gear (Overdrive)

The fifth-gear position causes the countershaft gear, which is driven by the input shaft, to rotate the fifth gear (Figure 4-31). The fifth-gear synchronizer sleeve is moved to engage with the dog teeth on the fifth gear. The power on the input gear transfers to the counter gear and then to the fifth gear. Power transfers through the synchronizer sleeve and hub to the main shaft for output power. As a result, the output shaft rotates at a higher speed than the input shaft.

Power Flow in Reverse Gear

Shop Manual
Chapter 4, page 153

In the reverse-gear position, if the transmission has a synchronized reverse gear, the reverse synchronizer sleeve moves to engage with the reverse gear. Power comes in through the input shaft, into the counter gear, through the reverse idler gear, and into the reverse gear on the main shaft. The reverse gear rotates the synchronizer sleeve, which rotates the hub and main shaft in a reverse direction.

If reverse is a nonsynchronized gear, a reverse-gear shift relay lever slides the reverse idler gear into contact with the counter shaft reverse gear and the reverse gear on the output shaft (Figure 4-32). The reverse idler gear causes the reverse output shaft gear to rotate counterclockwise. The counter shaft rotates counterclockwise and causes the output shaft to rotate clockwise to produce forward motion of the vehicle.

Shift rails are machined with interlock and **detent notches**. The interlock notches prevent the selection of more than one gear during operation. The detent notches give the driver a positive shift feel when engaging the speed gears.

Gearshift Linkages

There are two designs of gearshift linkages: Some are internal and others are external to the transmission. Internal linkages are located at the side or top of the transmission housing (Figure 4-33). The control end of the shifter is mounted inside the transmission, as are all of the shift controls. Movement of the shifter moves a **shift rail** and shift fork toward the desired gear and moves the synchronizer sleeve to lock the speed gear to the shaft.

As the rail moves, a detent ball moves out of its **detent notch** and drops into the notch for the selected gear. At the same time, an interlock pin moves out of its interlock notch and into the other shift rails (Figure 4-34).

External linkages function in much the same way except rods, external to the transmission, act on levers connected to the internal shift rails of the transmission.

Reverse

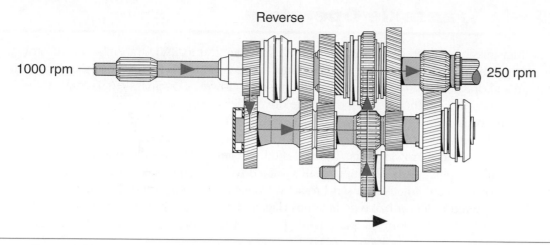

1000 rpm — 250 rpm

Figure 4-32 Power flow in reverse gear.

Figure 4-33 An external shifter assembly mounted to the transmission.

Right interlock plate is moved by the 1-2 shift rail into the 3-4 shift rail slot

The 3-4 shift rail pushes both the interlock plates outward into the slots of the 5-R and 1-2 shift rails

Right interlock plate is moved by lower tab of the left interlock plate into the 1-2 shift rail

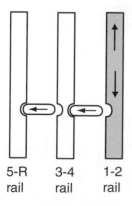

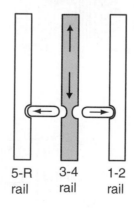

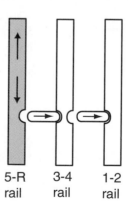

5-R rail	3-4 rail	1-2 rail
5-R rail	3-4 rail	1-2 rail
5-R rail	3-4 rail	1-2 rail

Left interlock plate is moved by lower tab of the right interlock plate into the 5-R shift rail slot

3-4 rail

Left interlock plate is moved by the 5-R shift rail into the 3-4 shift rail slot

Figure 4-34 Operation of interlock plates.

Basic Transaxle Operation

Shop Manual

Chapter 4, page 150

The transmission section of a transaxle is practically identical to RWD transmissions. It provides for torque multiplication, allows for gear shifting, and is synchronized. They also use many of the design and operating principles found in transmissions. However, a transaxle also contains the differential gear sets and the connections for the drive axles (Figure 4-35).

Transaxles normally use fully synchronized, constant mesh helical gears for all forward speeds and spur gears for reverse. To keep a transaxle compact, many designs use pressed-fit synchronizer hubs and narrower gears. Transaxles differ from RWD transmissions in that the cluster gear assembly is eliminated. The input shaft's gears drive the output shaft gears directly. The output shaft rotates according to each synchro-activated gear's operating ratio. The shafts usually ride on roller, tapered roller, or ball-type bearings (Figure 4-36). Shaft endplay is normally controlled by using thrust washers or by the placement of shims or spacers between the end plate and case.

A BIT OF HISTORY

Not many years ago, a new convenience for the driver was introduced: a column mounted gear shifter. For many years the shifter was mounted to the floor and took up valuable space in the interior of cars. By mounting the shifter on the column, that floor space was reclaimed. This trend was reversed in the 1960s and today column shifted manual transmission vehicles are a rarity.

The **input shaft** transmits the engine's power to the transmission or transaxle's gears.

The **output shaft** transmits torque from the transmission to the drive axles via a drive shaft.

Normally a transaxle has two separate shafts: an **input shaft** and an **output shaft**. The engine's torque is applied to the input shaft and the revised torque (due to the transaxle's

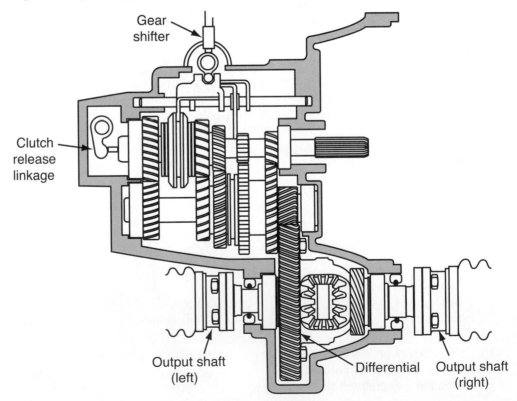

Figure 4-35 Typical transaxle assembly.

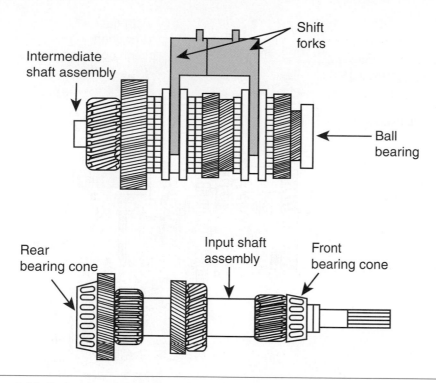

Figure 4-36 Typical transaxle input and output shafts.

gearing) rotates the output shaft. Normally the input shaft is located above and parallel to the output shaft. The main gears freewheel around the output shaft unless they are locked to the shaft by synchronizers. The main speed gears are in constant mesh with their mating gears on the input shaft and rotate whenever the input shaft rotates.

Some transaxles are equipped with an additional shaft designed to offset the power flow on the output shaft (Figure 4-37). Power is transferred from the output shaft to the third shaft using helical gears and by placing the third shaft in parallel with the output shaft and input shaft. The third shaft is added only when an extremely compact transaxle installation is required. Other transaxles with a third shaft use an offset input shaft that receives the engine's power and transmits it to a main shaft, which serves as an input shaft.

In a transaxle, the transmission and differential are both located in a lightweight housing bolted to the engine. A pinion gear is machined onto the end of the transaxle's output shaft and is in constant mesh with the differential ring gear. When the output shaft rotates, the pinion gear causes the differential ring gear to rotate. The resultant torque rotates the other differential gears, which in turn rotate the vehicle's drive axles and wheels.

Reverse is usually engaged by moving a sliding idler gear arrangement rather than a sliding synchronizer collar (Figure 4-38), although some five-speed units have synchronized gears in all forward gears and in reverse. The sliding gears have splines through their center bore and rotate at the speed of the shaft. There is no speed-matching action, so nonsynchronized gears tend to clash when shifted into mesh.

The driver changes gears by moving the shift lever to move the internal shift mechanisms of the transaxle. Movement of the shift lever is transmitted to the main shift control shaft and forks to select and engage the forward gears. An internal shift mechanism assembly transfers shift lever movements through the main shift control shaft assembly and the shifting forks (Figure 4-39). A reverse relay lever assembly is commonly used to engage the reverse idler gear with the reverse sliding gear and the input shaft.

The external shift mechanism normally consists of a floor-mounted shift lever that pivots through the shifter boot and is held in place by a stabilizer assembly and bushing (Figure 4-40).

Cable shifters are also used to decrease noise that is transmitted into the vehicle.

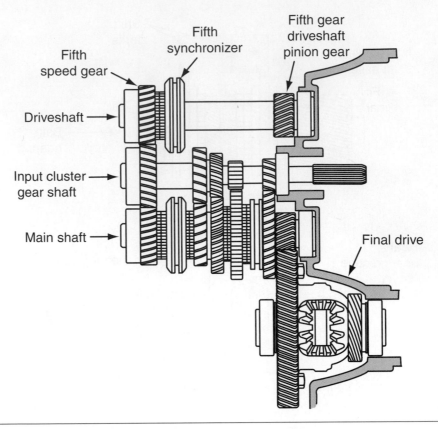

Figure 4-37 Typical five-speed transaxle with three gear shafts.

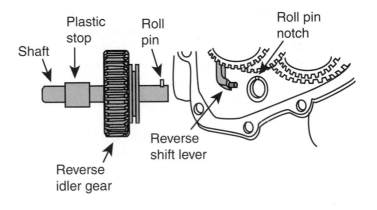

Figure 4-38 A slider-type reverse gear used in many transaxles.

Shift lever motion is transmitted to the internal shift mechanism by a shift rod that is connected to and operates the transaxle input shift shaft.

Some transaxles use a two-cable assembly to shift the gears (Figure 4-41). One cable is the transmission selector cable and the other is a shifter cable. The selector cable activates the desired shift fork and the shifter cable causes the engagement of the desired gear.

Power Flow in Neutral

When a transaxle is in its "neutral" position, no power is applied to the differential. Because the synchronizer collars are centered between their gear positions, the meshed drive gears are not

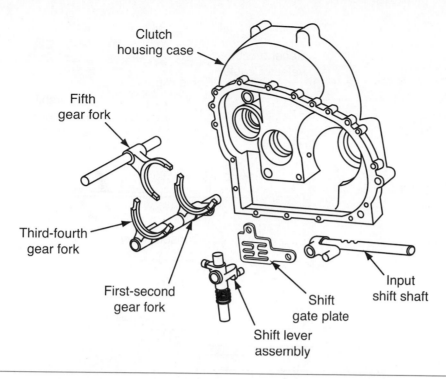

Figure 4-39 Internal shift mechanism for a typical transaxle.

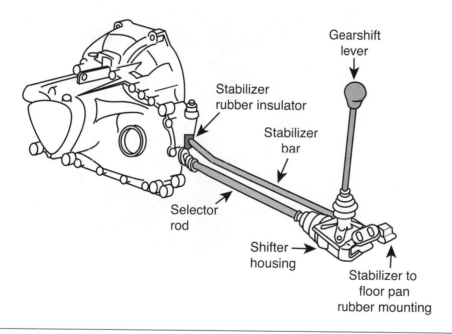

Figure 4-40 External gear shift linkage.

locked to the output shaft. Therefore, the gears spin freely on the shaft and the output shaft does not rotate.

Power Flow through Forward Gears

When first gear is selected (Figure 4-42), the first and second gear synchronizer engages with first gear. Because the synchronizer hub is splined to the output shaft, first gear on the input shaft drives its mating gear (first gear) on the output shaft. This causes the output shaft to rotate at the ratio of first gear, and to drive the differential ring gear at that same ratio.

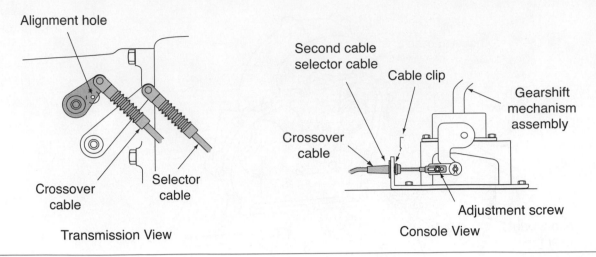

Transmission View

Console View

Figure 4-41 Typical two cable gear linkage.

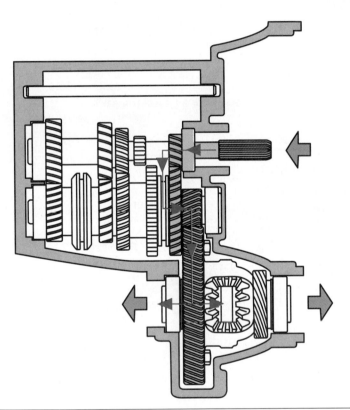

Figure 4-42 Power flow in first gear.

As the other forward gears are selected, the appropriate shift fork moves to engage the synchronizer with the gear. Because the synchronizer's hub is splined to the output shaft, the desired gear on the input shaft drives its mating gear on the output shaft. This causes the output shaft to rotate at the ratio of the selected gear and drive the differential ring gear at that same ratio (Figure 4-43).

Power Flow in Reverse

When reverse gear is selected on transaxles that use a sliding reverse gear (Figure 4-44), the shifting fork forces the gear into mesh with the input and output shafts. The addition of this third gear reverses the normal rotation of first gear and allows the car to change direction.

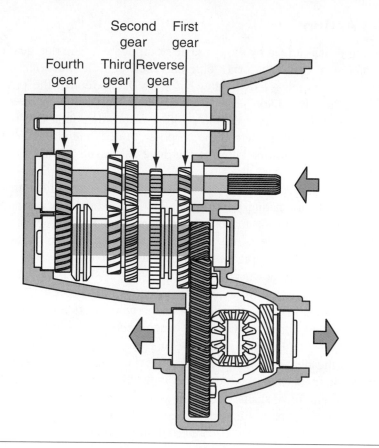

Figure 4-43 Power flow in fourth gear.

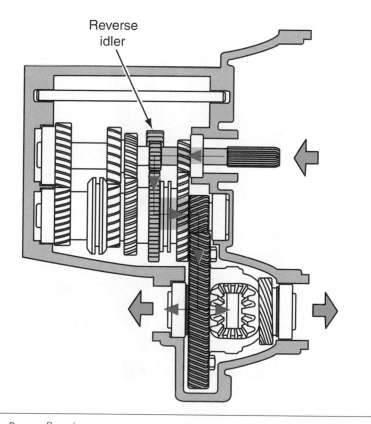

Figure 4-44 Power flow in reverse gear.

Some transaxles need the 90-degree-power-flow change in the differential. These units are used in rear-engine with RWD applications or in longitudinally positioned engines with FWD. Alfa Romeo, Audi, Subaru, Volkswagen, Chrysler LH, and Porsche are good examples of these types of transaxles.

Constant velocity (CV) joints are much like the universal joints used in the drive shaft of RWD cars. However, the speed of the driven side of the U-joint may vary with relation to the driving side, depending on the angle of the shaft. A CV joint maintains an equal speed on both sides of the joint, which helps reduce vibration and wear.

Shop Manual

Chapter 4, page 172

Differential Action

The final drive ring gear is driven by the transaxle's output shaft. The ring gear then transfers the power to the differential case. One major difference between the differential in a RWD car and the differential in a transaxle is power flow. In a RWD differential, the power flow changes 90 degrees between the drive pinion gear and the ring gear. This change in direction is not needed with most FWD cars. The transverse engine position places the crankshaft so that it already is rotating in the correct direction. Therefore, the purpose of the differential is only to provide torque multiplication and divide the torque between the drive axle shafts so that they can rotate at different speeds (Figure 4-45).

The final drive or differential case is driven by the output shaft of the transmission. The case holds the ring gear with its mating pinion gear. The differential side gears are connected to inboard constant velocity (CV) joints by splines and are secured in the case with circlips. The drive axles extend out from each side of the differential to rotate the car's wheels (Figure 4-46). The axles are made up of three pieces connected together to allow the wheels to turn for steering and to move up and down with the suspension. A short stub shaft extends from the differential to the inner CV joint. An axle shaft connects the inner CV joint and the outer CV joint. A short spindle shaft that fits into the hub of the wheels extends from the outer CV joint. To keep dirt and moisture out of the CV joints, a neoprene boot is installed over each CV joint assembly.

On many cars the right and left drive axles are different lengths. Because of this, drivers experience **torque steer**. Because the transaxle is typically placed to one side of the car, it is difficult to have equal length drive shafts. Therefore, some manufacturers use two-piece drive shafts on the longer side. The inner part of the shaft connects to the transaxle and to a bearing support. At the bearing support, the outer shaft is connected and is able to move with the suspension of the car. The outboard shaft is the same length as the shaft on the other side of the car. With the equal length drive shafts on both sides of the car, torque steer is minimized.

Torque steer is a term used to describe a pull to one side during hard acceleration. The pulling is often felt in the steering wheel.

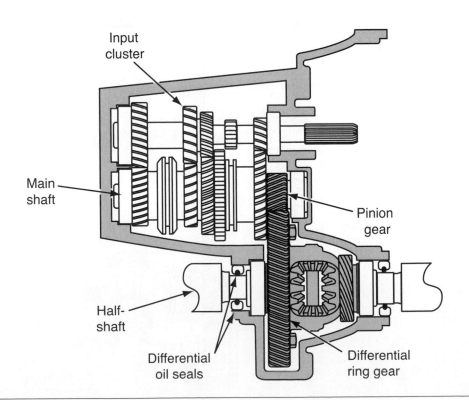

Figure 4-45 Transaxle final drive assembly.

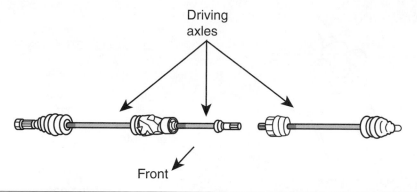

Driving
axles

Front

Figure 4-46 FWD drive axles.

A BIT OF HISTORY

The first mass-produced FWD passenger car was introduced on an Alvis Model 12/75 produced in England in 1928. The first series-produced FWD cars marketed in the United States were the Ruxton and the Cord in 1929.

Continuously Variable Transmissions (CVTs)

The **continuously variable transmission (CVT)**, which is an unconventional transmission design, may be regarded as an automatic transmission in that the driver does nothing to the gears. However, the design also is rather more mechanical than most automatic transmissions; therefore, it is sort of a manual transmission. Basically, a CVT is a transmission with no fixed forward speeds. The gear ratio varies with engine speed and load. These transmissions are, however, fitted with a one-speed reverse gear.

A CVT transmission can automatically select any desired drive ratio within its operating range. A CVT continuously selects the best overall ratio for all operating conditions. During the drive ratio changes, there is no perceptible shift. During maximum acceleration, the drive ratio is adjusted to maintain peak engine horsepower. At a constant vehicle speed, the drive ratio is set to obtain maximum fuel mileage while maintaining good driveability.

The transmissions do not have a torque converter; rather, they use a manual transmission–type flywheel with a start clutch that allows engine power to be transferred through the transmission when starting from a rest position. Instead of relying on planetary or helical gear sets to provide drive ratios, a CVT uses belts and pulleys (Figure 4-47). Several different designs have been used over the years. The torodial design uses a moveable roller between two curved metal plates. One plate is the input and the other plate is the output. By changing the angle and position at which the roller touches the plates, the drive ratio is varied. This design typically is used with engines that have a high torque output. Another CVT uses a rubber belt that transfers power between two variable width pulleys. The most common CVT uses a steel push belt running between two variable width pulleys.

Each pulley consists of a pair of cones that can be moved close together or further apart to adjust the diameter at which the belt operates (Figure 4-48). The pulley ratios are electronically controlled to select the best overall drive ratio based on throttle position, vehicle speed, and engine speed. Different speed ratios are available any time the vehicle is moving. Because the size of the drive and driven pulleys can vary greatly, vehicle loads and speeds can be changed without changing the engine's speed. With this type of transmission, attempts are made to keep the engine operating at its most efficient speed. This decreases fuel consumption and exhaust emissions.

Shop Manual
Chapter 4, page 130

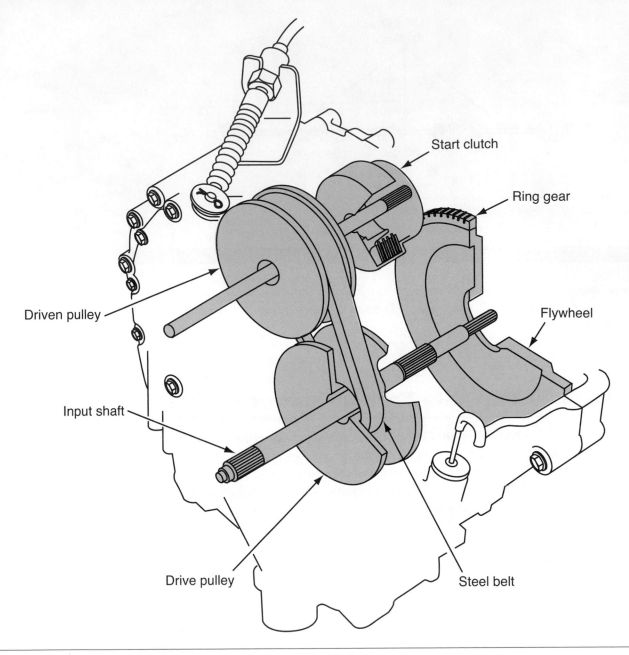

Start clutch

Ring gear

Driven pulley

Flywheel

Input shaft

Drive pulley

Steel belt

Figure 4-47 A CVT is not equipped with gears; rather, it uses pulleys and a belt.

One pulley is the driven member and the other is the drive member. Each pulley has a moveable face and a fixed face. When the moveable face moves, the effective diameter of the pulley changes. The change in effective diameter changes the effective pulley (gear) ratio. A belt links the driven and drive pulleys.

To achieve a low pulley ratio, high hydraulic pressure works on the moveable face of the driven pulley to make it larger. In response to this high pressure, the pressure on the drive pulley is reduced. Because the belt links the two pulleys and proper belt tension is critical, the drive pulley reduces just enough to keep the proper tension on the belt. The increase of pressure at the driven pulley is proportional to the decrease of pressure at the drive pulley. The opposite is true for high pulley ratios. Low pressure causes the driven pulley to decrease in size and the subsequent high pressure on the drive pulley causes it to increase in size.

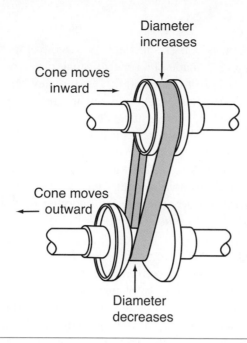

Figure 4-48 The action of the pulleys in a CVT.

Many late model CVTs are equipped with a feature that simulates the activity of a manual shifting automatic transmission. These transmissions have five or six predetermined areas that the pulleys stop in when the driver selects manual control. When in manual, these steps provide the feel and shift effect of distinct gear ratios.

CVT Controls

The operation of a CVT is controlled electronically. To illustrate these electronic controls, look at Honda's CVT. The system includes a transmission control module (TCM), various sensors, three linear solenoids, and an inhibitor solenoid. Pulley ratios are always controlled by the control system. Input from the various sensors determines which linear solenoid the TCM will activate. Activating the shift control solenoid changes the shift control valve pressure, causing the shift valve to move. This changes the pressures applied to the driven and drive pulleys, which change the effective pulley ratio. Activating the start clutch control solenoid moves the start clutch valve. This valve allows or disallows pressure to the start clutch assembly. When pressure is applied to the clutch, power is transmitted from the pulleys to the final drive gear set.

The start clutch allows for smooth starting. Because this transaxle does not have a torque converter, the start clutch is designed to slip just enough to get the car moving without stalling or straining the engine. The slippage is controlled by the hydraulic pressure applied to the start clutch. To compensate for engine loads, the TCM monitors the engine's vacuum and compares it to the measured vacuum of the engine when the transaxle was in Park or Neutral.

The TCM controls the pulley ratios to reduce engine speed and maintain ideal engine temperatures during acceleration. If the car is continuously driven at full throttle acceleration, the TCM causes an increase in pulley ratio. This reduces engine speed and maintains normal engine temperature while not adversely affecting acceleration. After the car has been driven at a lower speed or not accelerated for a while, the TCM lowers the pulley ratio. When the gear selector is placed into reverse, the TCM sends a signal to the PCM. The PCM then turns off the car's air conditioning and causes a slight increase in engine speed.

Other CVT Designs

Nissan's latest CVT uses a steel push belt. This technology was developed to enable CVTs to mate with high output engines. The belt is made of a series of small plates held in position by a cable. When torque is applied to the belt as it rotates off the drive pulley, the plates lock together and the belt becomes a solid link. As the belt begins to rotate around the driven pulley, torque is no longer applied to the belt and the belt becomes flexible.

To help in the activity of the belt, these CVTs use a special oil that helps lock the steel belt to the pulleys and lubricates and cools the transmission. Controlling the temperature of the fluid is important to this CVT; therefore, it is equipped with three transmission coolers.

This CVT has one clutch and a simple planetary gear set that are used to allow for neutral and reverse. Nissan has added a torque converter for smoother operation at very low speeds. Once the vehicle begins to move, the torque converter locks up to allow the CVT's belt and pulleys to provide all of the drive ratios.

Audi's stepless Multitronic CVT (Figure 4-49) is based on two variable pulleys and a steel belt. However, the chainlike belt is not a conventional multiplate chain with a free pivot pin design. Instead, this CVT uses adjacent rows of plates linked together with cradle type pressure pieces (oval-shaped pins). The cradle type pressure pieces are jammed between the tapered sides of the pulleys as they are pressed together. Torque is transmitted only by the frictional force between the ends of the cradle pieces and the contact faces of the pulleys.

To change the direction of the vehicle, two wet clutches are used, one for forward and one for reverse. A planetary gear set is used in conjunction with the clutches to change direction. These clutches are applied when the shifter is placed into Drive or Reverse.

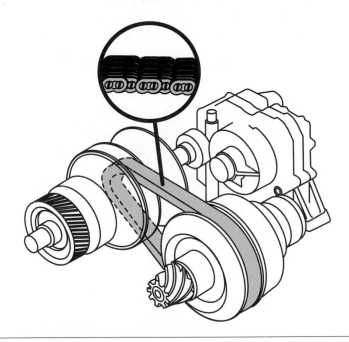

Figure 4-49 An Audi Multitronic transmission.

Summary

❏ The purpose of a transmission is to use various sized gears to keep the engine in an efficient operating range through all vehicle speeds and loads.

❏ A transmission is a system of gears that transfers the engine's power to the drive wheels of the car.

❏ A manual transaxle is a single unit composed of a transmission, differential, and final drive.

❏ Three major types of manual transmissions have been used by the automotive industry: the sliding gear, collar shift, and synchromesh designs.

❏ A synchromesh transmission is a constant mesh, collar shift transmission equipped with synchronizers.

❏ A synchronizer's primary purpose is to bring components that are rotating at different speeds to one synchronized speed.

❏ Block synchronizers consist of a hub, sleeve, blocking ring, and inserts.

❏ Synchronization occurs in three stages. In the first stage, the sleeve is moved toward the gear by the shift lever and engages the hub assembly. In the second stage, the movement of the sleeve causes the inserts to press the blocking ring onto the cone of the gear. In the third stage, the synchronizer ring completes its friction fit over the gear cone and the gear is brought up to the same speed as the synchronizer assembly. The sleeve slides onto the gear's teeth and locks the gear and its synchronizer assembly to the main shaft.

❏ Most current transmissions have five forward speeds.

❏ The front bearing retainer of a transmission serves to house an oil seal for the input shaft, holds the input shaft rigid, serves as a centering and holding fixture for the clutch release bearing, and limits the movement of the input shaft.

❏ The counter gear assembly is in constant mesh with the speed gears.

❏ The speed gears are located on the main shaft.

❏ Reverse gear is not meshed with the counter gear and is normally engaged by sliding the reverse gear into the reverse idler gear.

❏ Shift forks are moved by the gear shift and move the synchronizer sleeves to engage a gear.

❏ There are two basic types of gear shift linkages: internal and external.

❏ One of the primary differences between a transmission and a transaxle is the absence of a counter shaft and gear set in the transaxle.

❏ One major difference between the differential of a RWD car and the differential of a transverse FWD car is power flow. In a RWD differential, power flow changes 90 degrees between the drive pinion gear and the ring gear.

❏ CVTs have no fixed drive ratios and depend on pulleys and a belt to respond to operating conditions.

Terms to Know
Block synchronizers
Cluster gear
Clutch gear
Clutch hub
Collar shift
Constant mesh
Continuously variable transmission (CVT)
Counter gear assembly
Detent notches
Dog teeth
High gear
Input shaft
Main shaft
Output shaft
Reverse idler gear
Shift forks
Shift rails
Sliding gear
Synchromesh
Torque steer

Review Questions

Short Answer Essays

1. Define the purpose of a transmission.

2. What is the primary purpose of a synchronizer?

3. Describe the three stages of synchronization.

4. List the functions of a transmission's front bearing retainer.

5. How is reverse speed obtained by most transmissions?

6. Define the primary differences between a transmission and a transaxle.

7. Describe and explain the major difference between the differential of a RWD car and the differential of a FWD car.

8. Describe how a basic CVT works.

9. Describe the power flow through a transaxle when it is in first gear.

10. What is the purpose of fifth gear in most transaxles and transmissions?

Fill-in-the-Blanks

1. A transmission is a system of _____ that transfers the _____ to the _____ of the car.

2. A manual transaxle is a single unit composed of a _____ , _____ , and _____ .

3. Block synchronizers consist of a _____ , _____ , _____ _____ , and _____ .

4. The counter gear assembly is in constant mesh with the _____ _____ .

5. _____ _____ are moved by the gear shift and move the synchronizer sleeves to engage a gear.

6. The speed gears are located on the _____ .

7. Three major types of manual transmissions have been used by the automotive industry: the _____ , _____ , and _____ designs.

8. A synchromesh transmission is a _____ _____ , _____ _____ transmission equipped with synchronizers.

9. Instead of relying on planetary or helical gearsets to provide drive ratios, a CVT uses _____ and _____ .

10. Smooth and quiet shifting can only be possible when the _____ and the _____ are rotating at the same speed.

Multiple Choice

1. When discussing the operation of a synchronizer, *Technician A* says that the synchronizer inserts force the blocking ring against the conical face of the gear, which slows the speed of the gear and allows for engagement. *Technician B* says that by matching the speed of the blocking ring and the gear, the synchronizer sleeve is able to engage with the gear dog teeth.
Who is correct?
   A. A only
   B. B only
   C. Both A and B
   D. Neither A nor B

2. When discussing the construction of a transaxle, *Technician A* says that transaxles normally have two main gear shafts. *Technician B* says that most transaxles have a countershaft, an input shaft, and an output shaft.
Who is correct?
   A. A only
   B. B only
   C. Both A and B
   D. Neither A nor B

3. When discussing the causes of torque steer, *Technician A* says that it is caused by the position of the engine in the frame, which places more weight on one wheel than the other. *Technician B* says that it is caused by using right and left drive axles of different lengths.
Who is correct?
   A. A only
   B. B only
   C. Both A and B
   D. Neither A nor B

4. While discussing CVT units: *Technician A* says that the torodial design uses a moveable roller between two curved metal plates. One plate is the input and the other plate is the output. By changing the angle and position where the roller touches the plates, the drive ratio is varied. *Technician B* says that the most commonly used design has a steel push belt that runs between two variable width pulleys.
Who is correct?
   A. A only
   B. B only
   C. Both A and B
   D. Neither A nor B

5. When discussing the various types of transmissions used in automobiles, *Technician A* says that one type is the sliding gear transmission. *Technician B* says that another type uses sliding collars.
Who is correct?
   A. A only
   B. B only
   C. Both A and B
   D. Neither A nor B

6. *Technician A* says that most transmissions use two main shafts, the input and output shafts. *Technician B* says that most transmissions use three shafts, the third shaft is located below the input and output shafts.
Who is correct?
   A. A only
   B. B only
   C. Both A and B
   D. Neither A nor B

7. When discussing current trends in manual transmissions, *Technician A* says that five-speed transmissions are commonly used today. *Technician B* says that transmissions with three speeds and an additional overdrive gear are no longer used.
Who is correct?
   A. A only
   B. B only
   C. Both A and B
   D. Neither A nor B

8. *Technician A* says that in automotive transmissions, the commonly used type of synchronizer is the block type. *Technician B* says that in automotive transmissions, the commonly used type of synchronizer is the plain type.
Who is correct?
   A. A only
   B. B only
   C. Both A and B
   D. Neither A nor B

9. When discussing overdrive gears, *Technician A* says that the typical ratio is 0.70 to 0.90:1. *Technician B* says that the typical ratio is 1.0:1.
Who is correct?
   A. A only
   B. B only
   C. Both A and B
   D. Neither A nor B

10. When discussing the power flow through a five-speed transmission when it is in first gear, *Technician A* says that power enters in on the input shaft, which rotates the countershaft that is engaged with first gear. *Technician B* says that the first gear synchronizer engages with the clutching teeth of first gear and locks the gear to the main shaft, allowing power to flow from the input gear through the countershaft and to first gear and the main shaft.
Who is correct?
   A. A only
   B. B only
   C. Both A and B
   D. Neither A nor B

Front Drive Axles

Upon completion and review of this chapter, you should be able to:

❑ Explain the purposes of a FWD car's drive axles and joints.

❑ Understand and describe the different methods used by manufacturers to offset torque steer.

❑ Name and describe the different types of CV joints currently being used.

❑ Name and describe the different designs of CV joints currently being used.

❑ Explain how a ball-type CV joint functions.

❑ Explain how a tripod-type CV joint functions.

Introduction

The drive axle assembly transmits torque from the engine and transmission to drive the vehicle's wheels. FWD drive axles transfer engine torque from the transaxle differential to the front wheels. With the engine mounted transversely in the car, the differential does not need to turn the power flow 90 degrees to the drive wheels. However, on a few older FWD cars, the differential unit is a separate unit (like RWD cars) and the power flow must be turned. Both arrangements use the same type of drive axle (Figure 5-1). Some FWD cars have a longitudinally mounted transaxle.

> Between 1975 and 1979, only 4–5 percent of the total vehicle population had front wheel drive. In 2000, FWD accounts for approximately 70 percent of the vehicle population.

Drive Axle Construction

The basic FWD driveline consists of two drive shafts. The shafts extend out from each side of the differential to supply power to the drive wheels. Steel tubes or solid bars of steel are used as the drive axles on the front of four-wheel-drive and front-wheel-drive cars. In all FWD and some 4WD systems, the transaxle is bolted to the engine and the axles pivot on **CV joints** (Figure 5-2). The outer parts of the axles are supported by the steering knuckles that house the axle bearings. Steering knuckles serve as suspension components and as the attachment points for the steering gear, brakes, and other suspension parts.

> A **CV joint** is a constant velocity joint used to transfer a uniform torque and a constant speed while operating through a wide range of angles.

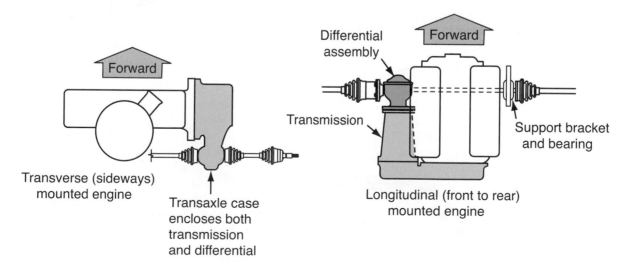

Figure 5-1 The two basic FWD engine and transaxle configurations.

FWD drive axles are also called **axle shafts**, drive shafts, and half-shafts.

The complete drive axle, including the inner and outer CV joints is typically called a half-shaft.

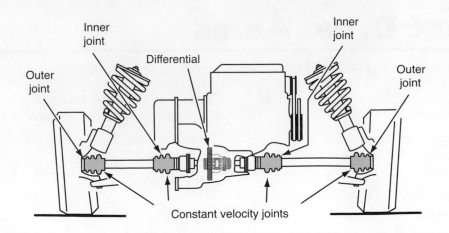

Figure 5-2 Typical FWD drive axle arrangement.

The drive axles used with FWD systems are actually made up of three pieces attached together in such a way as to allow the wheels to turn and move with the suspension. The **axle shaft** is connected to the differential by the inboard CV joint. The axle shaft then extends to the outer CV joint. A short spindle shaft runs from the outer CV joint to mate with the wheel assembly (Figure 5-3). The hub connects to the spindle shaft of the outer CV joint.

The drive shafts, on a FWD vehicle operate at angles as high as 40 degrees for turning and 20 degrees for suspension travel (Figure 5-4). Each shaft has two CV joints, an inboard joint that connects to the differential, and an outboard joint that connects to the steering knuckle. As a front wheel is turned during steering, the outboard CV joint moves with it around a fixed center. Up and down movements of the suspension system force the inboard joint to slide in and out.

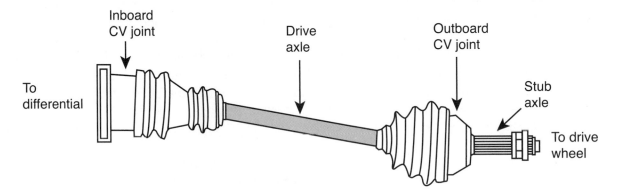

Figure 5-3 Typical FWD drive axle assembly.

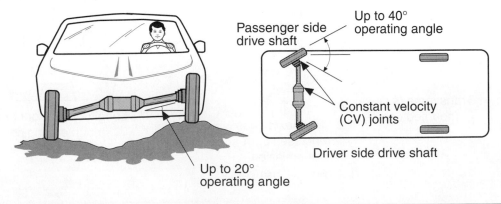

Figure 5-4 FWD drive shaft angles.

116

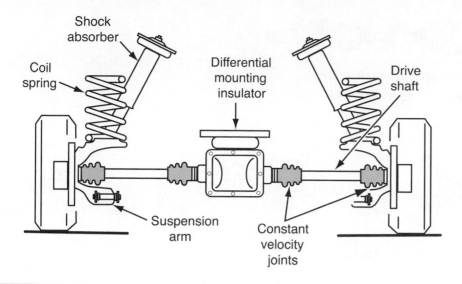

Figure 5-5 Typical IRS RWD power train with CV joints.

CV joints are used on the front axles of many 4WD vehicles as well. They have also been used on RWD buses and cars that have the engine mounted in the rear, such as Porsches. Mid engine cars with RWD, such as Pontiac Fieros and Toyota MR2s, also are equipped with CV joints. Some RWD cars with independent rear suspension (IRS), such as BMWs, Nissan 300ZXs, and Ford Thunderbirds, use CV joints on their drive axles (Figure 5-5).

The spindle shaft is commonly called the stub axle.

A BIT OF HISTORY

The first patent application in the United States for a front-wheel-drive automobile was made by George Selden in 1879. His gasoline buggy had a three-cylinder engine mounted over the front axle.

Drive Axles

With the engine mounted transversely, the transaxle sits to one side of the engine compartment. Because of this offset, one of the axles must be longer than the other (Figure 5-6). The unequal lengths result in torque steer.

Torque steer occurs when the CV joints on one drive shaft operate at different angles from those on the other shaft. The joints on the longer shaft almost always operate at less of an angle than those on the shorter drive shaft. Therefore, when the differential in a transaxle sends power out to the wheels, the longer shaft will have less resistance, and most of the engine's torque will be sent to that wheel. This is the same thing that happens when one tire has more traction than the other. The tire with the least traction will get the most power. This, of course, doesn't happen if the differential is equipped with limited-slip.

Torque steer is a term used to describe a condition in which the car tends to steer or pull in one direction as engine power is applied to the drive wheels.

Equal Length Shafts

Equal length shafts are used on most FWD cars to reduce torque steer. To provide for equal length shafts, manufacturers divide the longer shaft into two pieces (Figure 5-7). One piece comes out of the transaxle and is supported by a bearing. The other piece is made the same length as the

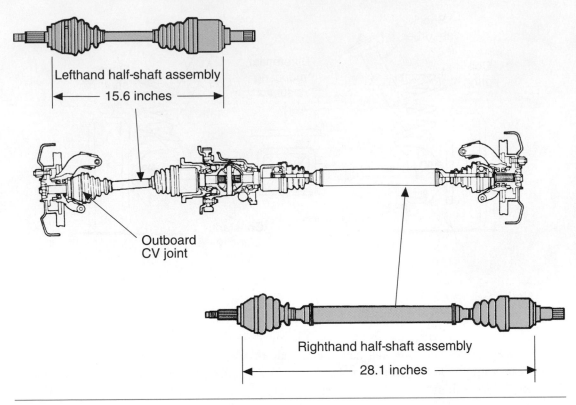

Figure 5-6 FWD drive axles of different lengths.

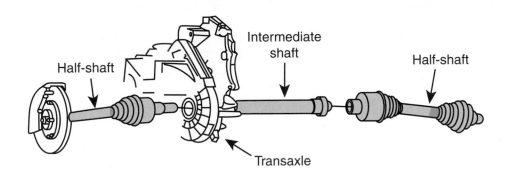

Figure 5-7 Equal length drive axles.

The inner piece of the axle shaft is normally called the **intermediate shaft** or linkshaft, which serves to equalize the lengths of the two outer drive shafts.

shorter axle on the other side of the car. The **intermediate shaft** is located between the right and left drive shafts and equalizes drive shaft length.

Vibration Dampers

A small damper weight is sometimes attached to the long half-shaft to dampen harmonic vibrations in the driveline and to stabilize the shaft as it spins, not to balance the shaft (Figure 5-8). This torsional vibration absorber is splined to the axle shaft and is often held in place by a snapring.

Unlike the drive shaft of RWD cars, FWD axle shaft balance is not very important because of the relatively slow rotational speeds. In fact, FWD drive shafts operate at about $1/3$ the rotational speed of a RWD drive shaft. This is because the shafts drive the wheels directly, with the final gear reduction already having taken place inside the transaxle's differential. The lower rotational speed has the advantage of eliminating vibrations that sometime result from high rotational speeds.

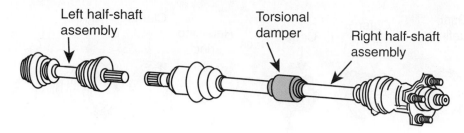

Figure 5-8 The long half shaft is sometimes fitted with a torsional damper.

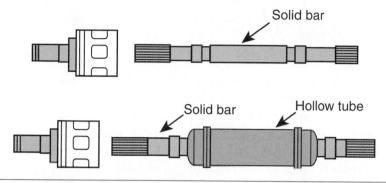

Figure 5-9 Solid and hollow drive axles.

Unequal Length Half-Shafts

If the **half-shafts** are not equal in length, the longer one is usually made thicker than the shorter one or one axle may be solid and the other tubular (Figure 5-9). These combinations would allow both axles to twist in the same amount while under engine power. If they twist unequal amounts, the car may experience torque steer.

Drive Axle Supports

The drive axles used with transaxles are the full-floating type. This type of axle does not support the weight of the vehicle; rather, all of the vehicle's weight is supported by the suspension. As the car goes over bumps, the axles move up and down, which changes their length. The inner CV joints let the axles slide in and out, changing their length, as they move up and down. The outer joints allow the steering system to turn the wheels, as well as allow for the up and down movement of the suspension.

CV Joints

One of the most important components of a FWD drive axle is the constant velocity joint (Figure 5-10). These joints are used to transfer a uniform torque and a constant speed, while operating through a wide range of angles. On FWD and 4WD cars, operating angles of as much as 40 degrees are common. The drive axles must transmit power from the engine to the front wheels that must also drive, steer, and cope with the severe angles caused by the up and down movement of the vehicle's suspension. CV joints are compact joints that allow the drive axles to rotate at a constant velocity, regardless of the operating angle.

CV joints do the same job as the **Universal joints** of front-engined RWD cars. The drive shaft of these cars is fitted with Universal joints at each end of the shaft. The joints allow the drive shaft to move with the suspension as it transfers power to the drive wheels. As the shaft rotates on an angle, the first U-joint sets up an oscillation in the drive shaft and then a second

Shop Manual
Chapter 5, page 203

A **Universal joint** is a mechanism that allows a shaft to rotate at a slight angle.

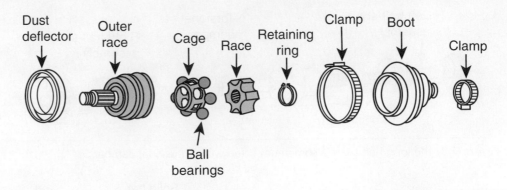

Figure 5-10 An exploded view of a CV joint.

Universal joints are commonly referred to as U-joints.

U-joint, at the other end of the shaft, cancels the oscillation before it reaches the axle. The ability of the joints to cancel the inherent oscillations lessens as the angle of the shaft increases. U-joints only work well if the shaft angle is 3 to 6 degrees. Two sets of U-joints are often used on drive shafts with greater operating angles. When the U-joints respond to the changes in the angle of the drive shaft, the speed of the shaft changes during each revolution. This change in speed causes the drive shaft to vibrate or pulse as it rotates. Constant velocity joints turn at the same speed during all operating angles and therefore can smoothly deliver power to the wheels.

A BIT OF HISTORY

In the 1920s an engineer from Ford, Alfred Rzeppa, developed a compact constant velocity joint using ball bearings between two bearing races. Through the years, this joint was further developed and improved. The improved design, similar to what is being used today, appeared on British and German FWD cars in 1959.

Shop Manual
Chapter 5, page 194

Bellows-type boots are rubber protective covers with accordion-like pleats used to contain lubricants and exclude contaminating dirt or water.

CV Boots

Bellows-type neoprene **boots** are installed over each CV joint to retain lubricant and to keep out moisture and dirt (Figure 5-11). These boots must be maintained in good condition. Each end of

The two commonly used boot clamps are the Band-it and Oetiker clamps.

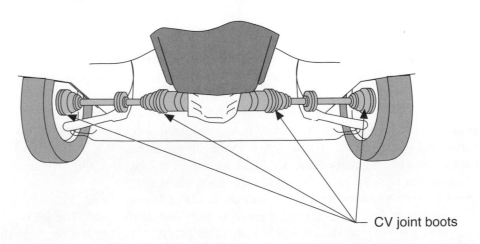

CV joint boots

If the CV joint boot keeps the joint properly sealed, the joint can last more than 100,000 miles.

Figure 5-11 Location of CV joint boots.

the boot is sealed tightly against the shaft or housing by a retaining clamp or strap (Figure 5-12). These straps may be metal or plastic and are available in many sizes and designs (Figure 5-13).

In some applications, the inboard CV joints operate very close to the engine's exhaust system. Special rubber boots of silicone or thermoplastic materials are required to withstand the temperatures. In these extremes, a special high temperature lubricant will be specified by the manufacturer.

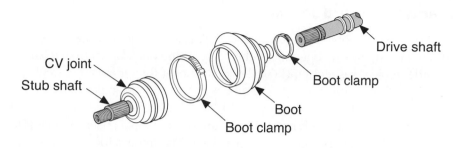

Figure 5-12 Clamps seal the boot against the CV joint and drive axle.

DESCRIPTION	APPEARANCE	TYPE
Large ladder clamp	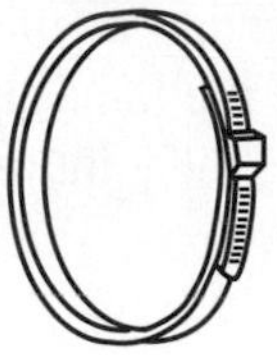	A.C.I. and G.K.N.
Small ladder clamp		A.C.I. and G.K.N.
Small strap and buckle clamp		Citroën
Large strap and buckle clamp		Citroën
Large spring clamp		A.C.I. and G.K.N.
Small rubber clamp		A.C.I. and G.K.N.

Nonpositive retention is used on the outer CV joints of all models of Ford and Chrysler cars, as well as some German-made cars.

Positive retention is used on most inner CV joints and on the outer joints of some European cars.

The outer CV joints of most Asian import cars use a modified single retention method for securing the CV joints to the shaft.

Figure 5-13 Various sizes and designs of boot clamps.

Types of CV Joints

To satisfy the needs of different applications, CV joints come in a variety of styles. The different types of joints can be referred to by position (inboard or outboard), by function (fixed or plunging), or by design (ball-type or tripod).

Inboard and Outboard Joints

In FWD drivelines, two CV joints are used on each half-shaft. The joint nearest the transaxle is the inner or **inboard joint**, and the one nearest the wheel is the outer or **outboard joint**. In a RWD vehicle with independent suspension (IRS), the joint nearest the differential can also be referred to as the inboard joint. The one closer to the wheel is the outboard joint.

The inboard joint is often called the plunge joint, and the outboard joint is called the fixed joint.

There are two basic types of outboard CV joints, the Rzeppa fixed CV joint and the fixed tripod joint. Three basic types of inboard CV joints are used: double-offset CV joint (DOJ), plunging tripod CV joint, and the cross groove plunge joint. The applications of these vary with car make and model (Figure 5-14).

CV joints are held onto the axle shafts by three different methods: nonpositive, positive, and single retention. Most inner joints use positive retention, whereas outboard joints may be held by any one of the three methods. Nonpositive retention is accomplished by the slight interference fit of the joint onto the shaft. Positively retained joints use a snapring to secure the joint to the shaft (Figure 5-15) or are splined to the shaft and retained with a nut. Single retention is accomplished by a very tight press-fit connection of the joint onto the shaft. Often, joints retained in this way cannot be removed without destroying the joint or the shaft.

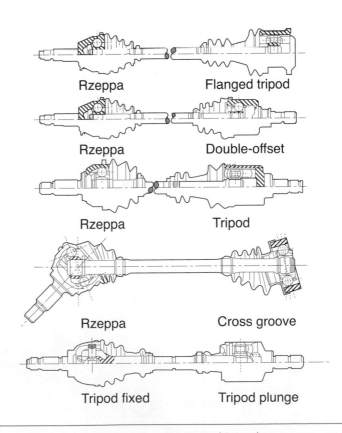

Rzeppa	Flanged tripod
Rzeppa	Double-offset
Rzeppa	Tripod
Rzeppa	Cross groove
Tripod fixed	Tripod plunge

Figure 5-14 Typical CV joint combinations on FWD drive axles.

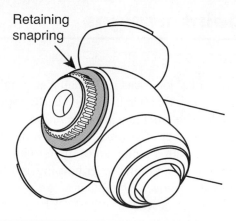

Retaining snapring

Figure 5-15 A snapring is often used to positively retain a CV joint to the drive shaft.

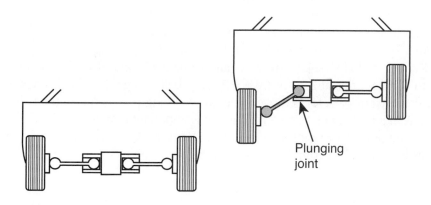

Plunging joint

Figure 5-16 Suspension movement and the resulting plunging action of an inboard CV joint.

Fixed and Plunging Joints

CV joints can also be categorized by function. They are either **fixed joints** (meaning they do not plunge in and out to compensate for changes in length) or **plunging joints** (one that is capable of in and out movement).

In response to the suspension of the car, the drive axles' effective length changes as the distance from the inboard to the outboard CV-joint changes. The inboard CV-joint must allow the drive shaft to freely move in and out of the joint housing as the front wheels go up and down (Figure 5-16).

The outboard joint is a fixed joint. Both joints do not need to plunge if one can do the job. The outboard joint must be able to handle much greater operating angles for steering than would be possible with a plunging joint.

In RWD applications with IRS, one joint on each axle shaft can be fixed and the other plunging, or both can be plunging joints. The operating angles are not as great because the wheels do not have to steer, thus plunging joints can be used at either or both ends of the axle shafts.

CV Joint Designs

CV joints are also classified by design. The two basic varieties are the ball-type and the tripod-type. Both are used as either inboard or outboard joints, and both are available in fixed or plunging designs.

Outboard CV Joint Designs

Ball-Type Joints

The most commonly used type of CV joint was named after its original designer, A. H. Rzeppa, and is based on a ball-and-socket principle (Figure 5-17). The **Rzeppa** type of outboard **joint** has its inner race attached to the axle. The inner race has several precisely machined grooves spaced around its outside diameter. The number of grooves equals the number of ball bearings used by the joint. These joints are designed with a minimum of three to a maximum of six ball bearings. The bearing cage is pressed into or is part of the outer housing and serves to keep the joint's ball bearings in place as they ride in the groove of the inner race.

When the axle rotates, the inner bearing race and the balls turn with it. The balls, in turn, cause the cage and the outer housing to turn with them (Figure 5-18). The grooves machined in the inner race and outer housing allow the joint to flex. The balls serve both as bearings between the races and the means of transferring torque from one to the other. This type of CV joint is used on almost every make and model of FWD car, except for most French designs.

If viewed from the side, the balls within the joint always bisect the angle formed by the shafts on either side of the joint regardless of the operating angle (Figure 5-19). This reduces the effective operating angle of the joint by a half and virtually eliminates all vibration problems. The cage helps to maintain this alignment by holding the balls snugly in their windows. If the cage windows become worn or deformed over time, the resulting play between the ball and window typically results in a clicking noise when turning. It is important to note that the opposing balls in a Rzeppa CV joint always work together as a pair. Heavy wear in the grooves of one ball almost always results in identical wear in the grooves of the opposing ball.

Named after its inventor, Alfred Rzeppa, Rzeppa joints are ball-type CV joints and are usually the outer joints on most FWD cars.

The term *bisect* means to divide by two.

The ball bearings are often referred to as driving balls or driving elements.

The machined grooves for the ball bearings are often referred to as tracks.

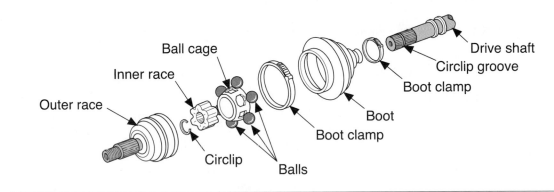

Figure 5-17 Rzeppa CV joint disassembled.

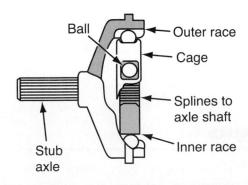

Figure 5-18 The inner race is splined to the axle shaft. The balls, placed between the ball groove and cage window, move the cage with the axle.

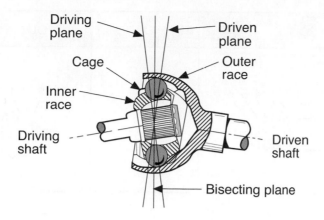

Figure 5-19 In a Rzeppa CV joint, the balls bisect the angle of the joint.

Another ball-type joint is the disc-style CV joint, which is used predominantly by Volkswagen as well as many German RWD cars. Its design is very similar to the Rzeppa joint.

Tripod-Type Joints

The fixed **tripod CV joint** uses a central hub or tripod that has three trunnions fitted with spherical rollers on needle bearings (Figure 5-20). These spherical rollers or balls ride in the grooves of an outer housing that is attached to the front wheels (Figure 5-21). Because the balls are not held

A **tripod** is the portion of an inner joint made up of rollers, needle bearings, and three arms.

A trunnion is a pivot or pin on which something can be rotated or tilted.

The outer housing of a tripod joint is called the tulip because of its three-lobed, flowerlike shape.

The hub of a tripod joint is called a spider.

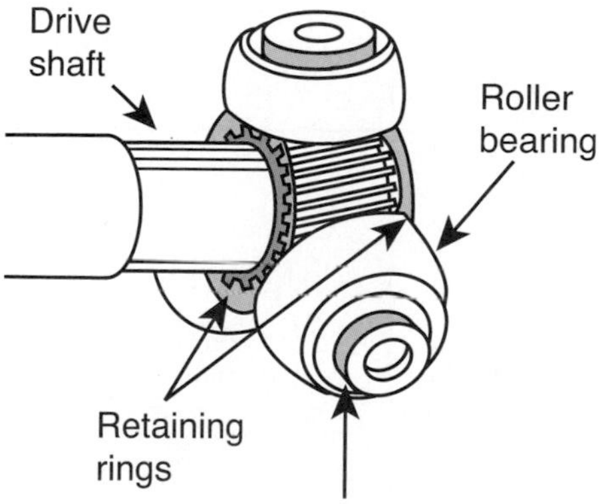

Figure 5-20 A tripod assembly

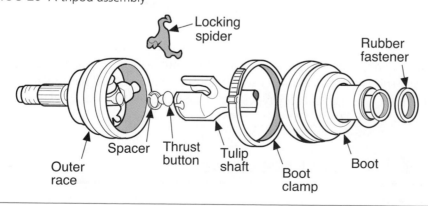

Figure 5-21 Typical tripod CV joint.

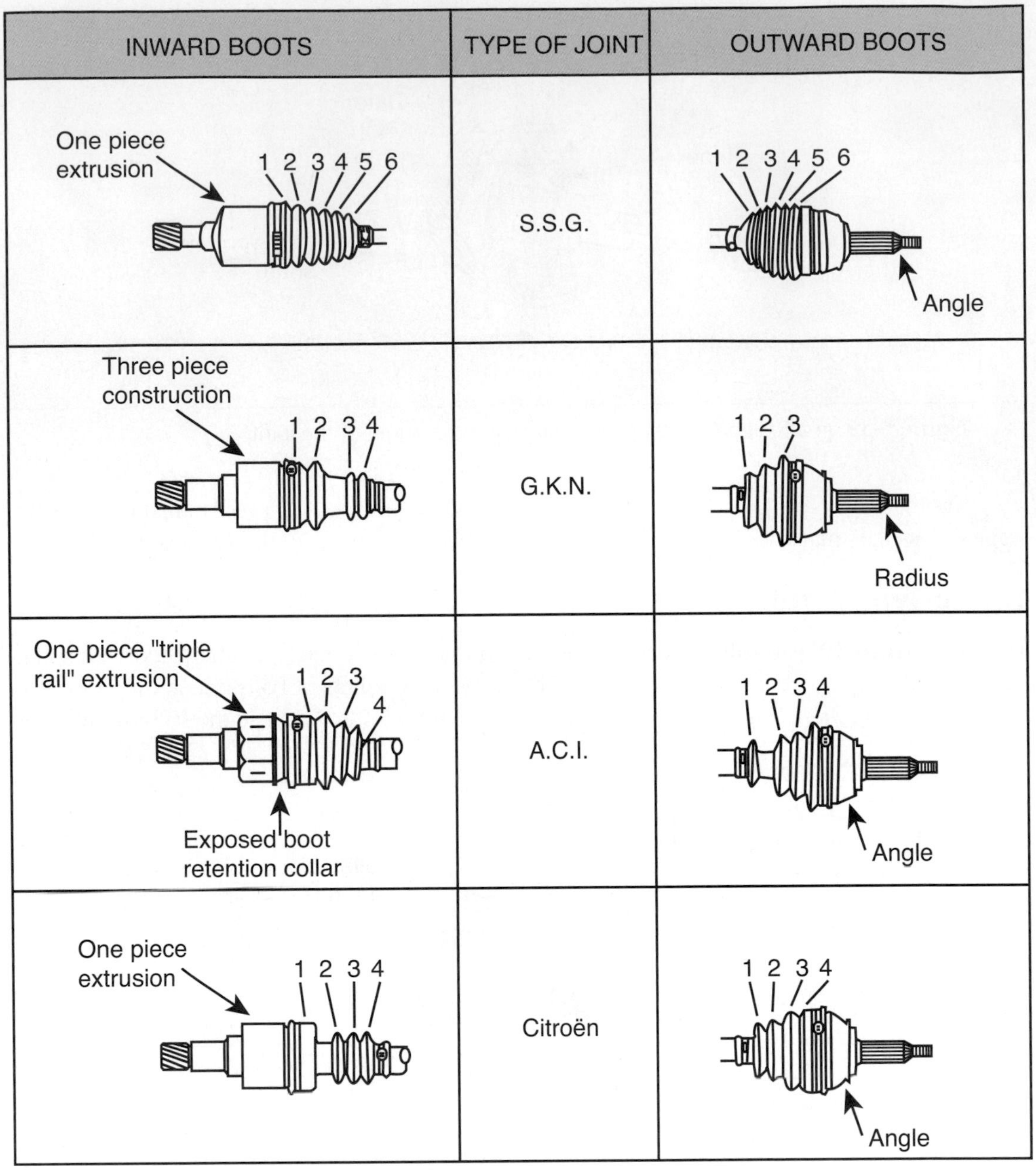

INWARD BOOTS	TYPE OF JOINT	OUTWARD BOOTS
One piece extrusion 1 2 3 4 5 6	S.S.G.	1 2 3 4 5 6 Angle
Three piece construction 1 2 3 4	G.K.N.	1 2 3 Radius
One piece "triple rail" extrusion 1 2 3 4 Exposed boot retention collar	A.C.I.	1 2 3 4 Angle
One piece extrusion 1 2 3 4	Citroën	1 2 3 4 Angle

Figure 5-22 The appearance of the joint's boot can identify the design of the joint.

in a set position in the hub, they are free to move back and forth within the hub. This allows for a constant velocity regardless of the movement of the hub as it responds to the steering or suspension system. This type of CV joint is used on most French cars and has great angular capability and is known by three subcategory types that define how they are retained: the Citroen, GKN, and ACI joints (Figure 5-22).

Inboard CV Joint Designs

Ball-Type Joints

A Rzeppa CV joint can be modified to become a plunging joint, simply by making the grooves in the inner race longer (Figure 5-23). Longer grooves in the inner race of a Rzeppa joint allow the

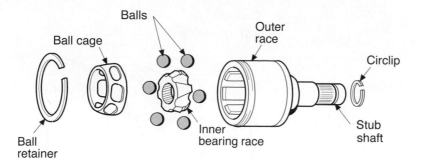

Figure 5-23 A double-offset CV joint.

bearing cage to slide in and out. This type of joint is called a **double-offset joint** and is typically used in applications that require higher operating angles (up to 25 degrees) and greater plunge depth (up to 2.4 inches). This type of joint can be found at the inboard position on some FWD half-shafts as well as on the drive shaft of some 4WD vehicles.

Like the Rzeppa joint, the **cross groove CV joint** uses six balls in a cage and inner and outer races (Figure 5-24). However, instead of the grooves in the races being cut straight, they are cut on an angle. The cross groove joint has a much flatter design than any other plunging joint. The feature that makes this joint unique is its ability to handle a fair amount of plunge (up to 1.8 inches) in a relatively short distance. The inner and outer races share the plunging motion equally so less overall depth is needed for a given amount of plunge. The cross groove joint can also handle operating angles of up to 22 degrees. Cross groove joints are commonly found on German-made cars and are used at the inboard position on FWD half-shafts or at either end of a RWD IRS axle shaft.

Tripod-Type Joints

A plunging tripod joint has longer grooves in its hub than a fixed joint allowing the spider to move in and out within the housing. On some tripod joints, the outer housing is closed, meaning the roller tracks are totally enclosed within it. On others, the tulip is open and the roller tracks are machined out of the housing (Figure 5-25). Plunging tripod-type joints are used on many American and European cars, including some Fords, Chryslers, General Motors, and French cars.

CAUTION: Many new vehicles equipped with ABS (antilock brake system) have special toothed rings fitted to the CV joint housing (Figure 5-26). These rings, called sensor rings, ABS rings, or tone wheels, provide individual wheel-speed information to the ABS computer. Careful inspection and handling procedures are required when CV service is performed to maintain proper ABS operation.

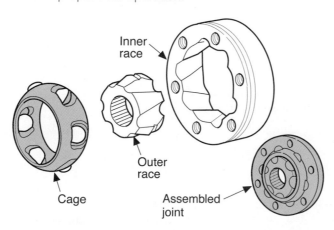

Figure 5-24 Cross groove CV joint.

The **double-offset joint** is often listed as a DOJ, and is another name for a plunging, inner CV joint.

Double-offset joints are commonly found on many Ford, General Motors, Honda, and Subaru cars.

A **cross groove CV joint** uses six balls in a cage and inner and outer races. The grooves in the races are cut on an angle instead of straight across as in the Rzeppa joint.

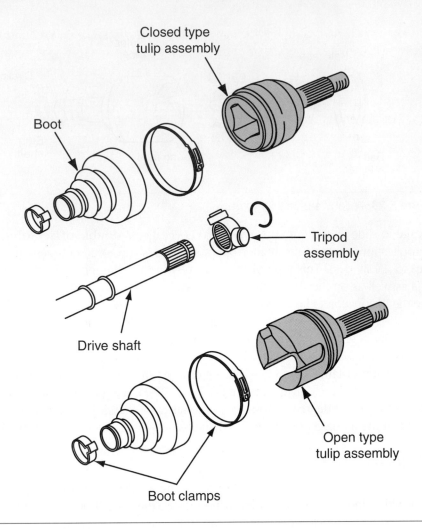

Figure 5-25 Open and closed tripod plunging joints.

The ABS speed
sensors are often
called reluctors.

Jounce is the
upward movement
of the car in
response to road
surfaces.

Rebound is the
downward
movement of the
suspension system as
it brings the car back
to normal heights
after jounce.

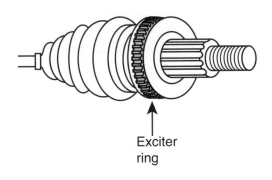

Figure 5-26 ABS speed sensor fitted to outboard joint.

CV Joint Wear

Regardless of the application, outer joints typically wear at a higher rate than inner joints, because of the increased range of operating angles to which they are subjected. Inner joint angles may change only 10 to 20 degrees as the suspension travels through **jounce** and **rebound**. Outer joints can undergo changes of up to 40 degrees in addition to jounce and rebound as the wheels are steered. That combined with more flexing of the outer boots, is why outer joints have a higher failure rate. On an average, nine outer joints are replaced for every inner joint. That does not mean you should overlook the inner joints. They wear, too. Every time the suspension travels through jounce and rebound, the inner joints must plunge in and out to accommodate the different arcs between the drive shafts and the suspension. Tripod inner joints tend to develop unique wear patterns on each of the three rollers and their respective tracks in the housing, which can lead to noise and vibration problems.

AUTHOR'S NOTE: Many students and techs forget that worn or damaged CV joints will cause handling problems. When a customer complains of a vibration, noise, or pulling, the first reaction is to check the tires, brakes, steering, and suspension systems. To save time and frustration, I suggest you include an inspection of the CV joints as part of your diagnosis of handling problems.

FWD Wheel Bearings

Shop Manual
Chapter 5, page 213

The drive axles are supported in the steering knuckle by wheel bearings (Figure 5-27). These bearings allow the axle to rotate evenly and smoothly and keep the axle in the center of the steering knuckle's hub. Basically there are two types of FWD wheel bearings. Most GM, Chrysler, and European cars use a double-row, angular-contact bearing. These units are simply two rows of ball bearings that are located next to each other. The races for these bearings are slightly offset to control radial loads during cornering. Ford and most Asian cars use opposed tapered roller bearings. Most FWD front-wheel bearings are one-piece sealed units. The unit is typically pressed into the wheel hub assembly or is held in the wheel hub by a snapring (Figure 5-28).

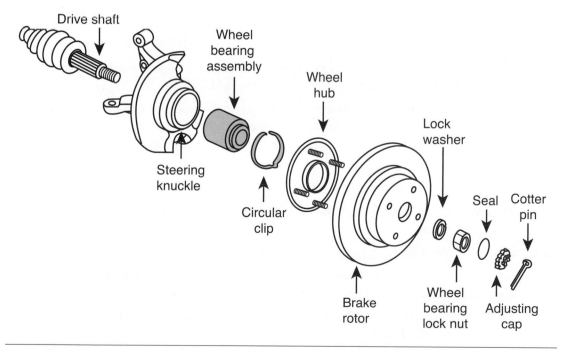

Figure 5-27 A typical FWD wheel bearing and hub assembly.

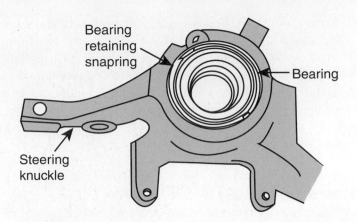

Figure 5-28 A sealed front wheel bearing that is held in the wheel hub/steering knuckle assembly by a snapring. On some cars the bearing is part of a wheel hub assembly, on others the bearing unit is pressed into the hub, and, on a few, the bearings are retained in much the same way as the front wheel bearings on a RWD car.

Summary

Terms to Know

Axle shaft

Bellows-type boots

CV joint

Cross groove CV joint

Double-offset joint

Fixed joint

Half-shaft

Inboard joint

Intermediate shaft

Jounce

Outboard joint

Plunging joint

Rebound

Rzeppa joint

Torque steer

Tripod CV joint

Universal joint

❏ The complete drive axle, including the inner and outer CV joints is typically called a half-shaft.

❏ A FWD car has two short drive shafts, one on each side of the engine, that drive the wheels and adjust for steering and suspension changes.

❏ CV joints are not only used on FWD cars; they are also used on the front axles of many 4WD vehicles. They have also been used on RWD buses and cars that have the engine mounted in the rear.

❏ *Torque steer* is a term used to describe a condition in which the car tends to steer or pull in one direction as engine power is applied to the drive wheels.

❏ Equal length shafts are used in some vehicles to help reduce torque steer. This is accomplished by making the longer side into two pieces. One piece comes out of the transaxle and is supported by a bearing. The other piece is made to the same length as the shorter side axle.

❏ A small damper weight is sometimes attached to one half-shaft to dampen harmonic vibrations in the driveline and to stabilize the shaft as it spins.

❏ If the half-shafts are not equal in length, the longer one is usually made thicker than the shorter one, or one axle may be solid and the other tubular. These combinations would allow both axles to twist the same amount when under engine power.

❏ Constant velocity joints are used to transfer a uniform torque and a constant speed when operating through a wide range of angles.

❏ Bellows-type neoprene boots are installed over each CV joint to retain lubricant and to keep out moisture and dirt.

❏ Each end of the boot is sealed tightly against the shaft or housing by a retaining clamp or strap.

❏ CV joints come in a variety of types that can be referred to by position (inboard or outboard), by function (fixed or plunging), or by design (ball-type or tripod).

❏ The CV joint nearest the transaxle is the inner or inboard joint, and the one nearest the wheel is the outer or outboard joint

❏ CV joints are held onto the axle shafts by three different methods: nonpositive, positive, and single retention.

- CV joints are either fixed (meaning they do not plunge in and out to compensate for changes in length) or plunging (one that is capable of in and out movement) joints.

- The most commonly used type of CV joint is a Rzeppa, which has its inner race attached to the axle. The inner race has several precisely machined grooves spaced around its outside diameter. The number of grooves equals the number of ball bearings used by the joint. These joints are designed with a minimum of three to a maximum of six ball bearings. The bearing cage serves to keep the joint's ball bearings in place as they ride in the groove of the inner race.

- The fixed tripod CV joint uses a central hub or tripod that has three trunnions fitted with spherical rollers on needle bearings. These spherical rollers or balls ride in the grooves of an outer housing that is attached to the front wheels.

- A Rzeppa CV joint can be modified to become a plunging joint by making the grooves in the inner race longer. Longer grooves in the inner race of a Rzeppa joint allow the bearing cage to slide in and out. This type is called a double-offset joint.

- Like the Rzeppa joint, the cross groove CV joint uses six balls in a cage and inner and outer races. However, the grooves in the races are cut on an angle rather than straight. The cross groove joint has a much flatter design than any other plunging joint.

- A plunging tripod joint has longer grooves in its hub than a fixed joint, allowing the spider to move in and out within the housing.

- Regardless of the application, outer joints typically wear at a higher rate than inner joints because of the increased range of operating angles to which they are subjected.

- The drive axles are supported in the steering knuckle by wheel bearings. These bearings allow the axle to rotate evenly and smoothly and keep the axle in the center of the steering knuckle's hub.

Review Questions

Short Answer Essays

1. Define the purpose of a CV joint.
2. Describe the difference between fixed and plunging CV joints.
3. Explain why CV joints are preferred over conventional universal joints.
4. Explain the different methods used by automobile manufacturers to offset torque steer.
5. Describe the purpose of the boot on a CV joint.
6. Describe purposes of the wheel bearing of a FWD car.
7. Describe the major differences between a Rzeppa and a tripod CV joint.
8. Explain why a fixed CV joint tends to wear much faster than a plunging joint.
9. Explain why some half-shafts are fitted with a vibration damper.
10. Explain how a ball-type CV joint is constructed.

Fill-in-the-Blanks

1. CV joints come in a variety of types that can be referred to by position

(_____ or _____), by function

(_____ or _____), or by design

(_____ or _____).

2. If the half-shafts are not equal in length, the longer one is usually made

_____ than the other, or one may be _____ and the

other _____ .

3. The major components of a Rzeppa joint are three to six _____ , an

inner _____ , and an outer _____ .

4. The major components of a tripod CV joint are the _____ , three

_____ , and an outer _____ .

5. _____ CV joints typically wear at a higher rate than the

_____ ones because of the _____ range of operating

_____ .

6. The type of joint that allows for changes in axle length is the _____ joint.

7. Half-shafts are also called _____ _____ and

_____ _____ .

8. CV joints are used to transfer _____ _____ and a

_____ _____ .

9. CV joints are held onto the axle shafts by three different methods: _____ ,

_____ , and _____ .

10. The use of an intermediate shaft allows for half-shafts of _____ .

Multiple Choice

1. When discussing CV joints, _Technician A_ says that they are called constant velocity joints because their rotational speed does not change with their operational angle. _Technician B_ says that all FWD and some RWD vehicles use CV joints on their rear drive axles.
Who is correct?
A. A only
B. B only
C. Both A and B
D. Neither A nor B

2. When discussing the types of CV joints used on FWD cars, _Technician A_ says that most use a fixed inboard joint. _Technician B_ says that all use at least one plunging joint on each axle.
Who is correct?
A. A only
B. B only
C. Both A and B
D. Neither A nor B

3. When trying to decide which is the most commonly used type of CV joint, _Technician A_ says that the ball-type is the most common. _Technician B_ says

that the Rzeppa type is the most common.
Who is correct?
A. A only
B. B only
C. Both A and B
D. Neither A nor B

4. When discussing the differences between RWD and FWD front-wheel bearings, _Technician A_ says that most FWD cars use one tapered roller bearing, whereas RWD cars use two. _Technician B_ says that most FWD front bearings are pressed onto the spindle end of the drive axle.
Who is correct?
A. A only
B. B only
C. Both A and B
D. Neither A nor B

5. When discussing Rzeppa CV joints, _Technician A_ says that the inner race of the joint is connected to the drive axle. _Technician B_ says that the ball bearings rotate on a trunnion within the outer housing.
Who is correct?

A. A only   **C.** Both A and B
B. B only   **D.** Neither A nor B

6. When discussing the types of CV joints used on FWD cars, *Technician A* says that plunging joints are most often used as outboard joints. *Technician B* says that plunging joints connect the axle with the stub axle.
Who is correct?
A. A only   **C.** Both A and B
B. B only   **D.** Neither A nor B

7. When explaining torque steer to a customer, *Technician A* says that joints on the longer axle shaft operate at less of an angle than the shorter axle shaft. *Technician B* says that the vibration damper on the half-shaft does not help correct torque steer problems.
Who is correct?
A. A only   **C.** Both A and B
B. B only   **D.** Neither A nor B

8. When discussing the purposes of the protective boot on a CV joint, *Technician A* says that the boot prevents contamination of the joint's lubricant.

Technician B says that the boot prevents the joint's lubricant from flying off when the shaft is rotating.
Who is correct?
A. A only   **C.** Both A and B
B. B only   **D.** Neither A nor B

9. *Technician A* says that a double-offset joint is similar to a Rzeppa joint. *Technician B* says that a cross groove joint is similar to a Rzeppa joint.
Who is correct?
A. A only   **C.** Both A and B
B. B only   **D.** Neither A nor B

10. When discussing the differences between the various designs of CV joints, *Technician A* says that most joints use the tripod design and all are basically interchangeable. *Technician B* says that the tripod-type joint uses needle bearings, not ball bearings like the ball type.
Who is correct?
A. A only   **C.** Both A and B
B. B only   **D.** Neither A nor B

Drive Shafts and Universal Joints

Upon completion and review of this chapter, you should be able to:

❏ Understand and describe the purpose and construction of common RWD drive shaft designs.

❏ Understand and describe the purpose and construction of common Universal joint designs.

❏ Explain the importance of drive shaft balance.

❏ Explain the natural speed variations inherent to a drive shaft.

❏ Describe the effects of canceling Universal joint angles.

CHAPTER 6

Introduction

On front engined rear-wheel-drive cars, the rotary motion of the transmission's output shaft is carried through the drive shaft to the differential, which causes the rear drive wheels to turn (Figure 6-1).

The **drive shaft** is normally made from seamless steel or aluminum tubing with Universal joint yokes welded to both ends of the shaft (Figure 6-2). To save weight, some manufacturers use epoxy-and-carbon fiber shafts. Some drive lines have two drive shafts and three Universal joints and use a center support bearing that serves as the connecting link between the two halves (Figure 6-3).

Four-wheel-drive vehicles use two drive shafts, one to drive the front wheels and the other to drive the rear wheels (Figure 6-4). FWD cars, 4WD vehicles equipped with an independent front suspension, and RWD vehicles equipped with independent rear suspension use an additional

The **drive shaft** is an assembly of one or two Universal joints connected to a shaft or tube used to transmit power from the transmission to the differential. It is also called the propeller shaft.

Although most late model cars are front-wheel drive, which are not equipped with a drive shaft and Universal joints, the best selling vehicles in America are pickup trucks, which do have a drive shaft and Universal joints. These, plus the many older cars on the road, give many opportunities to technicians trained in the diagnosis and repair of drive shafts and Universal joints.

The short drive shafts or axles used on FWD, 4WD, and RWD with IRS are called half-shafts.

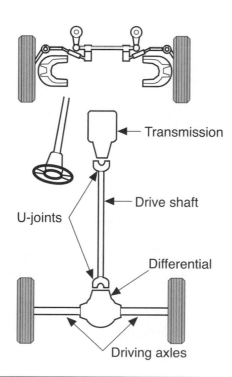

Figure 6-1 RWD drive train.

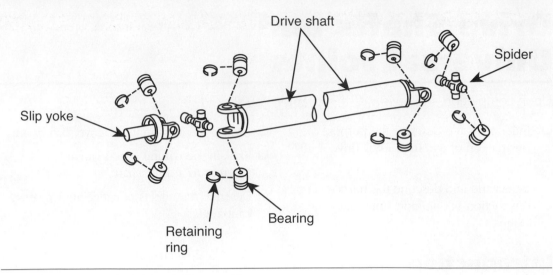

Figure 6-2 Typical drive shaft assembly.

Figure 6-3 A two-piece drive shaft is often used when there is a great distance between the transmission and the rear axle.

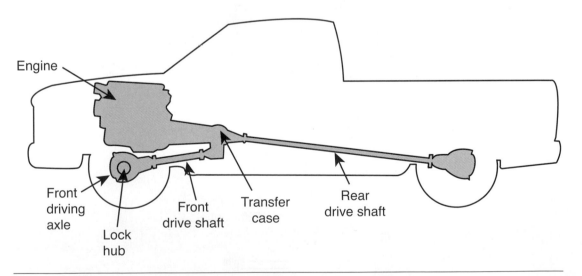

Figure 6-4 4WD truck with two drive shafts.

pair of short drive shafts. These shafts are actually the car's drive axles, which transmit the torque from the differential to each drive wheel.

The Autocar, in 1901, was the first car in the United States to use a drive shaft.

Drive Shaft Construction

Two facts must be considered when designing a drive shaft: the engine and transmission are more or less rigidly attached to the car frame and the rear axle housing, with the wheels and differential attached to the frame by springs. As the rear wheels encounter irregularities in the road, the springs compress or expand. This changes the angle of the driveline between the transmission and the rear axle housing. It also changes the distance between the transmission and the differential.

In order for the drive shaft to respond to these constant changes, drive shafts are equipped with two or more universal joints that permit variations in the angle of the shaft, and a slip joint that permits the effective length of the driveline to change.

A drive shaft is actually an extension of the transmission's output shaft, as its sole purpose is to transfer torque from the transmission to the drive axle assembly (Figure 6-5). It is usually made from seamless steel or aluminum tubing with a yoke welded or pressed onto each end, which provides a means of connecting two or more components together.

The drive shaft, like any other rigid tube, has a natural vibration frequency. This means that if one end of the tube were held tightly, the tube would vibrate at its own frequency when it is deflected and released or when it rotates. It reaches this natural frequency at its critical speed. The critical speed of a drive shaft depends on the diameter of the tube and the length of the drive shaft. Drive shaft diameters are as large as possible and shafts as short as possible to keep the critical speed frequency above the normal driving range.

Shop Manual
Chapter 6, page 236

In normal automotive terms, suspension spring compression is called jounce.

The expansion of a suspension spring is suspension rebound.

Critical speed is the rotational speed at which an object begins to vibrate as it turns. This is mostly caused by centrifugal forces.

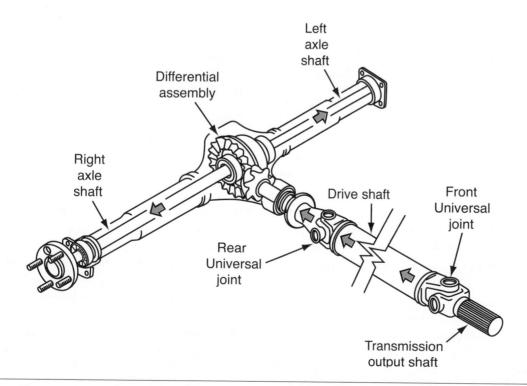

Figure 6-5 Power flows from the transmission to the rear axle.

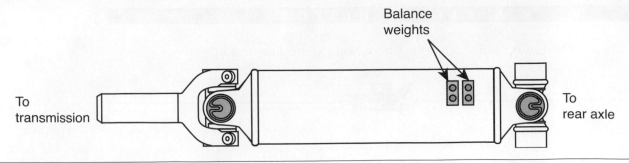

Balance weights

To transmission

To rear axle

Figure 6-6 Location of the balance weights on a drive shaft.

A drive shaft rotates at three to four times the speed of the car's tires.

Linear stiffness means the shaft will resist deflection regardless of length.

Because the drive shaft rotates at high speeds and at varying angles, it must be balanced to reduce vibration. As the drive shaft's length and the operating angle and speed increase, the necessity for balance also increases. Several methods are used to balance drive shafts. One of the most common techniques employed by manufacturers is to balance the drive shaft by welding balance weights to the outside diameter of the drive shaft (Figure 6-6).

To reduce the effects of vibrations and the resulting noises, manufacturers have used various methods to construct a drive shaft. An example of this is a drive shaft with cardboard liners inserted into the tube, which serve to decrease the shaft's vibration by damping the vibrations.

Drive shaft performance has also been improved by placing biscuits between the drive shaft and the cardboard liner. These biscuits are simply rubber inserts that reduce noise transfer within the drive shaft.

Another drive shaft design is the tube-in-tube, in which the input driving yoke has an input shaft that fits inside the hollow drive shaft. Rubber inserts are bonded to the outside diameter of the input shaft and to the inside diameter of the drive shaft. This design reduces the noise associated with drive shafts when they are stressed with directional rotation changes and also greatly reduces the vibration.

Recently, drive shafts have begun to be made with aluminum tubing or fiber composites. These composites give the shaft linear stiffness, while the positioning of the fibers provide for torsional strength. The advantages of a fiber composite drive shaft are weight reduction, torsional strength, fatigue resistance, easier and better balancing, and reduced interference from shock loading and torsional problems.

AUTHOR'S NOTE: Common causes of driveline vibrations are bad U-joints and missing balance pads on the drive shaft. While checking the U-joints, check the drive shaft for the balance pads. Nearly all drive shafts have them so if you don't find them, that is probably the reason for the vibration.

Types of Drive Shafts

The **torque tube** is a fixed tube over the drive shaft on some cars. It helps locate the rear axle and takes torque reaction loads from the drive axle so the drive shaft will not sense them.

Three types of drive shafts have been used in automobiles. The first type, and the most commonly used, is the **Hotchkiss drive**. This type of drive shaft is readily recognized by its external shaft and U-joints (Figure 6-7). These shafts are either one- or two-piece assemblies consisting of a shaft with U-joints attached to each end. A Hotchkiss drive system can be used with either leaf or coil springs. When it is used with coil springs, additional braces, called control arms, must be used to control the movement of the rear drive axle (Figure 6-8).

A two-piece drive shaft is used on many long wheelbase vehicles. It uses a third U-joint between the two shaft sections and a center bearing to support the middle of the shaft assembly (Figure 6-9).

The second type of drive shaft is called a **torque tube**. Vehicles with independent rear suspension and a rear-mounted transaxle—such as late model Corvettes and some Japanese RWD cars—use a torque tube. On these cars, the torque tube is rigidly connected at both ends. The

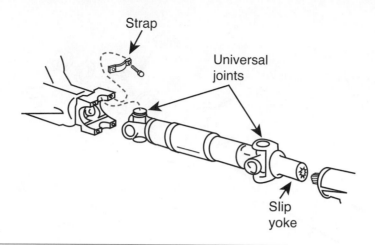

Figure 6-7 Hotchkiss drive.

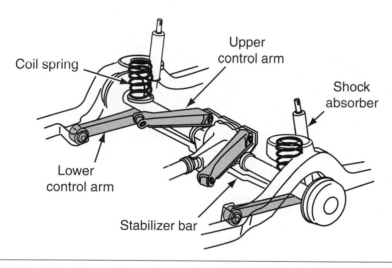

Figure 6-8 Hotchkiss drive with a coil spring suspension.

A **Hotchkiss drive** system actually describes the entire drive shaft and rear axle assembly. This system allows the rear axle to move with the suspension while allowing torque to be transferred from the transmission to the rear axle. It is commonly called an open drive shaft system.

A two-piece drive shaft is also referred to as a split drive shaft.

A torque tube system is commonly called an enclosed drive shaft system.

The joint's cross is often called its spider.

The arms of the joint are also called trunnions.

rotating inner drive shaft does not need universal joints because the transaxle location never changes relative to engine location (Figure 6-10). The Chevrolet Chevette used a two-piece drive shaft in which the forward section was a tubular shaft with universal joints at each end and the rear section was a torque tube that was rigidly mounted to the differential.

The third and least commonly used type of drive shaft is the flexible type. This shaft is actually a flexible steel rope, much like an oversized speedometer cable that does not use universal joints between the engine and the rear-mounted transaxle. The 1961–1963 Pontiac Tempest models used this type of drive shaft.

Universal Joints

A drive shaft must smoothly transfer torque while rotating, changing length, and moving up and down. The different designs of drive shafts all attempt to ensure a vibration-free transfer of the engine's power from the transmission to the differential. This goal is complicated by the fact that the engine and transmission are bolted solidly to the frame of the car while the differential is mounted on springs. As the rear wheels go over bumps in the road or changes in the road's

Shop Manual
Chapter 6, page 241

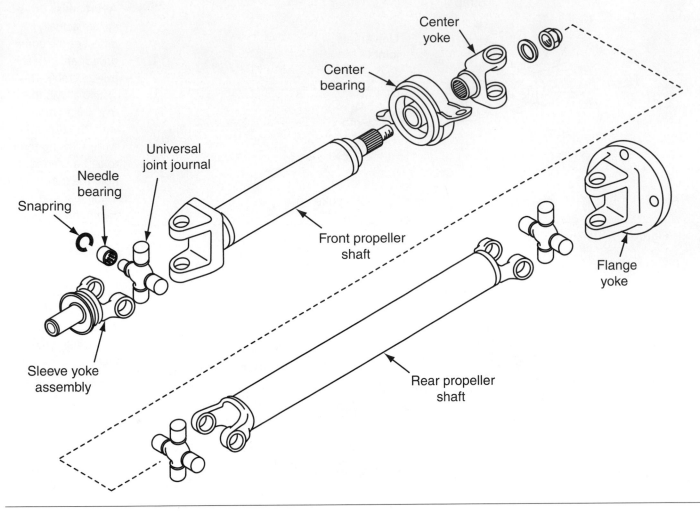

Figure 6-9 A two-piece drive shaft assembly.

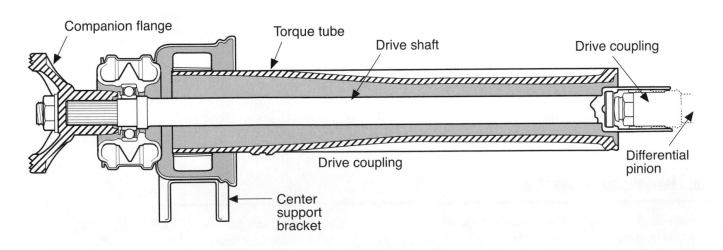

Figure 6-10 A typical torque tube assembly.

surface, the springs compress or expand. This changes the angle of the drive shaft between the transmission and the differential, as well as the distance between the two (Figure 6-11). To allow for these changes, the Hotchkiss-type drive shaft is fitted with at least two U-joints to permit variations in the angle of the drive and a slip joint that permits the effective length of the drive shaft to change. The Universal joint is basically a double-hinged joint consisting of two Y-shaped yokes, one on the driving or input shaft and the other on the driven or output shaft, plus a cross-shaped unit called the cross (Figure 6-12). A **yoke** is used to connect the U-joints together. The four arms of the cross are fitted with bearings in the ends of the two shaft yokes. The input shaft's yoke causes the cross to rotate, and the two other trunnions of the cross cause the output shaft to rotate. When the two shafts are at an angle to each other, the bearings allow the yokes to swing around on their trunnions with each revolution. This action allows two shafts, at a slight angle to each other, to rotate together.

The joint's **yoke** is a Y-shaped assembly into which two of the joint's arms fit.

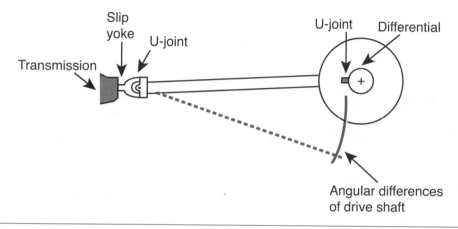

Figure 6-11 An illustration showing the changes in the length and angle of a drive shaft.

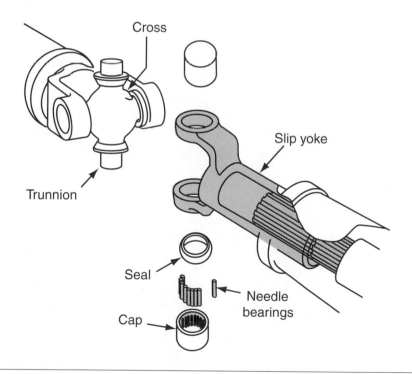

Figure 6-12 An exploded view of a Cardan U-joint.

Universal joints allow the drive shaft to transmit power to the rear axle through varying angles that are controlled by the travel of the rear suspension. Because power is transmitted on an angle, U-joints do not rotate at a constant velocity nor are they vibration free.

The original Universal joint was developed in the sixteenth century by a French mathematician named Cardan. In the seventeenth century, Robert Hooke developed a cross-type Universal joint, based on the Cardan design. Then in 1902, Clarence Spicer modified Cardan and Hooke's inventions for the purpose of transmitting engine torque to an automobile's rear wheels. By joining two shafts with Y-shaped forks to a pivoting cruciform member, the problem of torque transfer through a connection that also needed to compensate for slight angular variations was eliminated. Both names, Spicer and Hooke, are at times used to describe a Cardan U-joint.

Speed Variations

Although simple in appearance, a U-joint is more intricate than it seems. Its natural action is to cause the shaft it is connected to, to speed up and slow down twice during each revolution, while operating at an angle. The amount that the speed changes varies according to the steepness of the universal joint's angle.

As a U-joint transmits torque through an angle, its output shaft speed increases and decreases twice on each revolution. These speed changes are not normally apparent, but may be felt as torsional vibration due to improper installation, steep and/or unequal operating angles, and high speed driving.

If a U-joint's input shaft speed is constant, the speed of the output shaft accelerates and decelerates to complete a single revolution at the same time as the input shaft. In other words, the output shaft falls behind, then catches up with the input shaft during this revolution. The greater the angle of the output shaft, the more the velocity will change each shaft revolution.

U-joint **operating angle** is determined by the difference between the transmission **installation angle** and the drive shaft installation angle (Figure 6-13). When the U-joint is operating at an angle, the driven yoke speeds up and slows down twice during each drive shaft revolution. This acceleration and deceleration of the U-joint is known as speed variation.

These four changes in speed are not normally visible during rotation, but may be understood after examining the action of a U-joint. A universal joint serves as a coupling between two shafts that are not in direct alignment. It would be logical to assume that the entire unit simply rotates. This is only true of the joint's input yoke.

The output yoke's rotational path looks like an **ellipse** because it can be viewed at an angle instead of straight on. This same effect can be obtained by rotating a coin with your fingers. The height of the coin stays the same even though the sides seem to get closer together.

The **operating angle** of a drive shaft is the amount the drive shaft deviates from the horizontal plane.

The **installation angle** of an object describes how far the object is tilted away from the horizontal plane.

An **ellipse** is merely a compressed form of a circle.

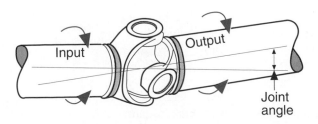

Figure 6-13 Universal joint angles.

This might seem to be merely a visual effect, however it is more than that. The U-joint rigidly locks the circular action of the input yoke to the elliptical action of the output yoke. The result is similar to what would happen when changing a clock face from a circle to an ellipse.

Like the hands of a clock, the input yoke turns at a constant speed in its true circular path. The output yoke, operating at an angle to the other yoke, completes the path in the same amount of time (Figure 6-14). However, its speed varies, or is not constant, compared to the input.

Speed variation is more easily visualized when looking at the travel of the yokes by 90-degree quadrants (Figure 6-15). The input yoke rotates at a steady or constant speed through a complete 360-degree rotation. The output yoke quadrants alternate between shorter and longer distances of travel than the input yoke quadrants. When one point of the output yoke covers the shorter distance in the same amount of time, it must travel at a slower rate. Conversely, when traveling the longer distance in the same amount of time, it must move faster.

Because the average speed of the output yoke through the four 90-degree quadrants equals the constant speed of the input yoke during the same revolution, it is possible for the two mating yokes to travel at different speeds. The output yoke is falling behind and catching up constantly. The resulting acceleration and deceleration produces fluctuating torque and torsional vibrations and is characteristic of all Cardan U-joints. The steeper the U-joint angle, the greater the speed fluctuations. Conversely, smaller angles produce less change in speed.

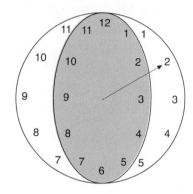

Figure 6-14 The face of a clock can be used to illustrate the elliptical action of the drive shaft's yokes.

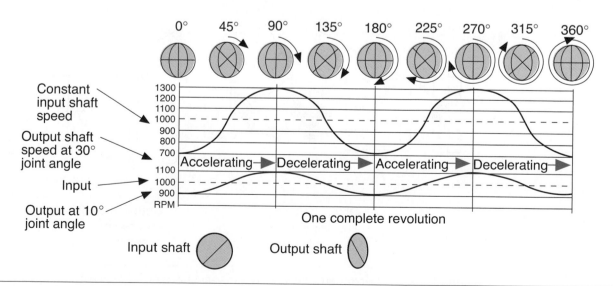

Figure 6-15 A chart showing the speed variations of a drive shaft's yoke.

Shop Manual
Chapter 6, page 253

In-phase describes the condition in which two events happen one after the other, regardless of speed.

Phasing of Universal Joints

The torsional vibrations set up by the changes in velocity are transferred down the drive shaft to the next U-joint. At this joint, similar acceleration and deceleration occurs. Because these speed changes take place at equal and reverse angles to the first joint, they cancel out each other whenever both occur at the same angle. To provide for this canceling effect, drive shafts should have at least two U-joints and their operating angles must be slight and equal to each other. Speed fluctuations can be canceled if the driven yoke has the same point of rotation, or same plane, as the driving yoke. When the yokes are in the same plane, the joints are said to be **in-phase** (Figure 6-16).

 SERVICE TIP: On a two-piece drive shaft, you may encounter problems if you are not careful. The center U-joint must be disassembled to replace the center support bearing. The center driving yoke is splined to the front drive shaft. If the yoke's position on the drive shaft is not indicated in some manner, the yoke could be installed in a position that is out of phase. Manufacturers use different methods of indexing the yoke to the shaft. Some use aligning arrows. Others machine a master spline that is wider than the others. When there are no indexing marks, the technician should always index the yoke to the drive shaft before disassembling the U-joint. This saves time and frustration during reassembly. Indexing requires only a light hammer and center punch to mark the yoke and drive shaft.

Shop Manual
Chapter 6, page 253

Canceling angles
occur when the opposing operating angles of two Universal joints cancel the oscillations developed by the individual Universal joint.

Canceling Angles

Oscillations, resulting from speed variations, can be reduced by using **canceling angles** (Figure 6-17). The operating angle of the front U-joint is offset by the one at the rear of the drive shaft. When the front U-joint accelerates, causing a vibration, the rear U-joint decelerates causing an equal but opposite vibration. These vibrations created by the two joints oppose each other and dampen the vibrations from one to the other. The use of canceling angles provides smooth drive shaft operation.

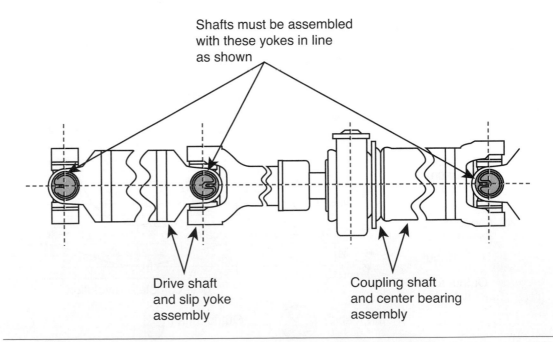

Figure 6-16 The U-joints are in the same plane and are in phase with each other.

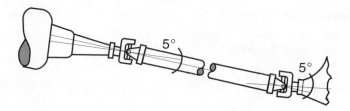

Figure 6-17 Equal U-joint angles reduce the vibrations of the shaft.

The correct operating angle of a U-joint must be maintained in order to prevent driveline vibration and damage. Shimming of leaf springs and the control arms on coil spring suspensions or adjusting the control arm eccentrics allow the operating angle of the drive shaft to be changed. Shimming at the transmission mount can also be done on some vehicles to change Universal joint angles.

Types of Universal Joints

There are three common designs of Universal joints: single Universal joints retained by either an inside or outside snapring, coupled Universal joints, (commonly called **double Cardan CV joints**) and Universal joints held in the yoke by U-bolts or lock plates.

Single Universal Joints

The single **Cardan/Spicer Universal joint's** (Figure 6-18) primary purpose is to connect the two yokes that are attached directly to the drive shaft. The joint assembly forms a cross, with four machined trunnions or points equally spaced around the center of the axis. Needle bearings used to reduce friction and provide smoother operation are set into bearing cups. The trunnions of the cross fit into the cup assemblies, which fit snugly into the driving and driven Universal joint yokes. U-joint movement takes place between the trunnions, needle bearings, and bearing cups. There should be no movement between the bearing cup and its bore in the Universal joint yoke. The bearings are usually held in place by snaprings that drop into grooves in the yoke's bearing bores. The bearing caps allow free movement between the trunnion and yoke. The needle bearing caps also may be pressed into the yokes, bolted to the yokes, or held in place with U-bolts or metal straps.

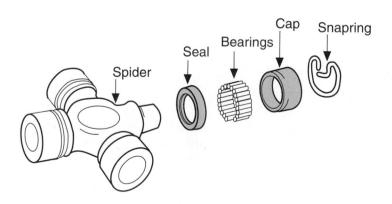

Figure 6-18 A Cardan U-joint.

Shop Manual
Chapter 6, page 241

Coupled Universal joints are commonly called **double Cardan CV joints**. A Double Cardan CV joint is actually two Cardan joints joined together by a yoke. By joining the two Cardan joints, rotational speed is not changed as the joint moves through its operating angles.

The **Cardan/Spicer Universal joint** is also known as the cross or four-point joint. Its primary purpose is to connect the two yokes that are attached directly to the drive shaft.

SERVICE TIP: There are many other methods used to retain the U-joint in its yoke, such as the use of a bearing plate, thrust plate, wing bearing, cap and bolt, U-bolt, and strap (Figure 6-19).

There are other styles of single Universal joints. The method used to retain the bearing caps is the major difference between these designs. The Spicer style uses an outside snapring that fits into a groove machined in the outer end of the yoke (Figure 6-20). The bearing cups for this style are machined to accommodate the snapring.

The mechanics style uses an inside snapring or C-clip that fits into a groove machined in the bearing cup on the side closest to the grease seal (Figure 6-21). When installed, the clip rests against the machined portion of the yoke. The snaprings are retained by spring tension against the retaining ring grooves. Some joints have nylon injected above the bearing cap to retain it. The nylon is heated to remove the joint.

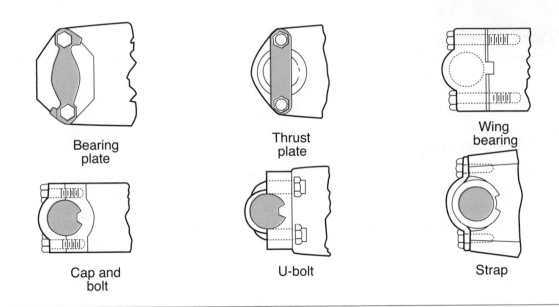

Bearing
plate

Thrust
plate

Wing
bearing

Cap and
bolt

U-bolt

Strap

Figure 6-19 Various methods used to retain U-joints in their yokes.

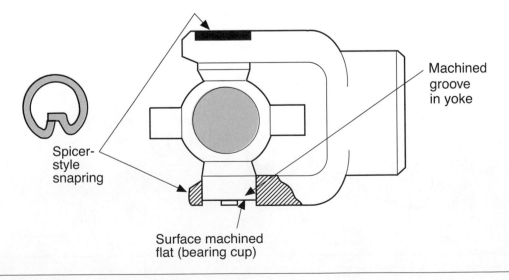

Spicer-
style
snapring

Machined
groove
in yoke

Surface machined
flat (bearing cup)

Figure 6-20 Spicer-style U-joint.

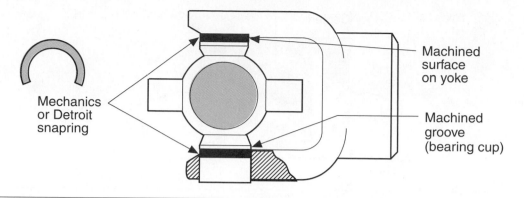

Figure 6-21 Mechanics-style U-joint.

The Cleveland style is an attempt to combine styles of Universal joints to obtain more applications from one joint. The bearing cups for this U-joint are machined to accommodate either Spicer- or mechanics-style snaprings. If a replacement U-joint comes with both style clips, use the clips that pertain to your application.

The mechanics-style U-joint is also called the Detroit/Saginaw style.

✔ **SERVICE TIP:** In order to get the correct parts, you should know the type of U-joint and the yoke span originally installed in the car (Figure 6-22).

Double Cardan Universal Joint

Shop Manual
Chapter 6, page 246

A double Cardan U-joint is used with split drive shafts and consists of two individual Cardan U-joints closely connected by a centering socket yoke and a center yoke that functions like a ball and socket. The ball and socket splits the angle of two drive shafts between two U-joints (Figure 6-23). Because of the centering socket yoke, the total operating angle is divided equally between the two joints. Because the two joints operate at the same angle, the normal fluctuations that result from the use of a single U-joint are canceled out. The acceleration or deceleration of one joint is canceled by the equal and opposite action of the other.

Most often installed in front-engined rear-wheel-drive luxury cars, the double Cardan Universal joint smoothly transmits torque regardless of the operating angle of the driving and driven members. It is therefore classified as a CV Universal joint (Figure 6-24). This joint is used when the U-joint operating angle is too large for a single joint to handle. On some vehicles, the double joint is used at both ends of the drive shaft. On other vehicles, it is used only on the drive end of the drive shaft.

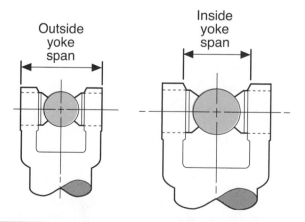

Figure 6-22 You should know both the outside (A) and inside (B) yoke span when ordering a new joint.

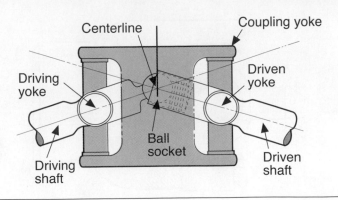

Figure 6-23 The ball and socket of a double Cardan joint splits the angle of the two shafts.

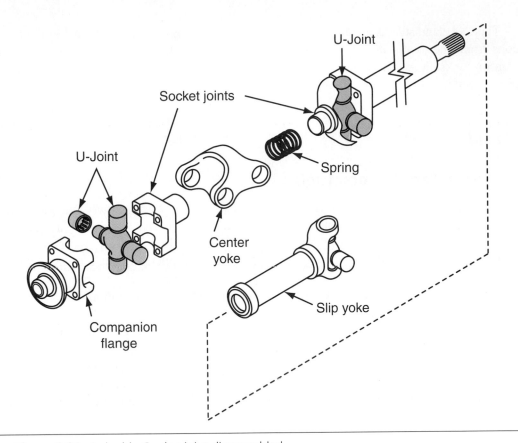

Figure 6-24 A double Cardan joint disassembled.

Constant Velocity Joints

CV joints are primarily used on FWD drive axles; however, some RWD cars with IRS—such as Ford Thunderbird, Corvette, BMW, and Porsche—use them. The commonly used types of CV joints are the Rzeppa and the tripod.

Slip Joints

Shop Manual
Chapter 6, page 237

As well as being equipped with universal joints to allow for angle changes, a drive shaft also must be able to change its effective length. As road surfaces change, the drive axle assembly moves up and down with the rear suspension. Because the transmission is mounted to the frame, it does not move with the movement of the suspension and the relative distance between the transmission

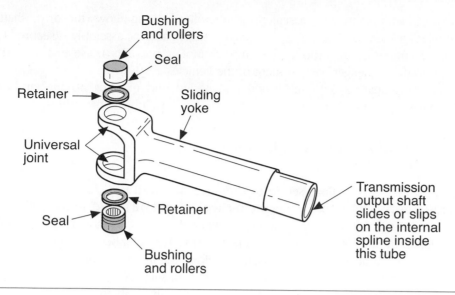

Figure 6-25 Typical slip joint assembly.

and rear drive axle changes. Drive shafts use a **slip joint** at one end of the drive shaft, which allows it to lengthen or shorten. The purpose of the slip joint is similar to the plunging CV joint used in FWD cars. The slip yoke is either positioned at the center of two-piece designs or at either end of the drive shaft, but is typically fitted to the front U-joint.

A slip joint assembly (Figure 6-25) includes the transmission's output shaft, the slip joint itself, a yoke, U-joint, and the drive shaft. The output shaft has external splines that match the internal splines of the slip joint. The meshing of the splines allows the two shafts to rotate together, but permits the ends of the shafts to slide along each other. This sliding motion allows for an effective change in the length of the drive shaft, as the drive axles move toward or away from the car's frame. The yoke of the slip joint is connected to the drive shaft by a U-joint.

Center Support Bearings

Vehicles with a long wheelbase are equipped with a propeller shaft that extends from the transmission or transfer case to a center support bearing (Figure 6-26) and they also have a propeller shaft that extends from the center support bearing to the rear axle. The center support bearing maintains the alignment of the two pieces. Pickup trucks and large SUVs commonly use a center support bearing.

A **slip joint** is a variable length connection that permits the drive shaft to change its effective length.

Some drive shafts are equipped with plunging-type CV joints in place of slip and universal joints.

A slip yoke is also called a sliding yoke.

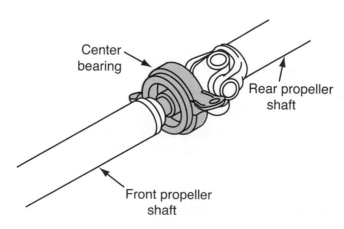

Figure 6-26 A center support bearing assembly.

In a center support bearing, a sealed ball or roller bearing allows the drive shaft to spin freely. The bearing is encased in rubber, or similar material, and the assembly is secured to a cross member of the frame or underbody. The rubber mount prevents noise and vibration from transferring into the passenger compartment of the vehicle.

The standard bearing is prelubricated and sealed and requires no further lubrication; however, some support bearings on heavy-duty vehicles have lubrication fittings.

Magneto-Rheological Fluid Center Bearing Brackets

In an attempt to reduce vibrations set up by the drive shaft as it revolves in the center bearing, manufacturers are beginning to use **magneto-rheological (MR) fluid** in the mount. The MR fluid is used to change the stiffness of the mount when the vehicle is moving. In the presence of a magnetic field, MR fluids rapidly change their viscosity. The fluid becomes more viscous when the intensity of the magnetic field is increased; therefore, the mount becomes more rigid. To provide vibration isolation, a bracket with a very soft material, such as elastomer, is used in conjunction with the fluid's damping capability. This allows drive shaft vibrations to be dampened and isolated during all operating conditions, while keeping the drive shaft pieces securely aligned.

MR fluid is a synthetic oil with soft magnetic particles, such as iron, suspended in it.

Summary

Terms to Know
Canceling angles
Cardan/Spicer
 Universal joint
Double Cardan CV
 joint
Drive shaft
Ellipse
Hotchkiss drive
In-phase
Installation angle
Magneto-
 rheological (MR)
 fluid
Operating angle
Slip joint
Torque tube
Yoke

❏ A drive shaft is normally made from seamless steel tubing with Universal joint yokes fastened to each end.

❏ U-joints are used to allow the angle and the effective length of the drive shaft to change.

❏ A drive shaft is actually no more than a flexible extension of the transmission's output shaft.

❏ Yokes on the drive shaft provide a means of connecting the shafts together.

❏ Balance weights are welded to the outside of a drive shaft to balance it and reduce its natural vibrations.

❏ Some drive shafts are internally lined with cardboard or cardboard with rubber inserts to help offset torsional vibration problems.

❏ A Hotchkiss drive system has an external drive shaft with at least two Universal joints.

❏ A torque tube consists of a tubular steel or small diameter solid shaft enclosed in a larger steel tube and is rigidly connected to the rear axle housing.

❏ The operating angle of a Universal joint is determined by the installation angles of the transmission and rear axle assembly.

❏ U-joints vibrate if the connecting shafts are not on the same plane because one shaft will be accelerating and decelerating at different speeds than the other.

❏ Drive shaft vibrations can be reduced by using canceling angles. The operating angle of the front joint is offset by the angle of the rear joint.

❏ Speed fluctuations can be canceled if the driven yoke has the same point of rotation, or same plane, as the driving yoke. When the two yokes are in the same plane, the joints are said to be in-phase.

❏ There are three common designs of Universal joints: single U-joints retained by either an inside or outside snapring, coupled U-joints, and U-joints held in the yoke by U-bolts or lock plates.

❏ A single Cardan joint, the most common type of joint, uses a spider, four machined trunnions, needle bearings, and bearing caps to allow the transmission of power through slight shaft angle changes.

❏ A double Cardan joint is called a CV joint because shaft speeds do not fluctuate, regardless of the shaft's angle. These joints are used on two-piece shaft assemblies and are actually two single Cardan joints joined together by a center bearing assembly.

❏ The methods used to retain a U-joint in its yoke are the use of a snapring, C-clip, bearing plate, thrust plate, wing bearing, cap and bolt, U-bolt, and/or strap.

❏ A center support bearing is used on all two-piece drive shaft.

Review Questions

Short Answer Essays

1. State the purposes of a Universal joint.

2. What methods are used by some auto manufacturers to reduce drive shaft torsional vibration problems?

3. What determines the operating angle of a Universal joint?

4. Explain why and when U-joints vibrate.

5. What effect do in-phase Universal joints have on the operation of a drive shaft?

6. What is meant by canceling angles?

7. Describe the construction and operation of a single Cardan joint.

8. Describe the construction and operation of a double Cardan joint.

9. Why are some cars equipped with a two-piece drive shaft?

10. What methods are used to retain a U-joint in its yoke?

Fill-in-the-Blanks

1. A drive shaft is normally made from _____ _____ tubing with _____ _____ _____ fastened to each end.

2. A drive shaft is actually an extension of the transmission's _____ _____ .

3. _____ on the drive shaft provide a means of connecting the shafts together.

4. _____ _____ are welded to the outside of a drive shaft to _____ it and reduce its natural _____ .

5. A slip joint is normally fitted to the _____ _____ _____ .

6. A Hotchkiss drive system has an external _____ _____ with at least _____ Universal joints.

7. A _____ _____ is made up of a small diameter solid shaft enclosed in a larger steel tube and is rigidly connected to the transmission and the rear axle assembly.

8. There are three common designs of Universal joints: _____ retained by either an inside or outside snapring, _____, and _____ held in the yoke by U-bolts or lock plates.

9. The basic styles of Cardan joints are: _____ and _____ .

10. The type of drive shaft with an input shaft bonded with rubber to the inside of a hollow drive shaft is a _____ .

Multiple Choice

1. When discussing drive shaft design, *Technician A* says that the shorter the shaft is, the less likely it is to become out of balance. *Technician B* says that cardboard is often inserted into the shaft to dampen the vibrations of the shaft.
   Who is correct?
   A. A only   **C.** Both A and B
   B. B only   **D.** Neither A nor B

2. When discussing the Hotchkiss design of drive shaft, *Technician A* says that it consists of a small diameter solid shaft enclosed in a larger tube. *Technician B* says that it has a Universal joint attached to each end of the shaft.
   Who is correct?
   A. A only   **C.** Both A and B
   B. B only   **D.** Neither A nor B

3. When discussing the phasing of Universal joints, *Technician A* says that this allows the speed changes at one joint to be canceled by the other joint. *Technician B* says that this means that both joints are positioned at opposite angles to each other.
   Who is correct?
   A. A only   **C.** Both A and B
   B. B only   **D.** Neither A nor B

4. When discussing the different types of Universal joints, *Technician A* says that the most commonly used type is the double Cardan. *Technician B* says that the Cardan type is also called the cross and bearing type.
   Who is correct?

   A. A only   **C.** Both A and B
   B. B only   **D.** Neither A nor B

5. When discussing slip joints, *Technician A* says that they allow the drive shaft to maintain its effective length. *Technician B* says that they normally consist of a shaft with external splines and internal splines on the other.
   Who is correct?
   A. A only   **C.** Both A and B
   B. B only   **D.** Neither A nor B

6. When discussing the different types of drive shafts that have been used to reduce vibrations, *Technician A* says that the flexible rope design is the most commonly used. *Technician B* says that the tube-in-tube design is not fitted with Universal joints.
   Who is correct?
   A. A only   **C.** Both A and B
   B. B only   **D.** Neither A nor B

7. *Technician A* says that joint speed fluctuations can be canceled if the driven yoke has the same plane of rotation as the driving yoke. *Technician B* says that joint speed fluctuations can be canceled by putting the joints in-phase.
   Who is correct?
   A. A only   **C.** Both A and B
   B. B only   **D.** Neither A nor B

8. When discussing slip joints, *Technician A* says that the yoke normally has external splines. *Technician B* says that the slip yoke is normally held in place on

the transmission's output shaft by a snapring.
Who is correct?

A. A only

C. Both A and B

B. B only

D. Neither A nor B

9. When discussing double Cardan joints, *Technician A* says that they are actually two single Cardan joints assembled together at a center bearing support. *Technician B* says that they are considered CV joints. Who is correct?

A. A only

C. Both A and B

B. B only

D. Neither A nor B

10. *Technician A* says that excessive joint operating angles can cause an increase in shaft vibration. *Technician B* says that a drive shaft will vibrate when the vehicle is driven at high speeds. Who is correct?

A. A only

C. Both A and B

B. B only

D. Neither A nor B

Differentials and Drive Axles

Upon completion and review of this chapter, you should be able to:

❏ Describe the purpose of a differential.

❏ Identify the major components of a differential and explain their purpose.

❏ Describe the various gears in a differential assembly and state their purpose.

❏ Describe the various methods used to mount and support the drive pinion shaft and gear.

❏ Explain the need for drive pinion bearing preload.

❏ Describe the difference between hunting, nonhunting, and partial nonhunting gear sets.

❏ Explain the purpose of the major bearings within a differential assembly.

❏ Describe the operation of a limited-slip differential.

❏ Describe the construction and operation of a rear axle assembly.

❏ Identify and explain the operation of the two major designs of rear axle housings.

❏ Explain the operation of a FWD differential and its drive axles.

❏ Describe the different types of drive axles and the bearings used to support each of them.

Introduction

The drive axle assembly of a RWD vehicle is mounted at the rear of the car. Most of these assemblies use a single housing to mount the differential gears and axles (Figure 7-1). The entire housing is part of the suspension and helps to locate the rear wheels.

Another type of rear drive axle is used with **IRS**. With IRS the differential is bolted to the chassis and does not move with the suspension. The axles are connected to the differential and drive wheel CV or U-joints. Because the axles move with the suspension and the differential is bolted to the chassis, a common housing for these parts is impossible.

On most RWD cars, the **final drive** is located in the rear axle housing. On most FWD cars, the final drive is located within the transaxle. Some current FWD cars mount the engine and transaxle longitudinally. These configurations use a differential that is similar to other FWD models. Some FWD cars have a longitudinally mounted engine fitted to a special transmission with a separate differential mounted to it.

A differential is needed between any two drive wheels, whether in a RWD, FWD, or 4WD vehicle. The two drive wheels must turn at different speeds when the vehicle is in a turn.

 AUTHOR'S NOTE: I began to learn about cars many years ago by tinkering on them and hanging out with guys who also tinkered. Some of what I learned was good, and some was not. One of the errors I learned back then has stayed with me; calling a final drive gearset the "diff." I know better now but it still stays in my mind. The reason for this is simply, back then, most rear axle work involved pulling the carrier from a removable carrier housing and changing the ring and pinion. We called the thing we pulled out the differential. Hopefully I used the term *differential* correctly throughout this chapter. If I did not, I apologize.

RWD final drives normally use a hypoid ring and pinion gearset that turns the powerflow 90 degrees from the drive shaft to the drive axles. A hypoid gearset allows the drive shaft to be

Shop Manual
Chapter 7, page 267

IRS stands for independent rear suspension.

Final drive is the final set of reduction gears the engine's power passes through on its way to the drive wheels.

Normally, rear axles on RWD vehicles are called live axles because they transmit power.

positioned low in the vehicle because the final drive pinion gear centerline is below the ring gear centerline (Figure 7-2).

On FWD cars with transversely mounted engines, the powerflow axis is naturally parallel to that of the drive axles. Because of this, a simple set of helical gears in the transaxle serve as the final drive gears.

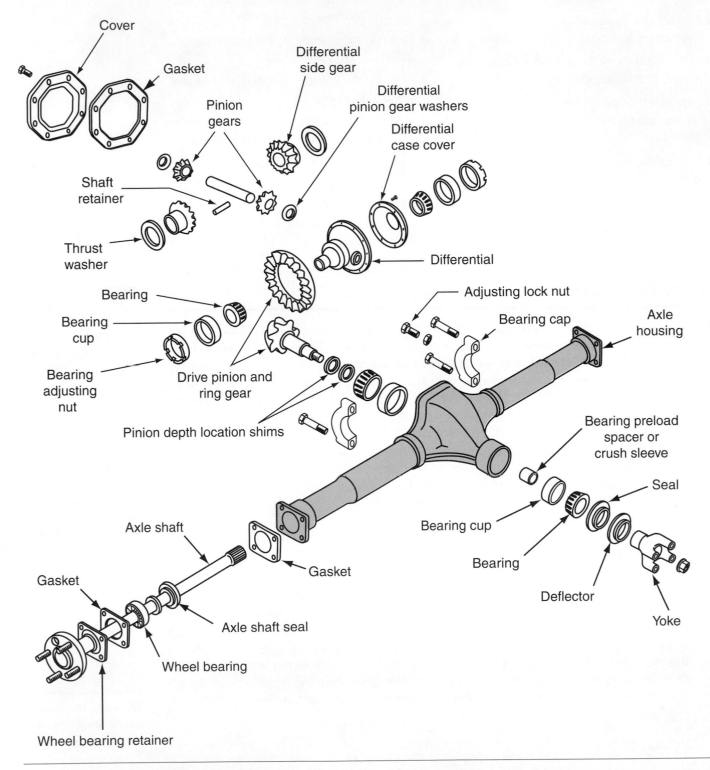

Figure 7-1 Typical RWD axle assembly

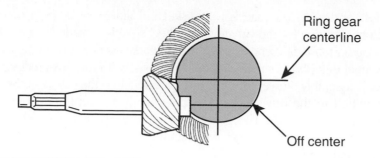

Figure 7-2 A hypoid gear set.

The differential is a geared mechanism located between the two driving axles. It rotates the driving axles at different speeds when the vehicle is turning a corner. It also allows both axles to turn at the same speed when the vehicle is moving straight. The drive axle assembly directs driveline torque to the vehicle's drive wheels. The gear ratio of the differential's ring and pinion gear is used to increase torque, which improves driveability. The differential serves to establish a state of balance between the forces or torques between the drive wheels and allows the drive wheels to turn at different speeds when the vehicle changes direction.

Function and Components

The differential allows for different speeds at the drive wheels when a vehicle goes around a corner or any time there is a change of direction. When a car turns a corner, the outside wheels must travel farther and faster than the inside wheels (Figure 7-3). If compensation is not made for this

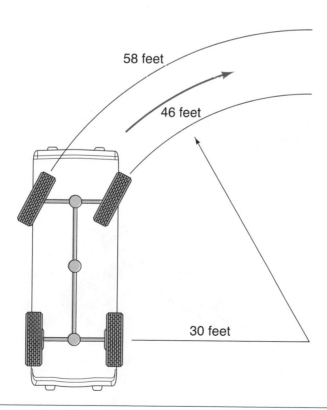

Figure 7-3 Travel of wheels when a vehicle is turning a corner.

Not too long ago, a differential was something that was in the rear axle assembly. Now, with the popularity of FWD vehicles, the differential is part of the transaxle and is most often called the final drive.

When engines are placed longitudinally in the car, they are said to have "north/south" placement.

Helical gears are gears on which the teeth are at an angle to the gear's axis of rotation.

When engines are mounted transversely in the car, they are said to have "east/west" or sideways placement.

difference in speed and travel, the wheels would skid and slide, causing poor handling and excessive tire wear. Compensation for the variations in wheel speeds is made by the differential assembly. While allowing for these different speeds, the differential also must continue to transmit torque.

The differential of a RWD vehicle is normally housed with the drive axles in a large casting called the rear axle assembly. Power from the engine enters into the center of the rear axle assembly and is transmitted to the drive axles. The drive axles are supported by bearings and are attached to the wheels of the car. The power entering the rear axle assembly has its direction changed by the differential. This change of direction is accomplished through the hypoid gears used in the differential.

A BIT OF HISTORY

Early automobiles were driven by means of belts and ropes around pulleys mounted on the driving wheels and engine shaft or transmission shaft. As there was always some slippage of the belts, one wheel could rotate faster than the other when turning a corner. When belts proved unsatisfactory, automobile builders borrowed an idea from the bicycle and applied sprockets and chains. This was a positive driving arrangement, which made it necessary to provide differential gearing to permit one wheel to turn faster than the other.

Power from the drive shaft is transmitted to the rear axle assembly through the pinion flange. This flange is the connecting yoke to the rear universal joint. Power then enters the final drive on the **pinion gear** (Figure 7-4). The pinion teeth engage the ring gear, which is mounted upright at a 90-degree angle to the pinion. Therefore, as the drive shaft turns, so do the pinion and ring gears.

The **ring gear** is fastened to the differential case with several hardened bolts or rivets. The differential case is made of cast iron and is supported by two tapered-roller bearings in the rear axle housing. Holes machined through the center of the differential housing support the differential pinion shaft. The pinion shaft is retained in the housing case by clips or a specially designed bolt. Two beveled differential pinion gears and thrust washers are mounted on the differential pinion shaft. In mesh with the differential pinion gears are two axle side gears splined internally to mesh with the external splines on the left and right axle shafts (Figure 7-5). Thrust washers are placed between the differential pinions, axle side gears, and differential case to prevent wear on the inner surfaces of the differential case.

In a **ring** and **pinion gear set**, the pinion is the smaller drive gear and the ring gear is the larger driven gear.

The term *differential* means relating to or exhibiting a difference or differences.

Other names commonly used to refer to the differential of RWD vehicles are rear axle, drive axle, third member, center section, or "pumpkin."

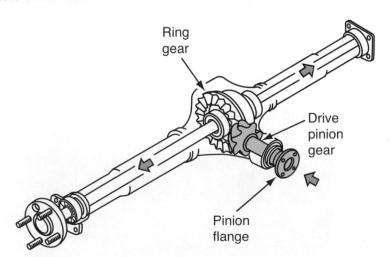

Figure 7-4 Main components of a RWD drive axle.

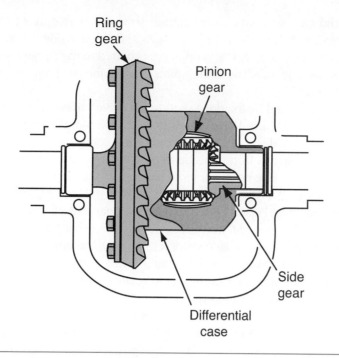

Figure 7-5 Components of a typical differential.

Differential Operation

The two drive wheels are mounted on axles that have a differential side gear fitted on their inner ends (Figure 7-6). To turn the power flow 90 degrees, as is required for RWD vehicles, the side gears are bevel gears.

Shop Manual
Chapter 7, page 268

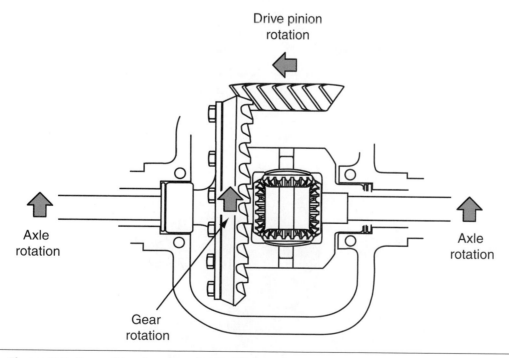

Figure 7-6 Power flow through a RWD differential.

When two beveled gears are meshed, the driving and driven shafts can rotate at a 90-degree angle.

The **differential case** is the metal unit that encases the differential side gears and pinion gears and to which the ring gear is attached.

The two side gears are placed on the side of the differential case, which is why they are called **side gears**.

The gear ratio in a differential is known as the axle ratio.

The **differential case** is mounted on bearings so that it is able to rotate independently of the drive axles. A pinion shaft, with small pinion gears, is fitted inside the differential case. The pinion gears mesh with the **side gears**. The ring gear is bolted to the flange of the differential case and the two rotate as a single unit. The drive pinion gear meshes with the ring gear and is rotated by the drive shaft (Figure 7-7).

Engine torque is delivered by the drive shaft to the drive pinion gear, which is in mesh with the ring gear and causes it to turn. Power flows from the pinion gear to the ring gear. The ring gear is bolted to the differential case, which drives the side gears, pinions, and axles as an assembly. The differential case extends from the side of the ring gear and normally houses the pinion gears and the side gears. The side gears are mounted so they can slip over splines on the ends of the axle shafts.

There is a gear reduction between the drive pinion gear and the ring gear, causing the ring gear to turn about one-third to one-fourth the speed of the drive pinion. The pinion gears are located between and meshed with the side gears (Figure 7-8), thereby forming a square inside the differential case. Differentials have two or four pinion gears that are in mesh with the side gears (Figure 7-9). The differential pinion gears are free to rotate on their own centers and can travel in a circle as the differential case and pinion shaft rotate. The side gears are meshed with the pinion gears and are also able to rotate on their own centers.

The small pinion gears are mounted on a pinion shaft that passes through the gears and the case. The pinion gears are in mesh with the axle side gears, which are splined to the axle shafts. In operation, the rotating differential case causes the pinion shaft and pinion gears to rotate end

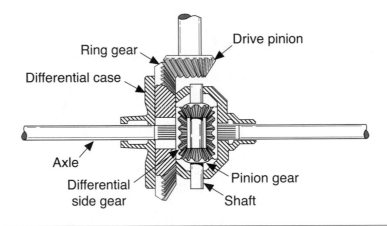

Figure 7-7 Basic differential.

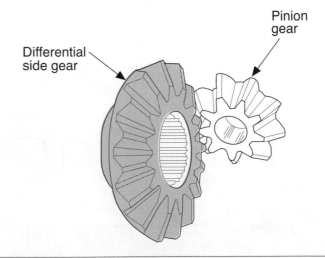

Figure 7-8 Pinion gears in mesh with the side gears.

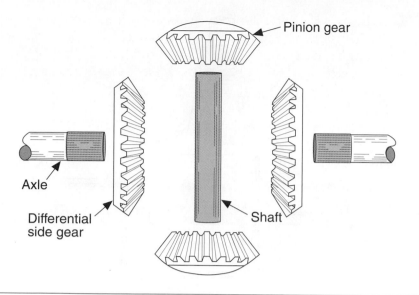

Figure 7-9 Position of side and pinion gears.

over end with the case (Figure 7-10). Because the pinion gears are in mesh with the side gears, the side gears and axle shafts are also forced to rotate.

When a car is moving straight ahead, both drive wheels are able to rotate at the same speed. Engine power comes in on the pinion gear and rotates the ring gear. The differential case is rotated with the ring gear. The pinion shaft and pinion gears are carried around by the ring gear and all of the gears rotate as a single unit. Each side gear rotates at the same speed and in the same plane as does the case and they transfer their motion to the axles. The axles are thus rotated, and the car moves. Each wheel rotates at the same speed because each axle receives the same rotation (Figure 7-11).

As the vehicle goes around a corner, the inside wheel travels a shorter distance than the outside wheel. The inside wheel must therefore rotate more slowly than the outside wheel. In this situation, the differential pinion gears will "walk" forward on the slower turning or inside side gear (Figure 7-12). As the pinion gears walk around the slower side gear, they drive the other side gear at a greater speed. An equal percentage of speed is removed from one axle and given to the other (Figure 7-13), however the torque applied to each wheel is equal.

Only the outside wheel rotates freely when a car is making a very sharp turn; therefore, only one side gear rotates freely. Because one side gear is close to being stationary, the pinion gears

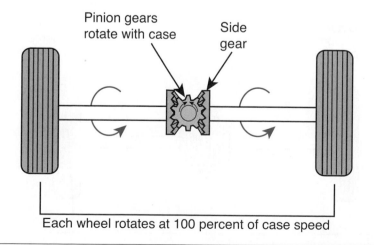

Figure 7-10 Position of pinion gears in the case causes the side gears to rotate.

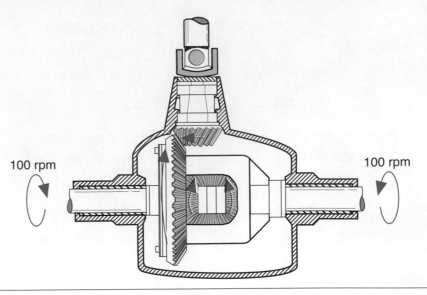

Figure 7-11 Differential action when the vehicle is moving straight ahead.

100 rpm   100 rpm

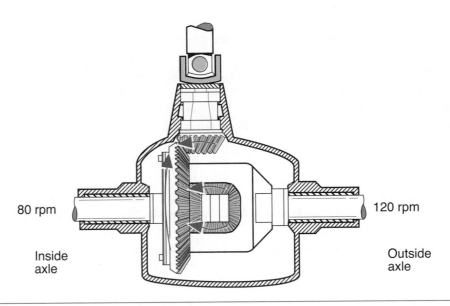

80 rpm   120 rpm

Inside   Outside
axle   axle

Figure 7-12 Differential action when the vehicle is turning a corner.

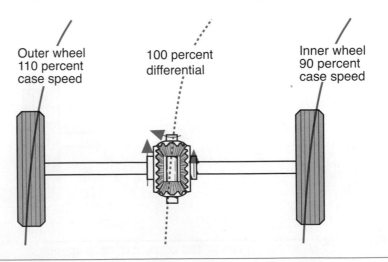

Outer wheel
110 percent
case speed

100 percent
differential

Inner wheel
90 percent
case speed

Figure 7-13 Speed differentiation when turning.

now turn on their own centers as they walk around that side gear. As they walk around that side gear, they drive the other side gear at twice their own speed. The moving wheel is now turning at twice the speed of the differential case, but the torque applied to it is only half of the torque applied to the differential case. This increase in wheel speed occurs because of these two actions: the differential pinion gears are rotating end over end with the pinion shaft and the action of the differential pinion gears rotating around the differential pinion shaft.

When one of the driving wheels has little or no traction, the torque required to turn the wheel without traction is very low. The wheel with good traction in effect is holding the axle gear on that side stationary. This causes the pinions to walk around the stationary side gear and drive the other wheel at twice the normal speed but without any vehicle movement. With one wheel stationary, the other wheel turns at twice the speed shown on the speedometer. Excessive spinning of one wheel can cause severe damage to the differential. The small pinion gears can actually become welded to the pinion shaft or differential case.

Axle Housings

Shop Manual
Chapter 7, page 273

Live rear axles use a one-piece housing with two tubes extending from each side. These tubes enclose the axles and provide attachments for the axle bearings. The housing also shields the parts from dirt and retains the differential lubricant.

In IRS (Figure 7-14) or FWD systems, the housing is in three parts. The center part houses the final drive and differential gears. The outer parts support the axles by providing attachments for the axle bearings. These parts also serve as suspension components and attachment points for the steering gear or brakes. In FWD applications, the differential and final drive are either enclosed in the same housing as the transmission or in a separate housing bolted directly to the transmission housing.

Rear Axle Housings. Based on their construction, rear axle housings can be divided into two groups, integral carrier or removable carrier. An **integral carrier** housing attaches directly to the rear suspension. A service cover, in the center of the housing, fits over the rear of the differential and rear axle assembly (Figure 7-15). When service is required, the cover must be removed. The components of the differential unit are then removed from the rear of the housing.

Integral carriers are commonly referred to as unitized or Salisbury-type differentials.

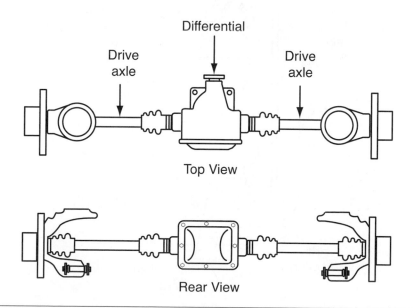

Figure 7-14 Drive axle assembly on a RWD vehicle with IRS.

The rear axle housing is sometimes called a banjo because of the bulge in the center of the housing. The bulge contains the final drive gears and differential gears.

In the rear, the outer sections of the housing may be called the uprights and in the front they are usually called the steering knuckle.

Removable carriers are often referred to as the third member, dropout carrier, or pumpkin.

In appearance the two designs of rear axle housing look similar except that the opening for the differential unit on a removable type is at the front and the rear of the housing is solid.

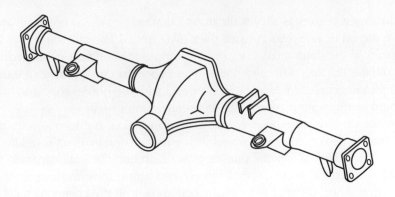

Figure 7-15 Typical integral carrier axle housing.

In an integral-type axle housing, the differential carrier and the pinion bearing retainer are supported by the axle housing in the same casting. The pinion gear and shaft is supported by two opposing tapered-roller bearings located in the front of the housing. The differential carrier assembly is also supported by two opposing tapered-roller bearings, one at each side (Figure 7-16).

The differential assembly of a **removable carrier** assembly can be removed from the front of the axle housing as a unit. The differential is serviced on a bench and then installed into the axle housing. The differential assembly is mounted on two opposing tapered-roller bearings

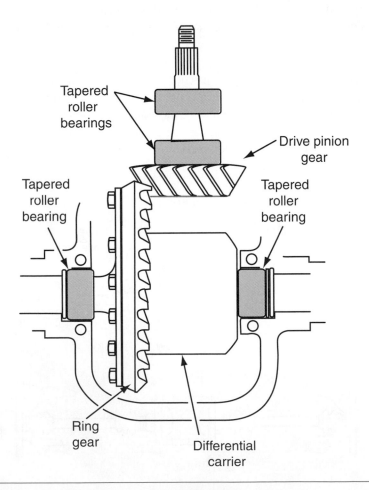

Figure 7-16 Location of bearings in a typical integral housing.

retained in the housing by removable caps. The pinion gear, pinion shaft, and the pinion bearings are typically assembled in a pinion retainer, which is bolted to the carrier housing (Figure 7-17).

A typical housing has a cast-iron center section with axle shaft tubes pressed and welded into either side. The rear axle housing encloses the complete rear-wheel driving axle assembly. In addition to housing the parts, the axle housing also serves as a place to mount the vehicle's rear

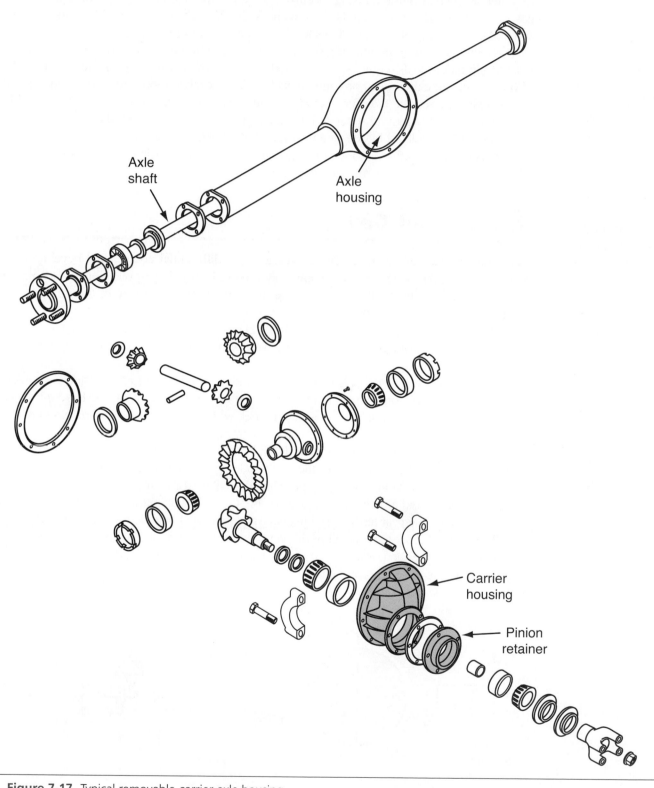

Figure 7-17 Typical removable carrier axle housing.

suspension and braking system. With IRS, the differential housing is mounted to the vehicle's chassis and does not move with the suspension.

Differential Fluids

On most nonremovable carrier-type rear axle assemblies, there is a differential inspection cover mounted to the rear of the housing. Normally, there is a fill plug near the center of the inspection cover. The opening for the plug is used to check the fluid level and to add fluid to the housing. When the housing is filled with oil, its level is at the bottom of the hole.

Hypoid gears require hypoid gear lubricant of the extreme pressure type and high viscosity. Limited slip differentials require special limited slip lubricant that provides the required coefficient of friction for the clutch discs or cones as well as proper lubrication. Transaxles and some RWD differentials may require a lower viscosity oil, such as automatic transmission fluid. In addition, some transaxles may require separate lubricants for the transmission and differential.

The oil is circulated by the ring gear and thrown over all the parts. Special troughs or gullies are used to return the oil to the ring and pinion area. The housing is sealed with gaskets and oil seals to keep the fluid in and dirt and moisture out.

Shop Manual
Chapter 7, page 273

The drive side of the teeth is the side that has engine power working on it when moving forward, whereas the coast side is the side of the teeth that has contact during deceleration.

When there is no torque applied either in drive or in coast, the condition is known as float.

Differential Gears

Two types of gears are currently being used as RWD differential gears: spiral bevel and hypoid (Figure 7-18). Spiral bevel gears are commonly used in heavy duty applications. In a spiral bevel gear set, the centerline of the drive pinion gear intersects the centerline of the ring gear. These designs are noisier than hypoid gears.

Hypoid gear sets are commonly used in RWD passenger car and light truck applications. The pinion gear in a hypoid gear set is mounted well below the centerline of the ring gear. Hypoid gears are quiet running.

This design allows for lower vehicle height and more passenger room inside the vehicle. By lowering the drive pinion gear on the ring gear, the entire drive shaft can be lowered. Lowering the drive shaft allows for a lower drive shaft tunnel, which in turn allows for increased passenger room and a lower ride height.

The teeth of a hypoid gear are curved to follow the form of a spiral, causing a wiping action when meshing. As the gears rotate, the teeth slide against each other. Because of this sliding action, the ring and pinion gears can be machined to allow for near perfect mating, which results in smoother action and a quiet-running gear set. Because this sliding action produces extremely high pressures between the gear teeth, only a hypoid-type lubricant should be used with hypoid gear sets.

The spiral-shaped teeth result in different tooth contacts as the pinion and ring gear rotate. The drive side of the teeth is curved in a convex shape, and the coast side of the teeth is concave (Figure 7-19). The inner end of the teeth on the ring gear is known as the toe and the outer end of the teeth is the heel (Figure 7-20).

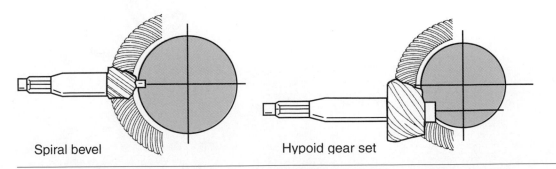

Spiral bevel   Hypoid gear set

Figure 7-18 Comparison of a spiral bevel and hypoid gear set.

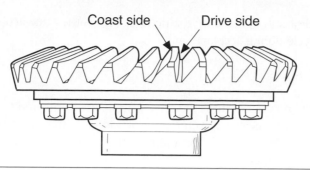

Figure 7-19 The drive and coast side of a ring gear.

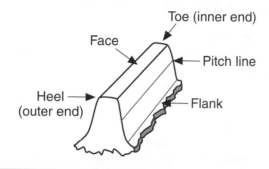

Figure 7-20 The toe and heel of a ring gear's tooth.

When engine torque is being applied to the drive pinion gear, the pinion teeth exert pressure on the drive side of the ring gear teeth. During coast or engine braking, the concave side of the ring gear teeth exerts pressure on the drive pinion gear.

Upon heavy acceleration, the drive pinion attempts to climb up the ring gear and raises the front of the differential. The suspension's leaf springs or the torque arm on coil spring suspensions absorb much of the torque to limit the movement of the axle housing (Figure 7-21).

This phenomenon actually always occurs; it is just more noticeable during heavy acceleration. It is caused by an "equal and opposite" reaction of the wheels pushing one way while

The climbing action of the drive pinion is sometimes called wind-up.

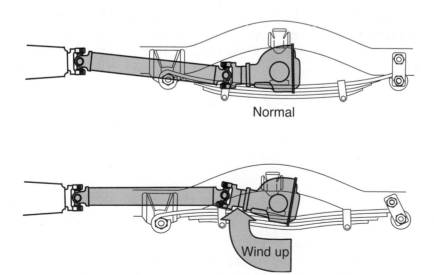

Figure 7-21 Great amounts of torque and good traction can cause the rear axle assembly to "wind up."

trying to push the axle housing in the other direction. This is the same action that lifts the front of a dragster or motorcycle during acceleration.

Gear Ratios

Gear ratios express the number of turns the drive gear makes compared to one turn of the driven gear it mates with. The ring gear is driven by the pinion gear, therefore causing torque multiplication. The ring gear is always larger than the pinion. This combination causes the ring gear to turn more slowly but with greater torque.

Many different final drive ratios are used. A final drive ratio of 2.8:1 is commonly used, especially on cars equipped with automatic transmissions. A 2.8:1 final drive ratio means the drive pinion must turn 2.8 times to rotate the ring gear one time. On cars equipped with manual transmissions, more torque multiplication is often needed, therefore a 3.5:1 final drive ratio is often used. To allow a car to accelerate more quickly or to move heavy loads, a final drive ratio of 4:1 can be used. Also, small engine cars with overdrive fourth and fifth gears often use a 4:1 final drive ratio, which allows them to accelerate reasonably well in spite of the engine's low power output. The overdrive in fourth and fifth gear effectively reduces the final drive ratio when the car is moving in those gears. Trucks also use a final drive ratio of 4:1 or 5:1 to provide more torque to enable them to pull or move heavy loads.

It is important to remember that the actual final drive or overall gear ratio is equal to the ratio of the ring and pinion gear multiplied by the ratio of the speed gear the car is operating in. For example, if a car has a final drive ratio of 3:1, the total final drive ratio for each transmission speed is as follows:

	Transmission Ratio	×	Final Drive Ratio	=	Total Final Drive Ratio
First gear	3:1		3:1		9:1
Second gear	2.5:1		3:1		7.5:1
Third gear	1.5:1		3:1		4.5:1
Fourth gear	1:1		3:1		3:1
Fifth gear	0.75:1		3:1		2.25:1

Notice that, in this example, the only time the total final drive ratio is the same as the ratio of the ring and pinion gear is when the transmission is in fourth gear, which has a speed ratio of 1:1.

Many factors are considered when a manufacturer selects a final drive ratio for a vehicle. Some of these factors are vehicle weight, engine rpm range, designed vehicle speed, frontal area of the body, fuel economy requirements, engine power output, and transmission type and gear ratios. Cars with final drive ratios around 2.5:1 will take longer to accelerate but will typically give a higher top speed. At the other end of the scale, a 4.11:1 ratio will give faster acceleration with a lower top speed. Since the 1970s there has been an emphasis on fuel economy, and most cars have been equipped with high gears to allow for lower engine speeds at normal driving speeds.

Determining Final Drive Ratio

To replace a ring and pinion gear set with one of the correct ratio, the ratio of the original set must be known. There are several ways to determine the final drive ratio of a ring and pinion gear set. If a shop manual is available, you can decipher the code found on the assembly or on a tag attached to it (Figure 7-22). Normally a table is given that lists the various codes and the ratios each represents.

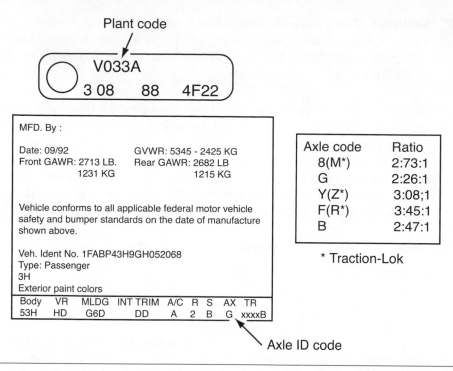

Figure 7-22 Deciphering differential codes from information given on the differential tag or from the VIN.

Most axles are shipped with an identification tag bolted to them. These tags contain all of the information needed to identify the axle for diagnosis and service. The tags are located under the housing-to-carrier stud nut or are attached by a cover-to-carrier bolt. Manufacturers also often stamp identification numbers into the axle housing. These codes are normally located on the front side of an axle tube (Figure 7-23). Always refer to your shop manual to locate and decipher the codes.

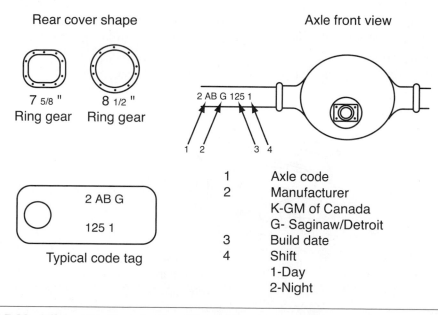

Figure 7-23 Different locations for the differential codes on an axle assembly.

SERVICE TIP: If the stamped numbers cannot be found or if the axle tag is not there, refer to the axle code letter or number on the vehicle identification number (VIN) plate (Figure 7-22). This will identify the ratio and type of axle with which the car was originally equipped.

Another way to determine the final drive ratio is to compare the number of revolutions of the drive wheels with those of the drive shaft. While turning both wheels simultaneously, note how many times the drive shaft turns to complete one revolution of the drive wheels. This count basically represents the ratio of the gears.

The gear ratio also can be determined when the differential is disassembled. Count the number of teeth on both the drive pinion and the ring gear. Divide the ring gear teeth number by the pinion drive number to calculate the final drive ratio.

Hunting and Nonhunting Gears

Shop Manual
Chapter 7, page 300

The alignment of nonhunting and partial nonhunting gears is often referred to as timing the gears.

Ring and pinion gear sets are usually classified as hunting, nonhunting, or partial nonhunting gears. Each type of gear set has its own requirements for a satisfactory gear tooth contact pattern. These classifications are based on the number of teeth on the pinion and ring gears.

A **nonhunting gear set** is one in which any one pinion tooth comes into contact with only some of the ring gear teeth. One revolution of the ring gear is required to achieve all possible gear tooth contact combinations. As an example, if the ratio of the ring gear teeth to the pinion gear teeth is 39 to 13 (or 3.00:1), the pinion gear turns three times before the ring gear completes one turn. One full rotation of the pinion gear will cause its 13 teeth to mesh with one third of the ring gear's teeth. On the next revolution of the pinion gear, its teeth will mesh with the second third of the ring gear's teeth and the third revolution will mesh with the last third of the ring gear. Each tooth of the pinion gear will return to the same three teeth on the ring gear each time the pinion rotates.

A **partial nonhunting gear set** is one in which any one pinion tooth comes into contact with only some of the ring gear teeth, but more than one revolution of the ring gear is required to achieve all possible gear tooth contact combinations. If the ratio of the ring gear teeth to the pinion gear teeth is 35 to 10 (or 3.5:1), any given tooth of the pinion will meet seven different teeth (seven complete revolutions of the pinion gear) of the ring gear before it returns to the space where it started.

When **hunting gear sets** are rotating, any pinion gear tooth will contact all the ring gear teeth. If the ring gear has 37 teeth and the pinion gear has 9, the gear set has a ratio of 37 to 9 (or 3.89:1). Any given tooth in the pinion gear meets all of the teeth in the ring gear before it meets the first tooth again.

During assembly the nonhunting and partial nonhunting gears must be assembled with the index marks properly aligned (Figure 7-24). When these gear sets were manufactured, they were

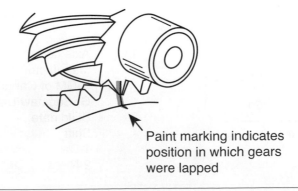

Paint marking indicates position in which gears were lapped

Figure 7-24 Index marks of a ring and pinion gear set.

probably **lapped** to ensure proper meshing and because specified teeth on the pinion will always meet specific teeth on the ring gear, a noisy gear will result if they are not properly aligned. Hunting gears do not need to be aligned because any tooth on the pinion may mesh with any tooth on the ring gear.

Differential Bearings

At least four bearings are found in all differentials. Two fit over the drive pinion shaft to support it and the other two support the differential case and are usually mounted just outboard of the side gears (Figure 7-25). The drive pinion and case bearings are typically tapered-roller bearings.

Different forces are generated in the differential due to the action of the pinion gear. As the pinion gear turns, it tries to climb up the ring gear and pull the ring gear down. Also, as the pinion gear rotates, it tends to move away from the ring gear and pushes the ring gear equally as hard in the opposite direction. Because of these forces, the differential must be securely mounted in the carrier housing. The bearings on each end of the differential case support the case and absorb the thrust of the forces (Figure 7-26). The pinion gear and shaft are mounted on bearings to allow the shaft to rotate freely without allowing it to move in response to the torque applied to it. All of these bearings are installed with a preload to prevent the pinion gear and ring gear from moving out of position.

Pinion Mountings

As torque is applied to a pinion gear, the pinion gear rotates, three separate forces are produced by its rotation and the torque applied to it. These forces make it necessary to securely mount the pinion gear.

Lapping is the process of using a grinding paste to produce a fine finish on the teeth of the two gears that will be in full contact with each other.

The rear Universal joint on the drive shaft attaches to the drive pinion flange or the companion flange.

Shop Manual
Chapter 7, page 304

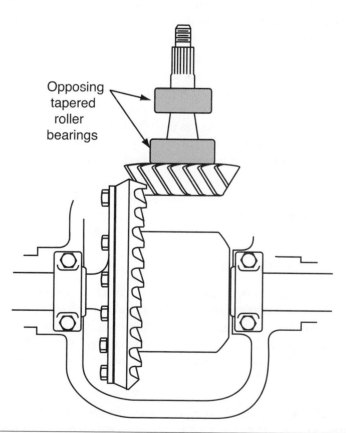

Figure 7-25 Position of bearings in a typical differential assembly.

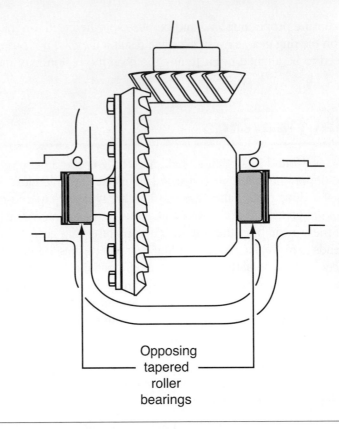

Figure 7-26 Position of differential case side bearings.

The **drive pinion flange** is splined to the rear axle's drive pinion gear. The drive pinion gear is placed horizontally in the axle housing and is positioned by one of two types of mounting, straddle or overhung. The straddle-mounted pinion gear is used in some removable carrier-type axle housings. The **straddle-mounted pinion** has two opposing tapered-roller bearings positioned close together with a short spacer between their inner races and ahead of the pinion gear. A third bearing, usually a straight roller bearing, is used to support the rear of the pinion gear (Figure 7-27).

The **overhung-mounted pinion** also uses two opposing tapered-roller bearings but does not use a third bearing. The two roller bearings must be farther apart than the opposing bearings of a straddle-mounted pinion because a third bearing is not used to support the pinion gear (Figure 7-28). This type of pinion gear mounting can be found on either the removable carrier or integral-type driving axle.

Some pinion shafts are mounted in a bearing retainer that is removable from the carrier housing. This type of pinion assembly utilizes a pilot bearing to support the rear end of the pinion and is equipped with two opposing tapered-roller bearings.

Drive Pinion Bearing Preload

A spacer is placed between the opposing tapered bearings to control the distance between them (Figure 7-29). This spacer also controls the amount of preload or loading pressure applied to the bearings. Preload prevents the pinion gear from moving back and forth in the bearing retainer.

Often a collapsible spacer is used between the two large tapered bearings to provide for proper pinion bearing preload. Some differentials use a solid noncollapsible spacer with selective thickness shims to adjust pinion bearing preload.

 SERVICE TIP: Collapsible spacers should never be reused. After they have been compressed once, they are not capable of maintaining preload when they are compressed again. The spacers should always be replaced when servicing the differential.

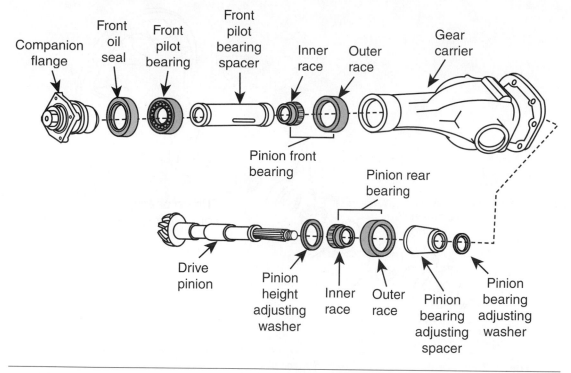

Figure 7-27 Typical straddle-mounted pinion gear.

When the pinion shaft nut is tightened to specifications, pressure is exerted by the pinion drive flange against the inner race of the front pinion bearing. This applies pressure against the spacer and the rear bearing, which cannot move because it is located against the drive pinion gear. This load on the two pinion bearings assures that there will be no pinion shaft end play. Any pinion shaft end play will result in rapid failure and noise.

End play is the amount of axial or end-to-end movement in a shaft due to clearance in the bearings.

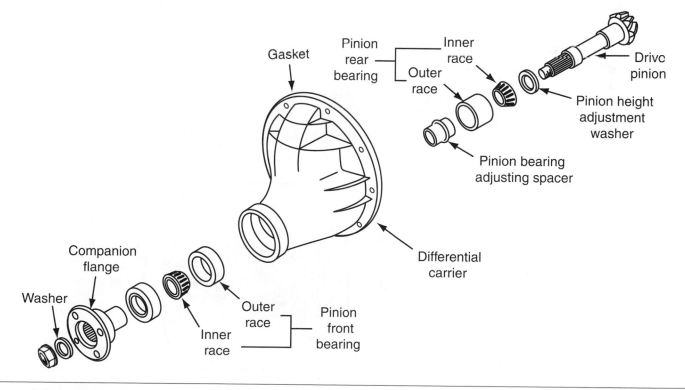

Figure 7-28 Typical overhung-mounted pinion gear.

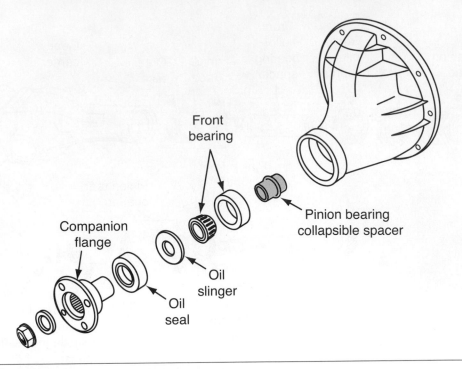

Figure 7-29 Location of pinion bearing spacer.

Shop Manual
Chapter 7, page 283

Differential Case

The differential case is supported in the carrier by two tapered-roller side bearings. This assembly can be adjusted from side to side to provide the proper backlash between the ring gear and pinion and the required side bearing preload. This adjustment is achieved by threaded bearing adjusters (Figure 7-30) on some units and the placement of selective shims and spacers (Figure 7-31) on others.

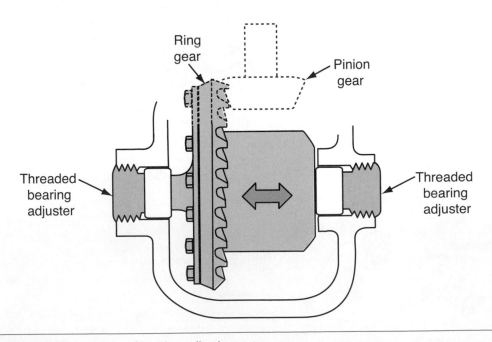

Figure 7-30 Location of bearing adjusting nuts.

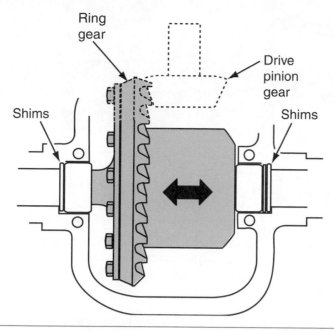

Figure 7-31 Location of bearing selective shims.

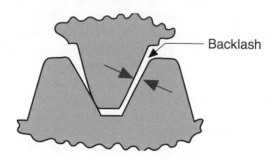

Figure 7-32 Gear backlash.

Backlash is the clearance or play between two gears in mesh. It is the amount one of the gears can be moved without moving the other (Figure 7-32).

Shop Manual
Chapter 7, page 285

Another name for collapsible spacer is crush sleeve.

The ring gear in a transaxle is sometimes referred to as the differential drive gear.

Transaxle Final Drive Gears and Differential

Transaxle final drive gears provide the means for transmitting transmission output torque to the differential section of the transaxle.

The differential section of the transaxle has the same components as the differential gears in a RWD axle and basically operates in the same way. The power flow in transversely mounted powertrains is in line with the wheels and therefore the differential unit does not need to turn the power 90 degrees.

The drive pinion and ring gears and the differential assembly are normally located within the transaxle housing of FWD vehicles. There are three common configurations used as the final drives on FWD vehicles: helical, planetary, and hypoid. The helical and planetary final drive arrangements are usually found in transversely mounted power trains. Hypoid final drive gear assemblies are used with longitudinal powertrain arrangements.

The drive pinion gear is connected to the transmission's output shaft and the ring gear is attached to the differential case. Like the ring and pinion gear sets in a RWD axle, the drive pinion and ring gear of a FWD assembly provide for a multiplication of torque.

The teeth of the ring gear usually mesh directly with the transmission's output shaft (Figure 7-33). However on some transaxles, an intermediate shaft is used to connect the transmission's output to the ring gear.

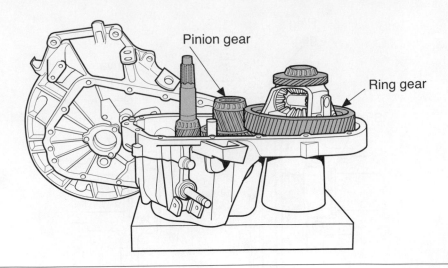

Figure 7-33 Typical ring and pinion gear set in a transaxle.

On some models, the differential and final drive gears operate in the same lubricant as the transmission section of the transaxle. On other designs, the differential section is separately enclosed and is lubricated by a different lubricant than the transmission section. These designs require positive sealing between the differential unit and the transmission to keep the different lubricants from mixing. All transaxles use seals between the differential and the drive axles to prevent dirt from entering the transaxle and to prevent lubricant from leaking past the attachment point of the drive axles.

Helical Final Drive Assembly

Helical gears are gears with teeth that are cut at an angle to the gear's axis of rotation.

Helical (Figure 7-34) final drive assemblies use helical gear sets that require the centerline of the pinion gear to be at the centerline of the ring gear. The pinion gear is cast as part of the main shaft and is supported by tapered-roller bearings. The pinion gear is meshed with the ring gear to provide the required torque multiplication. Because the ring is mounted on the differential case, the case rotates in response to the pinion gear.

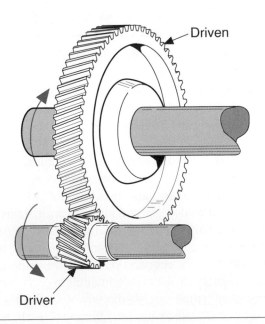

Figure 7-34 Helical gear set.

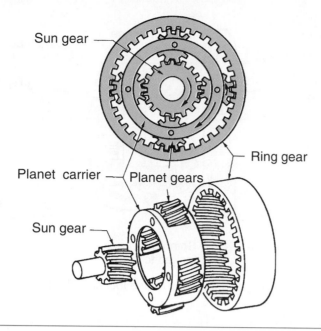

Figure 7-35 Planetary final drive gear set.

Planetary Final Drive Assembly

The ring gear of a planetary final drive assembly has lugs around its outside diameter. These lugs fit into grooves machined inside the transaxle housing. These lugs and grooves hold the ring gear stationary. The transmission's output shaft is splined to the planetary gearset's sun gear. The planetary pinions are in mesh with both the sun gear and ring gear and form a simple planetary gearset (Figure 7-35). The planetary carrier is constructed so that it also serves as the differential case.

In operation, the transmission's output drives the sun gear, which, in turn, drives the planetary pinions. The planetary pinions walk around the inside of the stationary ring gear. The rotating planetary pinions drive the planetary carrier and differential housing. This combination provides maximum torque multiplication from a simple planetary gear set.

Hypoid Final Drive Assembly

Hypoid gears have the advantage of being quiet and strong because of their thick tooth design. And due to their strength, hypoid-type gears can be used with large engines that are longitudinally mounted in vehicles. This type of final drive unit is identical to those used in RWD vehicles.

Limited-Slip Differentials

An open differential is built with a combination of interlocking gears that eliminates tire scrubbing, as the outer tire has further distance to travel during cornering. Although this differential is the easiest on the car's tires and suspension, it has one major disadvantage—a lack of traction. Power, is for the most part, transferred to the wheel with the least resistance or traction.

When a car is stuck in mud or snow, one drive wheel spins while the other remains stationary. In this example, the differential is transferring all of the torque and rotary motion to the drive wheel with the least resistance. Resistance, in this case, means traction. Applying torque to the wheel without traction does little good while trying to move the car.

A limited-slip differential assembly provides more driving force to the wheel with traction when one wheel begins to slip or spin freely. Friction material is often used to transfer the torque applied to a slipping wheel to the one with traction. Those that use a clutch pack (Figure 7-36)

An open differential is a standard-type differential, often called by racers the peg leg type.

Many names are used for limited-slip differentials, including Posi-Traction, Traction-Lok, and Posi-units.

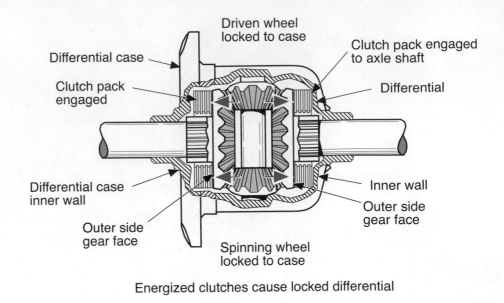

Figure 7-36 Action of clutches in a limited-slip differential.

have two sets (one for each side gear) of clutch plates and friction discs to prevent normal differential action. As long as the friction discs maintain their grip on the steel plates, the differential side gears are locked to the differential case. This allows the case and drive axles to rotate at the same speed and prevents one wheel from spinning faster than the other. There are other designs of limited-slip differentials that do not rely on a clutch assembly; these will be discussed later in this chapter.

Limited-slip differentials are used on high-performance and sports cars for increased traction while cornering and on off-road vehicles in which the drive wheels are constantly losing traction. Power flows in the same way as in an open differential. Most limited-slip differentials transfer at least 20 percent of the available torque to the wheel with traction. Limited-slip differentials merely limit the ease of differential action between the side gears through the use of these clutches.

Clutch Pack

The most common limited-slip differentials use two sets of multiple disc clutches to control differential action. Each **clutch pack** consists of a combination of steel plates and friction plates. The plates are stacked on the side gear hub and are housed in the differential case. A preload spring applies an initial force to the clutch packs (Figure 7-37).

The friction plates are splined to each side gear's hub. The ears of the steel plates are fitted into the case so that the clutch packs are always engaged. The discs rotate with the side gear and the plates with the differential case.

The clutch assembly consists of a multiple plate clutch, a center block, preload springs, and a preload plate (Figure 7-38). The clutch assembly is always engaged due to pressure constantly being applied to it by the preload springs. Under normal driving conditions, the clutch brake slips as the torque generated by differential action easily overcomes the capacity of the clutch assembly. This allows for normal differential action when the vehicle is turning. During adverse road conditions, where one or both wheels may be on a low friction surface such as snow, ice, or mud, the friction between the clutch plates will transfer a portion of usable torque to the wheel with the most traction.

The clutch packs are mounted behind each of the axle's side gears and springs between the side gears force the gears against the clutches. Although the springs allow enough slippage to permit driving around a curve, during slippery conditions they keep the side gears against the

Shop Manual
Chapter 7, page 325

A **clutch pack** consists of a complete set of alternating clutch plates and discs.

Common names for factory limited-slip differentials are Trac-Lok for Chrysler and Ford, Limited-Slip and Posi-Traction for GM, and Power-Lok and Sure-Grip for Chrysler.

178

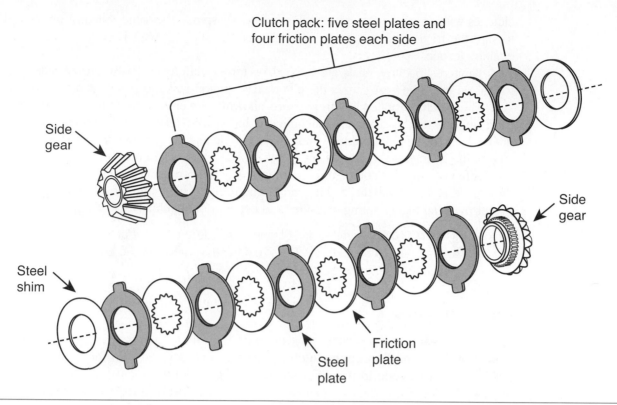

Figure 7-37 The clutch assembly in a typical limited-slip differential assembly.

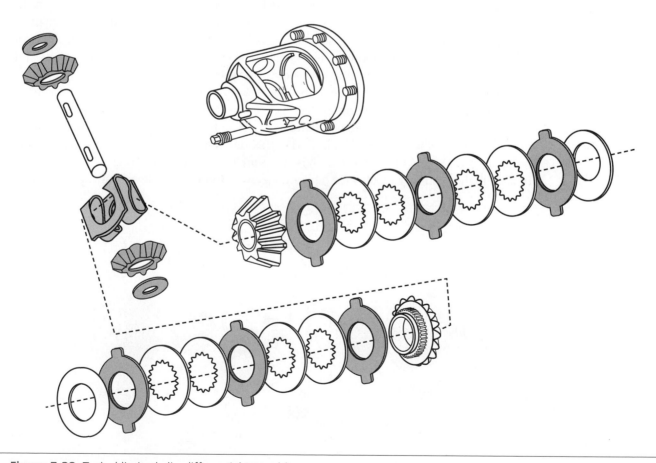

Figure 7-38 Typical limited-slip differential assembly.

clutches with enough pressure to make those gears spin at the same speed. If one wheel begins to slip, the friction of the clutches ensures that the slipping wheel does not receive all of the engine's torque.

Limited-slip differentials are applied by torque differences between the side gears. High torque on one side gear causes the differential's pinion gears to push against the opposite side gear. The clutch is applied by this pressure, allowing power to move to that axle. Preload springs (or a single spring) assist in applying the clutch. This provides enough pressure on the clutch to drive both axles when the drive wheels have an unequal amount of traction. However, the pressure of the springs is low enough to allow clutch slippage when the vehicle is turning a corner.

A few aftermarket companies make replacement differentials that use clutch plates much like those in a factory limited-slip differential. These aftermarket differentials offer greater holding power by using higher spring pressures and better gripping clutches.

■ **CAUTION:** A limited-slip equipped vehicle should not be driven with a mini-type spare tire on a drive axle. The size difference between the tires on that axle can cause excessive damage to the clutches.

Cone Clutches

Some vehicles are equipped with a limited-slip differential that uses cone clutches (Figure 7-39) preloaded by five springs. A **cone clutch** is simply a cone covered with frictional material that fits inside an internal cone in the differential case (Figure 7-40). When the two cones are pressed together, friction allows them to rotate as one. The cones' frictional surfaces have spiral grooves cut in them. These grooves allow lubricant to flow through the cones. When the vehicle is moving straight, spring pressure and the separating force created by the pinion gears pushes each clutch cone against the internal cone in the differential case. During cornering, normal differential action overcomes the pressure of the springs, releasing the clutches, and allows the inner axle to slip.

Sure-Grip (limited-slip) differentials are often called **ramp-type differentials**. BMW also uses a differential similar in design to the Sure-Grip.

Some cone-clutched limited-slip differentials, such as Chrysler's Sure-Grip, have beveled ends on the differential's pinion shafts and matching "ramps" cut in the shaft openings of the differential case. When torque is applied to the **ramp-type differential**, the ramps tend to force the side gears apart and apply pressure to the clutch assembly on the axle with the best traction. The cone clutches simultaneously grip the side gears and the inside of the differential case.

One aftermarket cone-clutched differential that is being offered as an option by automobile manufacturers is the Auburn (Figure 7-41). This differential uses interlocking cones to provide holding power when it can be used. These units are made with low spring pressures for street use or with higher spring pressures for higher breakaway torque, for racing only applications.

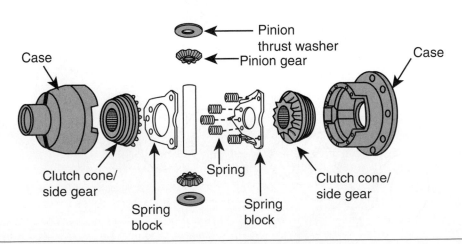

Figure 7-39 Typical cone clutched limited-slip differential.

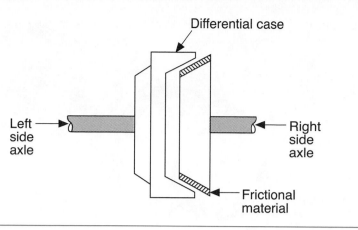

Figure 7-40 Basic construction of a cone-type clutch.

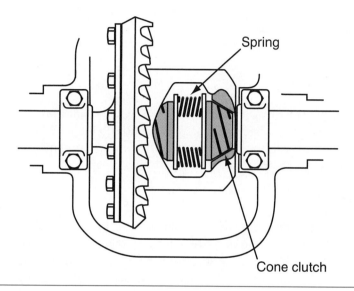

Figure 7-41 An Auburn cone clutched limited-slip differential.

Late-model, high-performance cars from GM may be equipped at the factory with an Auburn limited-slip differential.

When speed differential increases, the viscous torque also increases.

Locked differentials are often called "lockers."

The Auburn Gear limited-slip differential uses cone clutches coupled to beveled side gears. As torque is transmitted through the differential side gears to the axle shafts, the side gear separating forces and spring preload firmly seat the cones into the differential case. The cone design, along with the applied force, determines the torque transfer of the differential. When torque levels decrease—for example, when cornering—the gear separating forces also decrease, allowing the axle shafts to rotate independently.

Viscous Clutch

Some late-model vehicles use a viscous clutch in their limited-slip differentials (Figure 7–42). A viscous limited-slip differential has a **viscous coupling** with alternately positioned steel and frictional plates connected to the two drive axles. The application of the plates relies on the resistance generated by a high-viscosity silicone fluid. When there is no rotational difference between the left- and right-side axles, power is distributed evenly to both axles. When one wheel has less traction than the other, there is a difference in rotational speeds between the axles. This speed differential causes the silicone fluid to shear, generating viscous torque. This torque effectively reduces the difference in speed and reduces the spinning of the wheel with the least traction.

Viscous couplings are often found in four-wheel-drive systems but are also found in the differentials of some performance cars. In 2001, BMW released what it calls a Variable M Differential

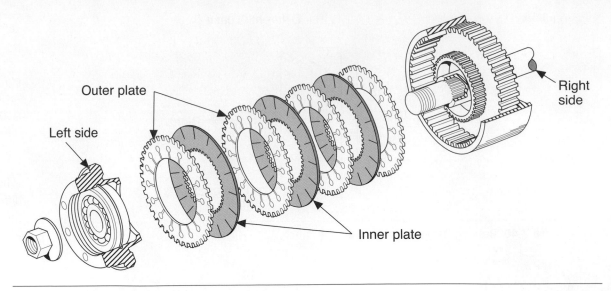

Figure 7-42 Typical viscous clutch assembly.

Lock. This unit is based on a viscous coupling. It uses silicone fluid pressurized by any relative motion between the two rear wheels to clamp a multidisc clutch. When clamped, the clutch directs torque to the wheel with the most traction. The unit has no limit to the ratio of torque it can send to one side or the other.

Gear-Based Units

Manufacturers are using a wide range of limited-slip designs beside the typical clutch-type. These designs were born out of the need to improve vehicle stability and tire traction. Some are torque sensing, such as a Torsen unit. Others are speed-sensitive, such as the Gerodisc differential. Many are gear-based and are commonly called torque-bias or torque-sensing (Torsen) units. This differential (Figure 7-43) automatically splits torque evenly between the left and right rear wheels for

Torsen style differentials may be found as standard equipment on some models from Audi, BMW, Mercedes, and the Honda S2000.

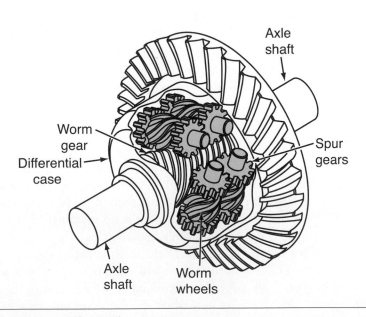

Figure 7-43 A Torsen differential.

maximum traction when needed, and changes modes in such a way that the driver never becomes aware of its actions. The basis of these units is a parallel-axis helical gear set. The Torsen differential multiplies the torque that is available from the wheel that is starting to spin or lose traction and sends it to the slower turning wheel with the better traction. This action is initiated by the resistance between the sets of gears in mesh. Helical geared units do not bind in turns nor do they lose their effectiveness with wear, as clutch-based units.

QUAIFE Automatic Torque Biasing Differential (ATB). The Quaife limited slip differential (Figure 7-44) is an automatic gear-operated torque biasing differential. This differential is an all-mechanical geared unit that does not use clutch packs or preloading to transfer torque from one axle to the other, and torque transfer occurs automatically when one wheel loses traction. Torque transfer also occurs gradually, without steps. It is particularly effective in reducing the effects of FWD torque steer.

Locked Differentials

Another type of special traction differential is the **locked differential**. This provides very limited differential action, if any. It is designed to provide both drive axles with nearly the same amount of power regardless of traction. Needless to say, this differential is designed only for off-road use and for racing applications.

Some trucks, off-the-road equipment, and cars use differentials that can be locked and unlocked by pressing a button. The button activates an air or hydraulic pump, which moves a sleeve that slides across between the drive gears to lock the drive gears or spider gears in place. This type of system gives the advantages of both an open and locked differential.

A commonly found, or at least much talked about, locked differential is the Detroit Locker. This unit is a ratcheting type of locking differential. It is very strong and will almost always provide equal torque application to each axle. It does not allow for much differential action; therefore, cornering is hampered. However, good drivers know when to lift off the throttle right before turning. This action allows time for the locker to unlock and provide some differential action during the turn. Detroit Lockers are primarily used in vehicles built for oval racing, such as NASCAR.

To eliminate all differential action, cars built for drag racing use a spool. A spool is basically a ring gear mounted to an empty differential case. Both the right and the left axles are splined to the case, providing for a solid connection between them. With a spool, even the slightest of turns causes the tires to scrub.

A **locked differential** is a type of differential with the side and pinion gears locked together.

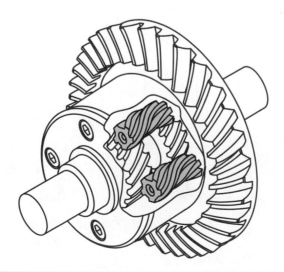

Figure 7-44 A Quaife-type differential.

Operation

When a vehicle is moving straight ahead, the axle shafts are linked to the differential case through the clutch and each wheel gets equal torque. While the vehicle is making a turn, depending on the direction the vehicle is turning, one clutch assembly slips a sufficient amount to allow a speed differential between the two axles. This is necessary because the wheels must move through two different arcs during a turn and must therefore spin at slightly different speeds. When one wheel has less traction than the other, a larger portion of the torque goes to the wheel with the most traction.

Normally, each axle gets an equal amount of torque through the differential. However when one wheel slips, some of that wheel's torque is lost through the pinion gears spinning on the pinion shaft. The clutch on the other wheel remains applied and some of the torque from the slipping side is applied to the wheel with traction. The amount of torque applied to the wheel with traction is determined by the frictional capabilities of its clutch assembly. Power is delivered to that wheel only until the torque overcomes the frictional characteristics of the clutch assembly, at which time it begins to slip. The friction between the clutch plates and discs will transfer a portion of the engine's torque to the wheel with the most traction. This action limits the maximum amount of torque that can be applied to the wheel with traction.

Gerodisc Differentials

Gerodisc differentials (Figure 7-45) are speed-sensitive, slip-limiting differential units. These units contain a clutch pack and a hydraulic pump. The pump is a gerotor-type whose pressure output depends upon rotational speed. This is the speed-sensitive part of the differential. The left axle shaft drives the pump. The output from the pump is fed to the clutch pack and the amount of pressure determines how tightly the clutches will be squeezed together. When the clutch pack is fully engaged, the two drive axles are locked together. This type of differential unit smoothly and progressively sends power to the drive axle that has the best traction.

Operation. When one axle spins faster than the other, the rotational speed of the hydraulic pump increases. This increase in speed increases the pressure from the pump. The pressure is applied to the clutch pack, which begins to lock the two axles together. The amount of tire slip determines the amount of pressure delivered by the pump. The pressure works on the clutch pack to lock the axles. When there is no slip, the pump is not delivering pressure and the differential functions as an open unit. When slippage is high, the axles are locked together. When there is some slippage, the axles are partially locked.

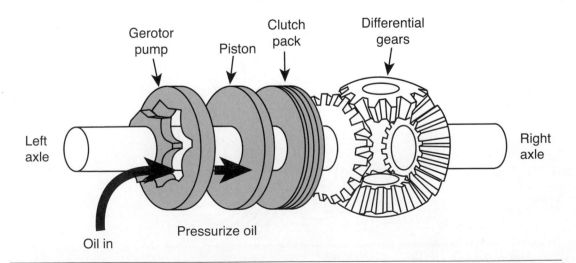

Figure 7-45 A Gerodisc differential setup.

Drive Axle Shafts and Bearings

Located within the hollow horizontal tubes of the axle housing are the axle shafts (Figure 7-46). The purpose of an axle shaft is to transmit the driving force from the differential side gears to the drive wheels. Axle shafts are heavy steel bars splined at the inner end to mesh with the axle side gear in the differential. The driving wheel is bolted to the wheel flange at the outer end of the axle shaft. The drive wheels rotate to move the vehicle forward or reverse.

The drive axles in a transaxle usually have two CV joints to allow independent front-wheel movement and steering of the drive wheels. These CV joints also allow for lengthening and shortening of the drive axles as the wheels move up and down.

The purpose of the axle shaft is to transfer driving torque from the differential assembly to the vehicle's driving wheels. There are two types of axles: the dead axle that supports a load and the **live axle** that supports and drives the vehicle.

There are basically three designs by which axles are supported in a live axle: full-floating, three-quarter floating, and semifloating. These refer to where the axle bearing is placed in relation to the axle and the housing. The bearing of a **full-floating axle** is located on the outside of the housing (Figure 7-47). This places all of the vehicle's weight on the axle housing with no weight on the axle.

Three-quarter and **semifloating axles** are supported by bearings located in the housing and thereby carry some of the weight of the vehicle (Figure 7-48). Most passenger cars are equipped with three-quarter or semi-floating axles. Full-floating axles are commonly found on heavy-duty trucks.

The axle shaft bearing supports the vehicle's weight and reduces rotational friction. With semifloating axles, radial and thrust loads are always present on the axle shaft bearing when the vehicle is moving. Radial bearing loads act at 90 degrees to the axle's center of axis. Radial loading is always present whether or not the vehicle is moving. Thrust loading acts on the axle bearing parallel with the center of axis. It is present on the driving wheels, axle shafts, and axle bearings when the vehicle turns corners or curves.

Three designs of axle shaft bearings are used on semifloating axles: ball-type, straight-roller, and tapered-roller bearings. The load on a bearing that is of primary concern is the axle's end thrust. When a vehicle moves around a corner, centrifugal force acts on the vehicle's body, causing it to lean to the outside of the curve. As the body leans outward, a thrust load is placed on the axle shaft and axle bearing. Each type of axle shaft handles axle shaft end thrust differently.

The end-to-end movement of the axle is controlled by a C-type retainer on the inner end of the axle shaft or by a bearing retainer and retainer plate at the outer end of the axle shaft.

Shop Manual
Chapter 7, page 323

Dead axles are found on trailers and are the type of axle found in the rear of FWD vehicles.

The **live axle** transmits power from the differential to the wheels.

Thrust loads are loads placed on a part that is parallel to the center of the axis of rotation.

Radial loads are loads applied at 90 degrees to an axis of rotation.

Body lean is also called body roll.

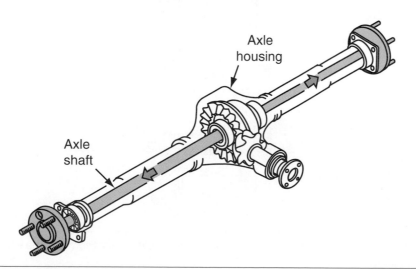

Figure 7-46 Position of drive axles within a RWD axle housing.

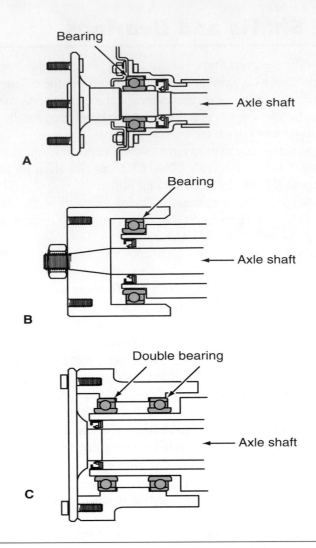

Figure 7-47 The types of rear axle shafts: (A) semifloating; (B) three-quarter floating; and (C) full-floating.

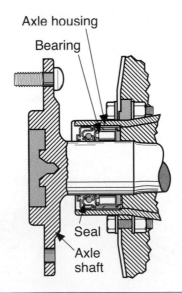

Figure 7-48 Locating the axle bearings in the axle housing places some of the vehicle's weight on the bearing.

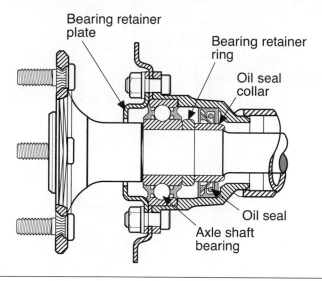

Figure 7-49 An axle shaft with a ball-type bearing.

Ball-Type Axle Bearings

An axle with ball-type axle bearings has the axle shaft and bearing held in place inside the axle housing by a stamped metal bearing retainer plate (Figure 7-49). The plate is bolted to the axle housing and is held in place on the axle shaft by a retaining ring, which is pressed onto the axle shaft.

The operation of the ball-type bearing is designed to absorb radial load as well as the axle shaft end thrust. Because both bearing loads are taken at the bearing, there is no axle shaft end thrust absorption or adjustment designed into the rear axle housing.

To seal in the lubricant, an oil seal collar and oil seal is used. The oil seal collar is a machined sleeve or finished portion of the axle on which the lips of the seal ride. The oil seal retains the gear lubricant inside the axle housing. The axle seal prevents the lubricant from leaking into the brakes.

Straight-Roller Axle Bearings

The straight-roller bearing uses the axle shafts as its inner race (Figure 7-50). The outer bearing race and straight rollers are pressed into the axle tubes of the rear axle housing. The inner end of the axle shaft at the differential has a groove machined around its outside diameter where the **C-type retainer** fits.

The bearing is lubricated by hypoid lubricant from the differential area of the axle housing. The grease seal, located outside the axle shaft bearing, prevents the lubricant from leaking out of the housing.

When the vehicle takes a turn, the body and axle housing move outward and the axle shaft moves inward on the bearings. The inner end of the axle shaft contacts the differential pinion shaft. The axle shaft end thrust exerted against the differential pinion shaft tries to move the differential housing and differential side bearing assembly against the integral housing axle tube. The axle housing absorbs the axle shaft end thrust. There is no end thrust adjustment designed into the rear axle housing.

Tapered-Roller Axle Bearings

The tapered-roller bearing and axle shaft assembly are held inside the axle housing by a flange, which is bolted to the axle housing (Figure 7-51). The inside of the flange may be threaded to receive an adjuster or it is machined to accept adjustment shims.

The axle shaft is designed to float, based on the slight in and out movements of the axles. As the axle shaft moves inward, it contacts a thrust block that separates both axle shafts at the center of the differential. The inward moving axle shaft contacts the thrust block, which passes the

Shop Manual
Chapter 7, page 324

A **C-type retainer** controls the end-to-end or side-to-side movement of the axle.

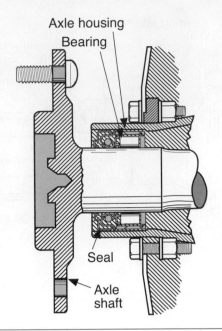

Axle housing
Bearing

Seal

Axle shaft

Figure 7-50 An axle shaft with a straight roller-type bearing.

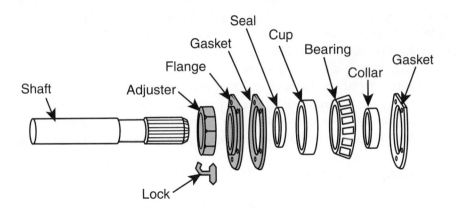

Seal
Gasket   Cup
Bearing
Flange   Gasket
Collar
Shaft   Adjuster

Lock

Figure 7-51 Tapered roller bearing assembly.

thrust force to the opposite axle shaft. There the axle shaft end thrust becomes an outward moving force, which causes the opposite tapered-roller axle bearing to seat further in its bearing cup. Axle end thrust adjustments can be made by a threaded adjuster or thin metal shims placed between the brake assembly plate and the axle housing.

The tapered-roller bearing is lubricated before installation in the axle housing. A seal and two gaskets keep the hypoid lubricant and foreign matter out of the bearing operating area. A collar holds the rear axle bearing in place on the axle shaft.

A BIT OF HISTORY

The use of axle half-shafts and a transverse tube to position the rear wheels of a vehicle was patented in 1894 by De Dion-Bourton. This design of axle is known as the De Dion axle.

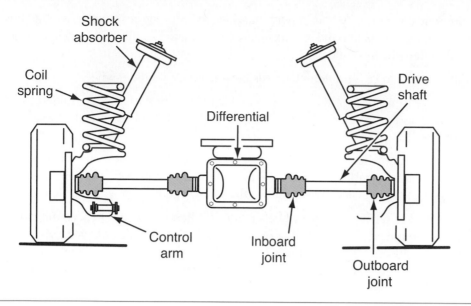

Figure 7-52 A typical IRS drive axle assembly.

IRS Axle Shafts

The drive axles on most newer IRS systems use two U- or CV joints per axle to connect the axle to the differential and the wheels (Figure 7-52). They are also equipped with linkages and control arms to limit camber changes. The axles of an IRS system are much like those of a FWD system. The outer portion of the axle is supported by an upright or locating member that is also part of the suspension.

Swing axles are a form of IRS. In a swing axle car, the differential is also bolted to the chassis but the axles are U-jointed only to where they meet the differential and not at the wheel end. This makes the wheels move in an arc as they move up and down with the suspension. The swinging of the wheels in an arc causes large camber changes and raises and lowers the rear of the car. These motions make the car difficult to drive in some situations, which is why swing axles are not found in many new cars. Swing axles were popular from the late 1940s through the 1960s because they were inexpensive to make and gave a softer ride than a solid live axle.

Camber is a suspension alignment term used to define the amount that the centerline of a wheel is tilted inward or outward from the true vertical plane of the wheel. If the top of the wheel is tilted inward, the camber is negative. If the top of the wheel is tilted outward, the camber is positive.

Summary

❑ The drive axle of a RWD vehicle is mounted at the rear of the car. It is a single housing for the differential gears and axles. It also is part of the suspension and helps to locate the rear wheels.

❑ The final drive is the final set of reduction gears the engine's power passes through on its way to the drive wheels.

❑ A differential is needed between any two drive wheels, whether in a RWD, FWD, or 4WD vehicle, because the two drive wheels must turn at different speeds when the vehicle is in a turn.

❑ RWD final drives use a hypoid ring and pinion gear set, which turns the power flow 90 degrees from the drive shaft to the drive axles. A hypoid gear set also allows the drive shaft to be positioned low in the vehicle.

❑ The differential rotates the driving axles at different speeds when the vehicle is turning and at the same speed when the vehicle is traveling in a straight line.

Terms to Know
C-type retainer
Clutch pack
Cone clutch
Differential case
Drive pinion flange
Final drive
Full-floating axle
Hunting gear set
Integral carrier
IRS
Lapping

❏ The differential is normally housed with the drive axles in the rear axle assembly. Power from the engine enters into the rear axle and is transmitted to the drive axles, which are attached to the wheels of the car.

❏ The differential allows for different speeds between the two drive wheels.

❏ The pinion gear meshes with the ring gear, which is fastened to the differential case. The pinion shafts and gears are retained in the differential case and mesh with side gears splined to the drive axles.

❏ When both driving wheels are rotating at the same speed, the differential pinions do not rotate on the differential pinion shaft; the differential assembly rotates as one and the driving wheels, axles, and axle side gears rotate at the same speed.

❏ When the vehicle turns, the drive wheels rotate at different speeds because the differential case forces the pinion gears to walk around the slow turning axle side gear. This action causes the outside axle side gear to reach a higher speed than the inside wheel. The amount of differential action taking place depends on how sharp the corner or curve is. Differential action provides control on corners and prolongs drive tire life.

❏ Live rear axles use a one-piece housing with two tubes extending from each side. These tubes enclose the axles and provide attachments for the axle bearings. The housing shields all parts from dirt and retains the differential lubricant.

❏ Rear axle housings can be divided into two groups, integral carrier or removable carrier.

❏ An integral carrier housing has a service cover that fits over the rear of the differential and rear axle assembly.

❏ The differential assembly of a removable carrier assembly can be removed from the front of the axle housing as a unit and is serviced on a bench and then installed into the axle housing.

❏ The types of gears currently used as final drive gears are the helical, spiral bevel, and hypoid gears.

❏ With hypoid gears, the drive side of the teeth is curved in a convex shape, whereas the coast side of the teeth is concave. The inner end of the teeth on a hypoid ring gear is known as the toe and the outer end of the teeth as the heel.

❏ The gear ratio of the pinion and ring gear is often referred to as the final drive ratio.

❏ Gear ratios express the number of turns the drive gear makes compared to one turn of the driven gear it mates with.

❏ Ring and pinion gear sets are usually classified as hunting, nonhunting, or partial nonhunting gears.

❏ A nonhunting gear set is one in which any one pinion tooth comes into contact with only some of the ring gear teeth and is identified by a .00 gear ratio.

❏ A partial nonhunting gear set is one in which any one pinion tooth comes into contact with only some of the ring gear teeth, but more than one revolution of the ring gear is required to achieve all possible gear tooth contact combinations. These gears are identified by a .50 ratio.

❏ When hunting gear sets are rotating, any pinion gear tooth will contact all the ring gear teeth.

❏ During assembly the nonhunting and partial nonhunting gears must be assembled with the timing marks properly aligned. Hunting gears do not need to be aligned because any tooth on the pinion may mesh with any tooth on the ring gear.

❏ At least four bearings are found in all differentials. Two fit over the drive pinion shaft to support it while the other two support the differential case.

❑ The drive pinion gear is placed horizontally in the axle housing and is positioned by one of two types of mounting, either straddle or overhung.

❑ The straddle-mounted pinion gear is used in most removable carrier-type axle housings and uses two opposing tapered-roller bearings positioned close together with a short spacer between their inner races and a third bearing to support the rear of the pinion gear.

❑ The overhung-mounted pinion uses two opposing tapered-roller bearings but not a third bearing.

❑ A spacer is placed between the opposing pinion shaft bearings to control the amount of preload applied to the bearings. Preload prevents the pinion gear from moving back and forth in the bearing retainer.

❑ The differential section of the transaxle has the same components as the differential gears in a RWD axle and basically operates in the same way, except that the power flow in transversely mounted powertrains does not need to turn 90 degrees.

❑ The final drive gears and differential assembly are normally located within the transaxle housing of FWD vehicles.

❑ There are three common configurations used as the final drives on FWD vehicles: helical, planetary, and hypoid. The helical and planetary final drive arrangements are usually found in transversely mounted powertrains. Hypoid final drive gear assemblies are normally used with longitudinal powertrain arrangements.

❑ A limited-slip unit provides more driving force to the wheel with traction when one wheel begins to spin by restricting the differential action.

❑ Most limited-slip differentials use either a clutch pack, a cone clutch, or a viscous clutch assembly.

❑ There are few variations of limited-slip differentials used by various manufacturers. These are torque or speed sensing.

❑ A locked differential provides very limited differential action and is designed to provide both drive axles with nearly the same amount of power regardless of traction.

❑ Gerodisc differentials are speed-sensitive slip-limiting differential units that have a clutch pack and a hydraulic pump.

❑ The purpose of the axle shaft is to transfer driving torque from the differential assembly to the vehicle's driving wheels.

❑ There are basically three ways drive axles are supported by bearings in a live axle: full-floating, three-quarter floating, and semifloating.

❑ There are three designs of axle shaft bearings used on semifloating axles: ball-type, straight-roller, and tapered-roller bearings.

Review Questions

Short Answer Essays

1. Define the term final drive.

2. Explain why a differential prolongs tire life.

3. List the reasons why hypoid gears are used in nearly all RWD final drives.

4. List the three major functions of a typical RWD differential.

5. Describe the main components of a differential and state their locations.

6. Explain the major differences between an integral carrier housing and a removable carrier housing.

7. Explain the differences between hunting, nonhunting, and partial nonhunting gears.

8. What do limited-slip differential units with a viscous coupling rely on to send power to the axle with the most traction?

9. What applies the clutch pack in a limited-slip differential assembly?

10. List the different ways drive axles are supported in an axle housing and explain the major characteristics of each one.

Fill-in-the-Blanks

1. The differential's _____ gear meshes with the _____ gear, which is fastened to the differential case, which houses the _____ shafts and gears, which are in mesh with the _____ gears, which are splined to the drive axles.

2. Rear axle housings can be divided into two groups: _____ carrier and _____ carrier.

3. The types of gears currently used as differential gears are the _____, _____ _____ , and _____ .

4. The drive side of a hypoid's gear teeth is curved in a _____ shape, and the coast side of the teeth is _____ . The inner end of the teeth on a hypoid ring gear is known as the _____ and the outer end of the teeth as the _____.

5. Gear ratios express the number of turns the _____ gear makes compared to one turn of the _____ gear it mates with.

6. Ring and pinion gear sets are usually classified as _____ , _____ , or _____ gears.

7. A straddle-mounted pinion gear is usually mounted on _____ bearings, whereas an overhung-mounted pinion is mounted on _____ .

8. The three common configurations used as the final drives on FWD vehicles are: _____ , _____ , and _____ .

9. Most limited-slip differentials use either a _____ , a _____ , or a _____ to limit the action of the differential.

10. _____ differentials are speed-sensitive slip-limiting differential units that contain a clutch pack and a hydraulic pump.

Multiple Choice

1. *Technician A* says that when a car is making a turn, the outside wheel must turn faster than the inside wheel. *Technician B* says that a locked differential may cause the car to slide around a turn. Who is correct?
   A. A only
   B. B only
   C. Both A and B
   D. Neither A nor B

2. When discussing the torque multiplication factor of the differential, *Technician A* says that all of the gears in a differential affect torque multiplication. *Technician B* says that there is a gear reduction as the power flows from the pinion to the ring gear. Who is correct?
   A. A only
   B. B only
   C. Both A and B
   D. Neither A nor B

3. *Technician A* says that when a car is moving straight ahead, all differential gears rotate as a unit. *Technician B* says that when a car is turning a corner, the inside differential side gear rotates slowly on the pinion, causing the outside side gear to rotate faster. Who is correct?
   A. A only
   B. B only
   C. Both A and B
   D. Neither A nor B

4. When discussing the mounting of the drive pinion shaft, *Technician A* says that drive pinion shafts may be mounted in a long bushing. *Technician B* says that drive pinion shafts may be held by two tapered-roller bearings. Who is correct?
   A. A only
   B. B only
   C. Both A and B
   D. Neither A nor B

5. When discussing gear ratios, *Technician A* says that they express the number of turns the driven gear makes compared to one turn of the drive gear. *Technician B* says that the gear ratio of a differential unit expresses the number of teeth on the ring gear compared to the number of teeth on the pinion gear. Who is correct?
   A. A only
   B. B only
   C. Both A and B
   D. Neither A nor B

6. When discussing final drive gear ratios, *Technician A* says that lower gear ratios allow for better acceleration. *Technician B* says that higher gear ratios allow for improved fuel economy but lower top speeds. Who is correct?
   A. A only
   B. B only
   C. Both A and B
   D. Neither A nor B

7. When discussing different types of ring and pinion gear sets, *Technician A* says that with a nonhunting gear set, each tooth of the pinion will always return to the same few teeth on the ring gear each time the pinion rotates. *Technician B* says that when a hunting gear set rotates, any pinion gear tooth is likely to contact each and every tooth on the ring gear. Who is correct?
   A. A only
   B. B only
   C. Both A and B
   D. Neither A nor B

8. When discussing limited-slip differentials, *Technician A* says that these differentials improve handling on slippery surfaces. *Technician B* says that these differentials limit the amount of differential action between the side gears. Who is correct?
   A. A only
   B. B only
   C. Both A and B
   D. Neither A nor B

9. When discussing the different designs of rear axles, *Technician A* says that the bearings for full-floating shafts are located within the axle tubes of the rear axle housing. *Technician B* says that the names used to classify the different designs actually define the amount of vehicle weight that is supported by the axles. Who is correct?
   A. A only
   B. B only
   C. Both A and B
   D. Neither A nor B

10. *Technician A* says that limited-slip differentials require a special lubricant. *Technician B* says that all differential units require a special hypoid compatible lubricant. Who is correct?
    A. A only
    B. B only
    C. Both A and B
    D. Neither A nor B

Four-Wheel-Drive Systems

Upon completion and review of this chapter, you should be able to:

- ❏ Explain the advantages and disadvantages of four-wheel drive.
- ❏ Use the correct terminology when discussing four-wheel-drive systems.
- ❏ Describe the different designs of four-wheel-drive systems and their applications.
- ❏ Compare and contrast the components of part- and full-time four-wheel-drive systems.

- ❏ Describe the operation of various transfer case designs and their controls.
- ❏ Identify the differences in operation and construction between manual and automatic locking front-wheel hubs.
- ❏ Identify the suspension requirements of vehicles equipped with four-wheel drive.

With the popularity of SUVs and pickup trucks, the need for technicians that can diagnose and service four-wheel-drive systems has drastically increased. This type of vehicle has topped the sales charts because it offers utility and comfort. Although all-wheel-drive passenger cars are available, most prospective buyers for all-wheel- and four-wheel-drive vehicles are opting for truck-based SUVs and pickups (Figure 8-1).

A BIT OF HISTORY

The first known gasoline-powered four-wheel-drive automobile was the Spyker, built in the Netherlands in 1902.

Introduction

The primary focus of this chapter is on the transfer cases and related systems used for four-wheel-drive (4WD) on light trucks (Figure 8-2), automobiles, and off-the-road vehicles. Although the principles of operation are the same for all 4WD units, the components, location, and controls of the various systems differ according to manufacturer and application.

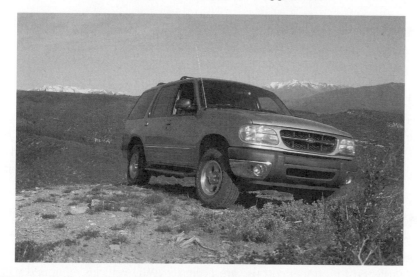

Figure 8-1 SUVs with 4WD are very popular.

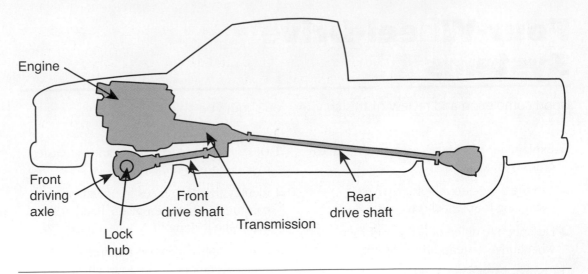

Figure 8-2 Typical arrangement of 4WD components.

The common acronym for Four-Wheel Drive is 4WD.

The designation of 4 × 4 is used with vehicles to indicate the number of wheels on the ground and the number of wheels that can be driven. For example, the typical car has four wheels, two of which can deliver power; therefore they could be designated as 4 × 2 vehicles. A 4WD vehicle is designated as a 4 × 4 because it has four wheels and power can be applied to all four wheels.

Four-wheel-drive and all-wheel-drive (AWD) systems can dramatically increase a vehicle's traction and handling ability in rain, snow, and off-road driving. Consider that the vehicle's only contact with the road is the small area of the tires. Driving and handling is vastly improved if the workload is spread out evenly among four wheels rather than two.

Factors such as the side forces created by cornering and wind gusts have less effect on vehicles with four driving wheels. The increased traction also makes it possible to apply greater amounts of energy through the drive system. Vehicles with 4WD and AWD can maintain control while transmitting levels of power that would cause two wheels to spin either on take off or while rounding a curve. The improved traction of 4WD and AWD systems allows them to cut through snow and water rather than hydroplane over it.

Both 4WD and AWD systems add initial cost and weight. With most passenger cars, the weight problem is minor. A typical 4WD system adds approximately 170 pounds to a passenger car. An AWD system adds even less weight. The additional weight in larger 4WD trucks can be as much as 400 pounds or more.

Vehicles equipped with 4WD and AWD require special service and maintenance not performed on 2WD vehicles. However, the slight disadvantages of 4WD and AWD are heavily outweighed by the traction and performance these systems offer. Their popularity is increasing at a rapid rate and technicians must be prepared to diagnose and repair these systems.

SERVICE TIP: Whenever diagnosing, servicing, or repairing a 4WD performance car, light truck, or off-the-road vehicle, refer to the appropriate service manual for specific information and service procedures for that vehicle.

Where the engine's torque is sent depends on the components of the 4WD system. If the vehicle is equipped with an open differential, power is applied to two drive wheels (one in the front and one in the rear). If the vehicle has a locking differential, it is possible for torque to be delivered to all four wheels. However, it is also possible for AWD vehicles to drive only one wheel. Understanding how each of the components works will allow you to determine if the system is working properly.

Four-Wheel-Drive Design Variations

Four-wheel drive is most useful when a vehicle is traveling off the road or on any slippery surface. 4WD vehicles designed for off-the-road use are normally RWD vehicles equipped with a **transfer case** (Figure 8-3), a front drive shaft, and a front differential and drive axles. Many 4WD vehicles use

Shop Manual
Chapter 8, page 345

three drive shafts. One short drive shaft connects the output of the transmission to the transfer case. The output from the transfer case is then sent to the front and rear axles through separate drive shafts.

Some high-performance cars are equipped with 4WD to improve the handling characteristics of the car. Nearly all of these cars are FWD models converted to 4WD. Normally, FWD cars are modified by adding a transfer case, a rear drive shaft, and a rear axle with a differential (Figure 8-4). Although this is the typical modification, some cars are equipped with a **center differential** (Figure 8-5) in place of the transfer case. This differential unit allows the rear and front wheels to turn at different speeds and with different amounts of torque.

The **center differential** is commonly referred to as the interaxle differential. It allows the front and rear wheels to turn at different speeds and with different amounts of torque.

Figure 8-3 A typical transfer case.

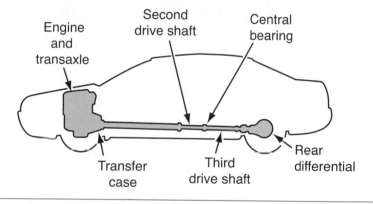

Figure 8-4 An AWD system based on a FWD platform.

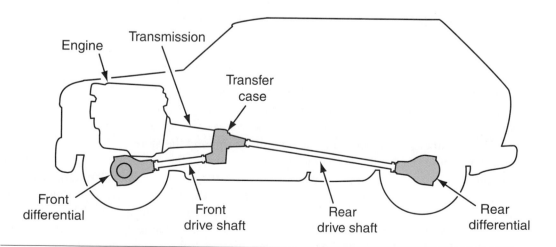

Figure 8-5 Location of front, rear, and center differentials.

The first 4WD car produced in the United States was the F.W.D. in 1908.

Terminology

As other terms commonly used by the automotive industry, the term "all-wheel-drive" (AWD) can mean many different things. The most common understanding of AWD is that it is a system that constantly provides power to all four wheels (Figure 8-6). Furthermore, these systems do not have a "low" operating range and are designed for light off-the-road use. On the contrary, 4WD usually is understood as a system that needs to be engaged by the driver in order to drive all four wheels. In addition, these systems have a "low" range and are suited for heavy off-the-road use. Often, "full-time 4WD" means AWD, or vice versa.

Other terms also are used to help in the understanding of the drive system. But for many, this only confuses the concept. This is especially true when manufacturers are not very technically careful about how they market their vehicles. Of course, as they develop new technologies for 4WD or AWD, they want to make sure the public knows a change or advance has been made. Selling automobiles is their business, so who can blame them? To help clear this up, let us look at what the authorities say.

The Society of Automotive Engineers (SAE) has the role of recommending specifications and practices for the automotive industry. Most are totally accepted by the manufacturers and some are included in government safety or emission control mandates. SAE defines all-wheel-drive as every system that sends power to four wheels. Based on that definition, all of the systems discussed in this and the previous chapters would be called all-wheel-drive systems.

SAE does attempt to clear this up by offering other terms to label the various systems. These terms are what will be used in this chapter, unless the manufacturer calls it something else. When this happens, hopefully you will recognize what type of system it really is so you can better understand how the system should work and how to diagnose it properly. According to SAE, part-time 4WD systems are manually controlled systems. These are the systems that were discussed in Chapter 7. They operate as a 2WD until the driver does something to activate or engage the other axle. Then torque is divided between the two axles and all four wheels can be driven. These systems have a fixed torque split between the front and rear. They also do not corner well at speed when 4WD is engaged because tire slip is the only way these systems accommodate for the differences in travel distance between the front and rear tires as the vehicle goes through a turn.

SAE says all-wheel-drive and full-time 4WD means the system is permanently AWD. Basically, these systems provide torque to four wheels regardless of conditions and the driver cannot select 2WD.

On-demand 4WD systems are those that are automatically controlled. Typically, these systems drive one axle until some slip is detected. Some systems send torque to the opposite axle when slip is anticipated. When there is slip, torque is sent to the other axle. The amount of torque transmitted to the other axle depends on the amount of slip and the system.

Some 4WD systems give the driver the option of selecting full-time 4WD or 2WD. In many cases, the full-time mode is really on-demand 4WD. When the driver selects 4WD Auto, nearly all of the power is sent to one drive axle. As soon as that axle experiences some slip, some torque is transferred to the other axle. An example of this is Ford's Control-Trac system. When the 4WD Auto mode is selected, 96 percent of the engine's torque is sent to the rear wheels until slippage or wheel spin is detected. In response to the slippage, the control computer energizes an electromagnetic clutch in the transfer case. The action of the clutch transmits torque toward the front

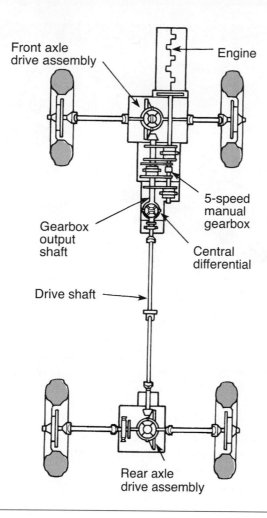

Figure 8-6 A typical AWD powertrain.

axle. The amount of torque sent to the front axle depends on the amount the rear wheels are slipping. Up to 96 percent of the torque can be sent to the front axle. This type of system is best described as a nonpermanent full-time system.

Components

The typical 4WD system consists of a front-mounted, longitudinally-positioned engine; either an automatic or manual transmission; front- and rear-drive shafts; front- and rear-drive axle assemblies; and a transfer case (Figure 8-7).

The transfer case is usually mounted to the side or rear of the transmission. When a drive shaft is not used to connect the transmission to the transfer case, a chain or gear drive (Figure 8-8), within the transfer case receives the engine's power from the transmission and transfers it to the drive shafts leading to the front and rear drive axles.

The transfer case itself is constructed similarly to a standard transmission. It uses shift forks to select the operating mode, plus splines, gears, shims, bearings, and other components found in manual and automatic transmissions. The outer case of the unit is made of cast iron or aluminum and is filled with lubricant (oil) that cuts friction on all moving parts. Seals hold the lubricant in the case and prevent leakage around shafts and yokes. Shims set up the proper clearance between the internal components and the case.

It must be kept in mind that vehicles with two drive axles may have different gear ratios between the front- and rear-drive axles, resulting in a pull-push action. The result of having the two axle ratios is a phenomenon called **driveline windup**. Driveline windup can be explained

Driveline windup is a reaction that takes place as a result of the transfer of torque to the rear wheels.

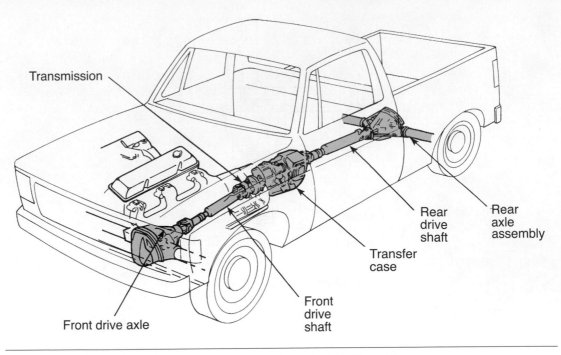

Figure 8-7 Location of major components of a typical 4WD vehicle.

by associating the driveline to a torsion bar. The driveline twists up when both driving axles are rotating at different speeds, pushing and pulling the vehicle on hard, dry pavement. Also remember that neither the front- nor rear-axle has any compensating factor for speed and gear ratio differences between the front- and rear-drive axles.

Driveline windup can cause handling problems, particularly when rounding turns on dry pavement. This is because the front-axle wheels must travel farther than the rear-axle wheels when rounding a curve. On wet or slippery roads, the front and rear wheels slide enough to prevent damage to the driveline components. However, this may not be the case on dry surfaces. This is why many older 4WD systems that do not include components to dissipate driveline windup can only be safely driven on wet or slippery surfaces.

2WD is an acronym commonly used for two-wheel-drive vehicles.

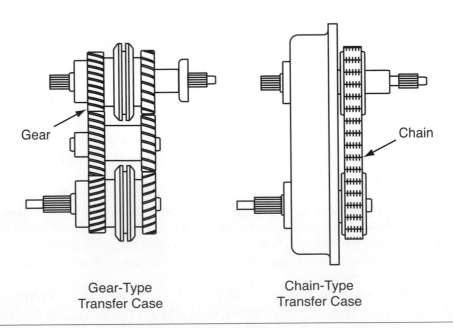

Figure 8-8 Typical gear and chain drive assemblies.

The drive shafts from the transfer case shafts connect to differentials at the front and rear drive axles. As on 2WD vehicles, these differentials are used to compensate for road and operating conditions by altering the speed of the wheels connected to the axles. This is important when the vehicle is turning a corner and the outside wheel must travel a farther distance than the inner wheel.

Universal joints are used to connect the drive shafts to the differential and the transfer case. The rear axles are either connected directly to the hub of the wheels or are connected to the hubs by U-joints. U-joints are also normally used to connect the front axles to the wheel hubs on heavy-duty trucks. Light-duty vehicles and 4WD passenger cars generally use half shafts and CV joints in their front-drive axle assembly (Figure 8-9).

An electric switch or shift lever, located in the passenger compartment, controls the transfer case so that power is directed to the axles selected by the driver. Power can typically be directed to all four wheels, two wheels, or none of the wheels. On many vehicles, the driver also can select a low-speed range for extra torque while traveling in adverse conditions.

Although most 4WD trucks and utility vehicles are design variations of basic RWD vehicles, most passenger cars and small SUVs equipped with 4WD are based on FWD designs. These modified FWD systems consist of a transaxle and differential to drive the front wheels, plus some type of mechanism for connecting the transaxle to a rear driveline. In many cases this mechanism is a simple clutch or differential. On some models, a transfer case that transfers power to the rear axle is fitted to the transaxle (Figure 8-10).

> **AUTHOR'S NOTE:** Customers should be made aware that wear on 4WD systems is much greater than on a 2WD transaxle or transmission. This is especially true if the driver leaves the vehicle engaged in 4WD on dry pavement. Some manufacturers will not warranty the parts of the 4WD system if there is evidence of abuse, such as operation of 4WD on dry surfaces.

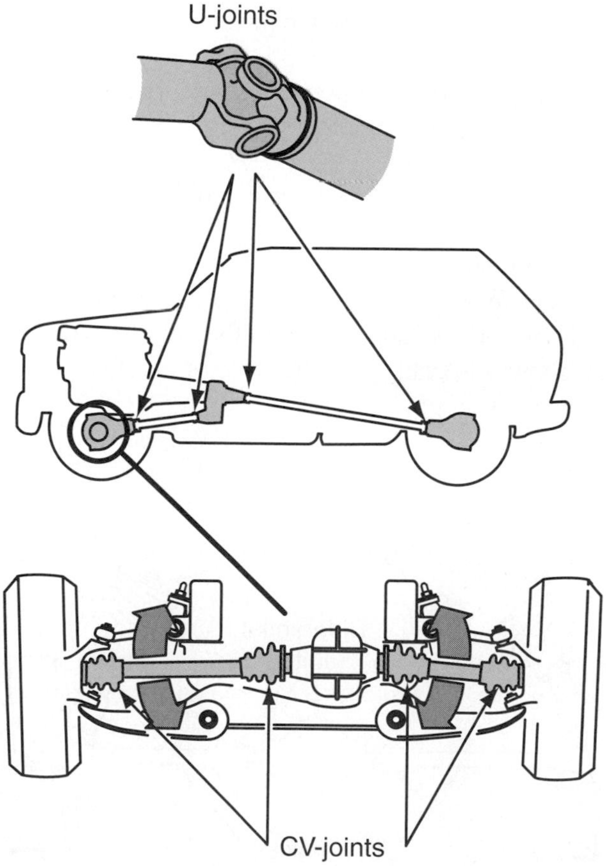

Integrated full-time 4WD systems use computer controls to enhance full-time operation, adjusting the torque split depending on which wheels have traction.

Typically, on-demand 4WD systems power a second axle only after the first begins to slip.

Figure 8-9 Location of CV and U-joints on a typical 4WD vehicle.

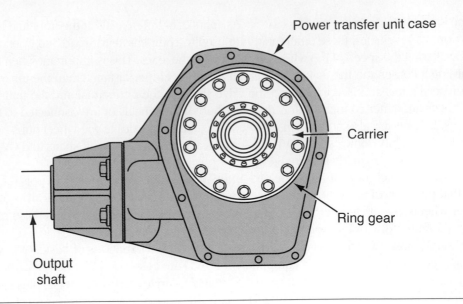

Figure 8-10 A power transfer unit (transfer case).

4WD Systems

A locking differential is one that under certain circumstances will allow no speed difference between two wheels or drive shafts. This can aid in traction on loose or slippery surfaces.

Shop Manual
Chapter 8, page 345

The rear drive axle of a 4WD vehicle is identical to those used in 2WD vehicles. The front drive axle is also like a conventional rear axle, except that it is modified to allow the front wheels to steer (Figure 8-11). Further modifications are also necessary to adapt the axle to the vehicle's suspension system. The differential units housed in the axle assemblies are similar to those found in a RWD vehicle.

The front differential is generally an open differential because of the great differences in wheel rotational speed when steering around corners. Some off-the-road vehicles have lockable front differentials for extreme conditions.

The rear differential is necessary to allow for speed differences when turning corners and to avoid stresses on the drivetrain. Open differentials split torque evenly between both axles or output shafts. This action can be modified by the inclusion of a limited-slip differential.

Although 4WD offers increased traction, it also has disadvantages. The additional axle, differential, drive shaft, and transfer case add weight to the vehicle and therefore decrease its fuel economy. Also, less horsepower is available to the wheels because of the power lost in turning the additional axle assembly.

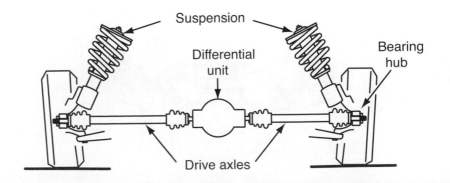

Figure 8-11 Typical front drive axle assembly.

Most 4WD units are equipped to allow the driver to select in and out of 4WD. The systems capable of operating in both 2WD and 4WD are called **part-time 4WD** systems. **Full-time 4WD** systems cannot be selected out of 4WD. The selection of 2WD or 4WD is controlled by a shifter, an electric switch (Figure 8-12), or locking axle hubs (Figure 8-13).

Part-Time Systems

Part-time 4WD systems are designed to be used only when driving off the road or on slippery surfaces. When 4WD is engaged, a part-time system locks the front and rear axles together. Therefore, the system does not allow for axle speed differences. When the vehicle is turning a corner on a slippery surface this doesn't present a problem, as the tires can easily skid or spin across the slick surface to

Figure 8-12 Typical 2WD/4WD selector switch.

Figure 8-13 Typical locking hub.

Part-time 4WD systems can be shifted in and out of 4WD and are designed to be used only when driving off the road or on slippery or wet surfaces.

Full-time 4WD systems use a center differential or similar device that accommodates speed differences between the two axles, which is necessary for on-highway operation.

accommodate the speed differences. However, on dry surfaces, the required speed differentials cause the tires to scrub against the pavement. A part-time 4WD system delivers equal power to each axle while in 4WD. Therefore, the driver must shift the transfer case out of 4WD when driving on a dry surface.

A transfer case is equipped with a gear or **chain drive** to transmit power to one or both of the drive axles. Some transfer cases are equipped with two speeds: a high range for normal driving and a low range for especially difficult terrain, such as deep sand or a steep grade. The low range of the transfer case lowers all of the transmission's ratios, enabling the engine's power to be used more effectively in demanding situations. Most part-time systems and some full-time systems are equipped with a two-speed transfer case.

Two-speed transfer cases are controlled by a shift lever that typically has four positions (Figure 8-14): 2WD High, which engages only the rear axle and is used for all dry-road driving, 4WD High, which engages both axles and is used at any speed on slippery surfaces (Figure 8-15), NEUTRAL which disengages both axles, and 4WD Low, which engages both axles, and lowers the ratios of the entire driveline. 4WD Low should only be used at low speeds and on very demanding terrain.

Other transfer cases are a single speed and only allow the driver to select between 2WD and 4WD modes. This switching is accomplished by an electric switch or a shift lever. The vehicle speed at which this changeover is permitted depends on the design of the system. Some systems require the vehicle to be at a stop before shifting into or out of 4WD; others allow the change at any speed.

Some FWD vehicles are fitted with 4WD and use a compact transfer case bolted to the front-drive transaxle. A drive shaft assembly carries the power to the rear differential. The driver can switch from 2WD to 4WD by pressing a dashboard switch, which activates a solenoid vacuum valve that applies vacuum to a diaphragm unit in the transfer case. The linkage of the diaphragm unit locks the output of the transaxle to the input shaft of the transfer case.

Transfer Cases

The transfer case delivers power to both the front- and rear-drive axle assemblies and is constructed much like a conventional transmission (Figure 8-16). It uses shift forks to select the operating mode, plus splines, gears, shims, bearings, and other components commonly found in transmissions. The outer case is made of magnesium, aluminum, or cast iron and is filled with lubricant. Seals are used to hold the lubricant in and keep dirt out. Shims are used to maintain proper clearances or preloads between the parts of the transfer case.

The purpose of a transfer case is to transfer torque from the output of the transmission to the vehicle's front and rear axles. The transfer case is normally connected to each axle by two

"Shift on the fly" 4WD is simply a system that can be shifted from two- to four-wheel drive while the vehicle is moving.

Shop Manual
Chapter 8, page 357

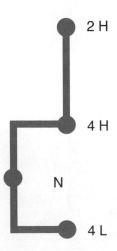

Figure 8-14 Typical shift lever positions.

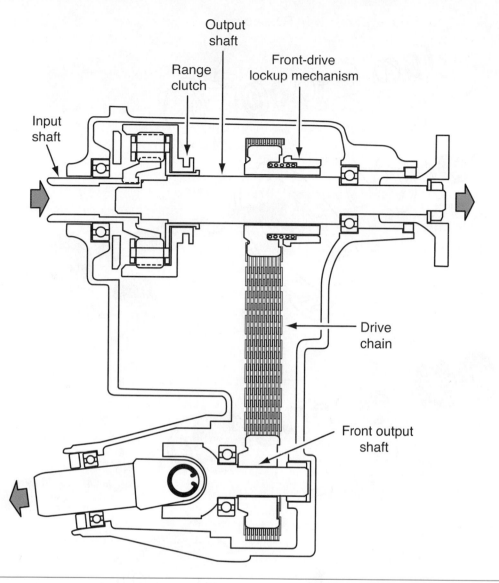

Figure 8-15 Power flow through a typical transfer case when it is in 4WD High.

drive shafts—one between the transfer case and the front axle and the other between the transfer case to the rear axle. Each axle has its own differential unit, which then turns the wheels. Torque is then multiplied by the gear reduction of the differential units and sent to the wheels.

Modes of Operation

A transfer case is an auxiliary transmission mounted to the side, or in the back of, the main transmission (Figure 8-17). There are basically three classifications of transfer cases: part-time, which provides the following ranges—NEUTRAL, 2WD HIGH, 4WD HIGH, and 4WD LOW; full-time, which provides 2WD HIGH, 4WD HIGH; and 4WD LOW; and part-time/full-time, which provides 2WD HIGH, full-time 4WD HIGH, and part-time 4WD LOW.

The change from the high-speed range to the low-speed range is made by moving the shift lever of the transfer case, which moves a gear on the transfer case's main drive shaft from the high-speed drive gear to the low-speed drive gear. High speed provides direct drive, whereas low speed usually produces a gear ratio of about 2:1.

Low speed is a torque multiplication gear that allows for increased power to the wheels and is an addition to the gear reductions of the transmission and final drive gears. When the transfer

Shop Manual
Chapter 8, page 357

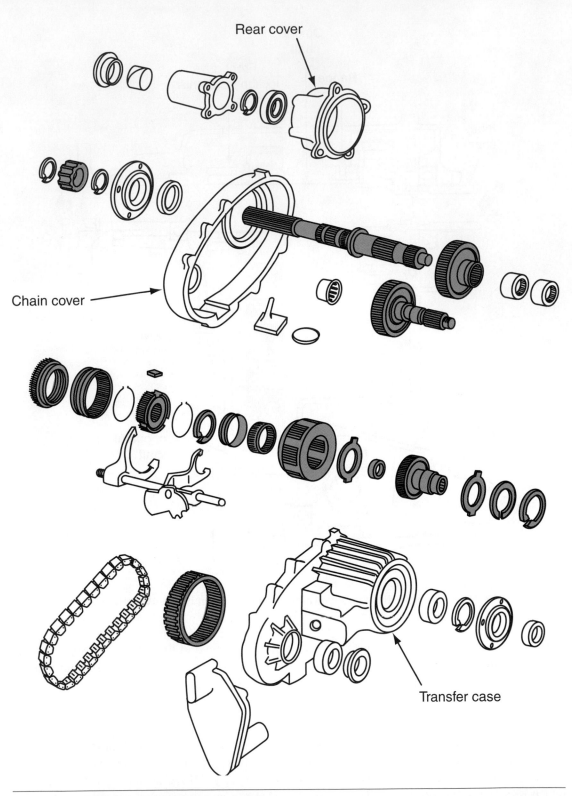

Figure 8-16 Typical transfer case.

case is in high speed, torque is only multiplied by the transmission and final drive gears. In addition to these two speeds, transfer cases may also have a neutral position. When the transfer case is in neutral, regardless of which speed gear the transmission is in, no power is applied to the wheels. Some 4WD vehicles are equipped with a single-speed transfer case and do not have a low range.

Most transfer cases use a planetary gear set (Figure 8-18) to provide for the different gear positions. Although the shifting mechanisms vary with the different models of transfer cases, the

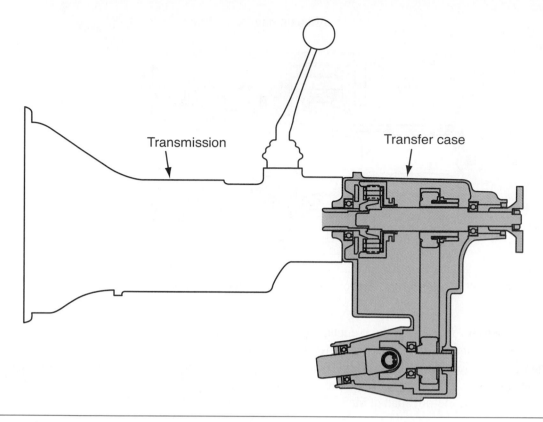

Figure 8-17 Location of transfer case on a transmission.

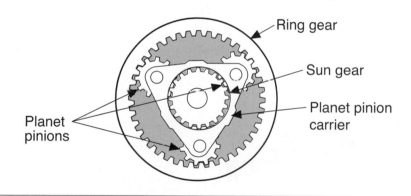

Figure 8-18 A simple planetary gear set.

power flow through nearly all transfer cases is the same. The shift mechanism moves the planetary gear set carrier and the **sun gear** along a splined section of the input shaft. This movement engages and disengages various gears, which results in the different gear selections. A description of the power flow through a typical two-speed transfer case follows.

When in neutral, the transfer case is driven directly by the main output shaft of the transmission. Power is not transmitted to the driving axles when the transfer case is in neutral, regardless of transmission gear position (Figure 8-19). With the transfer case in neutral, the sun gear turns the planetary gears, which drive the ring gear (often called the **annulus gear**). This gear set rotates with the input shaft. The planetary gear case, which is splined to the rear output shaft, remains stationary because the gear set is positioned away from and not engaged with the case. Therefore, power is not transmitted to the rear output shaft.

The **sun gear** is the central gear in a planetary gear set around which the rest of the gears rotate.

The ring gear of a planetary gear set is often called the **annulus gear**.

207

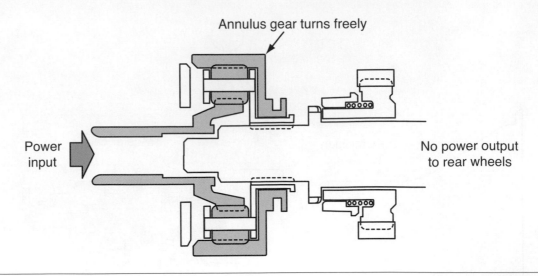

Annulus gear turns freely

Power input

No power output to rear wheels

Figure 8-19 Power flow through a transfer case when it is in neutral.

When in 2WD HIGH, the planetary gear set moves rearward and the clutch shift fork holds the sliding clutch in the 2WD position. As the planetary gear set moves rearward, it locks into the planetary gear case. This prevents the rotation of the planetary gears on their axes, causing the planetary gears, planetary case, and the ring gear to rotate as a single unit. As the gear set rotates, it drives the rear output shaft at the same speed as the input shaft (Figure 8-20). This provides for direct drive.

When in 4WD HIGH, the planetary gear set is in the same position as in 2WD HIGH, but the 4WD clutch shift fork is moved forward. This movement releases the sliding clutch, which is splined to the driving chain **sprocket** carrier gear (Figure 8-21). The clutch shift spring pushes the sliding clutch into engagement with the rear output shaft. This causes the chain to drive the front output shaft at the same speed as the rear output shaft, thus sending power to both axles and providing 4WD.

When in 4WD LOW, the 4WD lockup shift collar is positioned to provide for 4WD. However, the shifter moves the sun gear and planetary gear set assembly rearward, causing the ring gear to engage with a locking ring that is part of the bearing retainer assembly (Figure 8-22) that

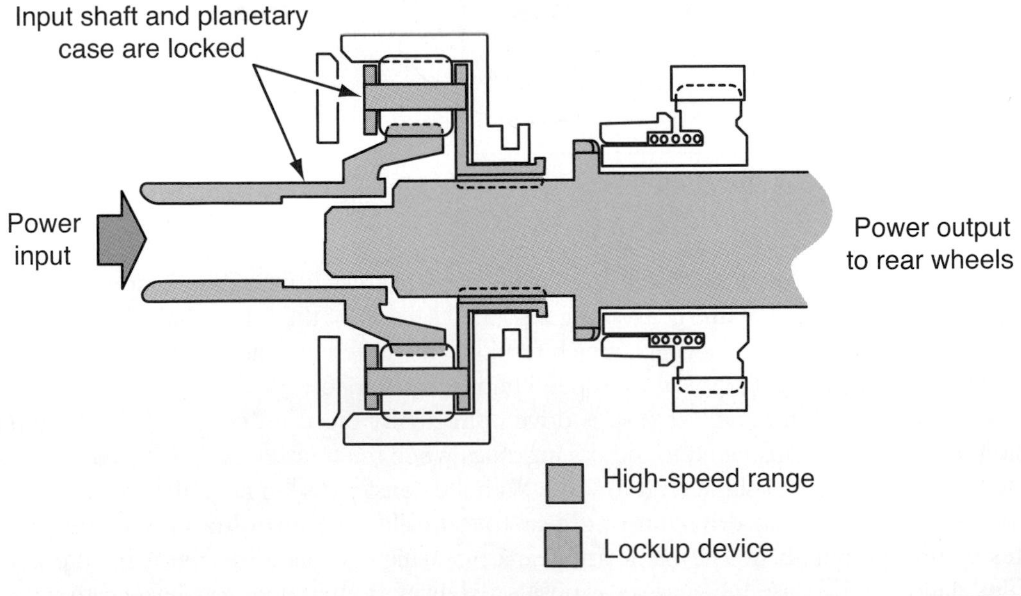

A **sprocket** is a projecting tooth on a wheel or cylinder that engages with the links of a chain to make it move forward.

Input shaft and planetary case are locked

Power input

Power output to rear wheels

High-speed range

Lockup device

Figure 8-20 Power flow through a transfer case when it is in 2WD High.

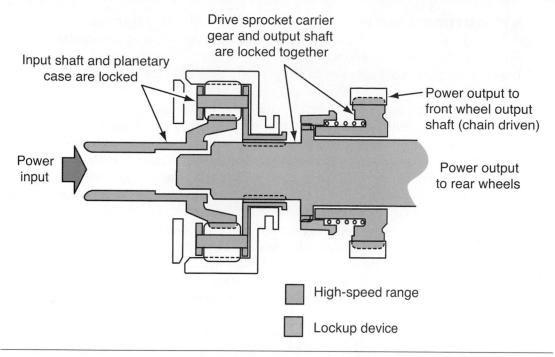

Input shaft and planetary
case are locked

Drive sprocket carrier
gear and output shaft
are locked together

Power output to
front wheel output
shaft (chain driven)

Power
input

Power output
to rear wheels

High-speed range

Lockup device

Figure 8-21 Power flow through a typical transfer case when it is in 4WD High.

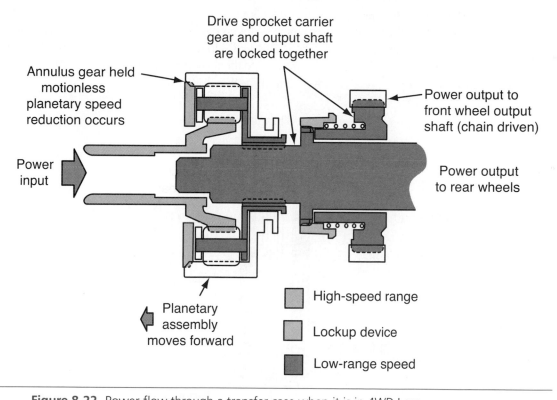

Drive sprocket carrier
gear and output shaft
are locked together

Annulus gear held
motionless
planetary speed
reduction occurs

Power output to
front wheel output
shaft (chain driven)

Power
input

Power output
to rear wheels

Planetary
assembly
moves forward

High-speed range

Lockup device

Low-range speed

Figure 8-22 Power flow through a transfer case when it is in 4WD Low.

holds the ring gear stationary and allows the planetary gears to "walk around" the inside of the ring gear. This causes the planetary gears to drive the planetary case around more slowly than the input shaft speed. Because the planetary case hub is splined directly to the output shaft, the output shaft rotates at the slower speed. This speed reduction results in increased torque at the drive wheels.

AUTHOR'S NOTE: Planetary gear sets can be difficult to figure out. The best way to get a good feel for how all of the potential gear combinations are possible is to play with a planetary unit. Hold and turn various parts of the gear set and watch what happens. Then duplicate the above positions to verify them.

Shop Manual
Chapter 8, page 358

Some transfer cases use a spur or helical gear set to provide for the speed ranges and for engagement and disengagement of 4WD. The front axle is engaged by shifting a sliding or clutching gear inside the transfer case into engagement with the driven gear on the drive shaft for the front wheels (Figure 8-23). The sliding gears and clutches are driven through splines on the shafts.

A sliding gear on the main shaft locks either the low-speed gear or the high-speed gear to the main shaft. In many transfer cases, this shift cannot be made unless the transmission is in neutral. Because the transfer case main shaft and the sliding gear, which is splined to it, will be turning at a different speed than the low- or high-speed gear, gear clash will occur.

To engage and disengage the front axle, another sliding gear or clutch is used to lock the front-axle drive gear to the front-axle drive shaft. On many transfer cases, the front axle can be engaged and disengaged when the vehicle is moving. This is done by releasing the accelerator pedal to remove the torque load through the gears and moving the shift lever. However, both the front and the rear wheels must be turning at the same speed. If the rear wheels have lost traction and are spinning, or if the brakes are applied and either the front or the rear wheels are locked and sliding, gear clashing will occur when engagement of the front axle is attempted. On some 4WD systems, the vehicle must be stopped before the change can be made. Also, the shift into 4WD Low mode normally requires that the vehicle be stopped first.

Some transfer cases are single speed and only allow for a change between 2WD and 4WD. An integral main drive gear in the transfer case, which is driven by the output shaft of the transmission, provides power to the rear driveline at all times. This main drive gear is in mesh with an idler gear that, in turn, meshes with a front axle drive gear to rotate the front driveline when it is in 4WD. When the vehicle is in 2WD, the idler gear is moved out of mesh with the front axle drive gear.

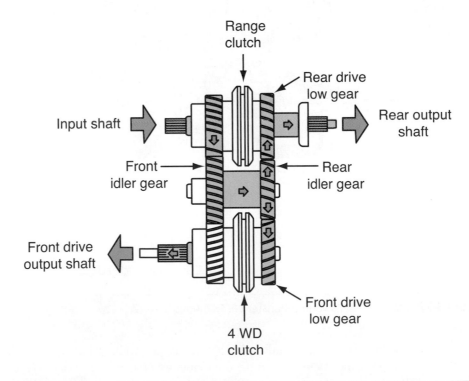

Figure 8-23 Power flow through a gear-driven transfer case when it is in 4WD Low.

Some of the newer "shift-on-the-fly" systems use a magnetic clutch in the transfer case to bring the front drive shaft, differential, and drive axles to the same speed as the transmission. When the speeds are synchronized, an electric motor in the transfer case (Figure 8-24) completes the shift. The system will not shift until the speeds are synchronized.

Transfer Case Designs

Some transfer cases rely entirely on gear sets to transfer power. Other designs use a combination of gears and a chain. Using a chain to link the drive axles, instead of a gear set, reduces the weight of the transfer case, thereby improving fuel economy.

Shop Manual
Chapter 8, page 362

Drive Chains

Drive chains are often used to link the input and output shafts in a transfer case. The chain only serves as a link and does not influence gear ratios. Chains are commonly used with planetary gear sets because they are very efficient and quiet and allow for flexible positioning of the transfer case's components. Two basic designs of chains are used: the round-pin style and the pin-and-rocker joint style.

In the **round-pin** design (Figure 8-25), a single pin is inserted into mating holes at each end of the link plate. Load is distributed over a large area, allowing greater chain tensions and higher dynamic loads. The round-pin joint is widely used in transfer cases on part-time 4WD vehicles.

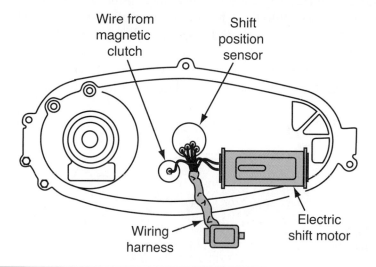

Figure 8-24 Location of electric shift motor on a transfer case.

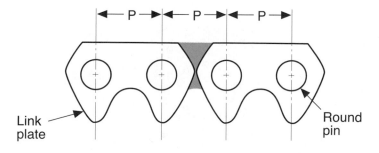

Figure 8-25 A round-pin-design drive chain.

The **pin-and-rocker joint** design (Figure 8-26) uses two convex joints that roll against one another as the chain moves. This type of chain is very efficient at continuous high speeds and is used on full-time 4WD systems.

Planetary Gear Drives

A typical part-time transfer case uses an aluminum case, a chain drive, and a planetary gear set for reduced weight and increased efficiency. Further efficiency is gained because the internal components of the transfer case do not rotate while the vehicle is operating in 2WD.

A simple **planetary gear set** is made up of three gears. In the center of the gear set is the sun gear. All other gears in the set revolve around the sun gear. Meshing with the sun gear are three or four planetary pinion gears. The pinions are held together by the planetary pinion carrier or the planetary carrier. The carrier holds the gears in place while allowing them to rotate around their shafts. On the outside of the planetary pinions is the ring gear. The ring gear has teeth around its inside circumference that mesh with the teeth of the planetary pinions. Each planetary pinion gear is mounted to the planetary carrier by a pin or shaft. The carrier assembly and the sun gear are mounted on their own shafts.

When the transfer case is in neutral, the rotation of the input shaft spins the planetary pinion gears and the ring gear around them. With both the pinion gears and the ring gear spinning freely, no power is transmitted through the planetary gear set.

Low- and high-range speeds are available from the planetary gear set. Speed reduction provides the low range and direct drive provides the high range. Speed reduction is provided when the ring gear is held stationary (Figure 8-27). When the transfer case's shift lever is moved to low range, the planetary gear assembly slides forward on its shaft and engages with a locking plate. The locking plate is bolted to the case and engages with the teeth of the ring gear to hold it stationary.

The driving gear of the gear set is the sun gear, which is connected to the output of the transmission. When the ring gear is held stationary and the sun gear is driving, the planetary pinions

In any set of two or more gears, the smallest gear is often called the pinion gear.

Shop Manual
Chapter 8, page 367

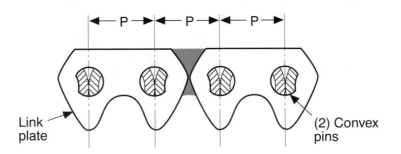

Figure 8-26 A pin-and-rocker-type drive chain.

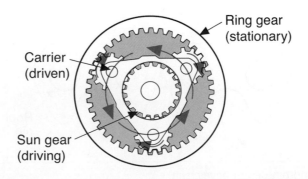

Figure 8-27 Speed reduction results from holding the ring gear and driving the sun gear.

rotate on their pins. As the pinions rotate, they must walk around the ring gear because they are in mesh with it. This action causes the pinion carrier to rotate in the same direction as the sun gear.

However, the planetary carrier turns more slowly than the sun gear because of the size difference between the pinions and the sun gear. As the pinions move around the inside of the ring gear, the shaft attached to the planetary carrier is driven in the same direction as the sun gear, but at a lower speed. This action causes speed reduction and torque multiplication, thereby providing the transfer case with low range.

The transfer case can be selected into the high range. In this operating mode, the ring gear and planetary carrier are locked together. As a result, the pinion gears cannot rotate on their pins and the entire assembly turns as a solid unit. Therefore, the output shaft rotates at the same speed as the input shaft, providing for direct drive.

Electronically Controlled Planetary Gear Sets. Some vehicles are equipped with an electronically controlled 4WD system that couples the front and rear axles through a planetary gear set that allows each axle to run at its own speed (Figure 8-28). The planetary gear set is not used for speed reduction; rather, it is used as a differential. The transmission output shaft drives the planetary carrier. The ring gear is connected to the driveline for the rear wheels and the sun gear is connected by a chain to the front driveline.

This type of system uses a speed sensor at each side of the planetary gear, in which power is split one third to the front axle and two thirds to the rear axle under normal driving conditions. When the sensors detect a difference between the speed of the front and rear wheels—a sign that one of the wheels has lost traction—an electronic controller sends current to an **electromagnetic clutch** that locks the planetary gear.

Half of the clutch's plates are splined to the sun gear and the other half are splined to the ring gear. By locking the sun gear to the ring gear, the axles are locked together and torque is split equally to the two driving axles. After a few seconds, the AWD electronic controller disengages the clutch and rechecks the speed of the axles. If slippage is still present, the controller reengages the clutch.

Shop Manual
Chapter 8, page 358

The **electro-magnetic clutch** locks the planetary gear when wheel sensors detect a difference between the speed of the front and rear wheels. It is also used on some models to lock up the center differential when power is needed at all four wheels.

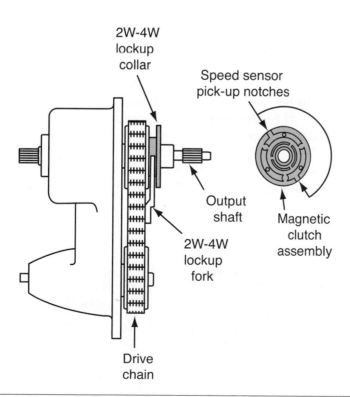

Figure 8-28 A magnetic clutch that synchronizes the shaft speeds before allowing a shift.

An electromagnetic clutch is also used on some models to lock up the center differential when power is needed at all four wheels. The clutch may be activated by a switch on the dash or it may be controlled electronically by a computer in response to wheel speed sensor signals.

A few vehicle models use a magnetic clutch to match the speed of the front axle with that of the rear axle. This speed matching takes very little time and allows a shift to be made from 2WD High to 4WD High at any speed. As soon as the front- and rear- output shafts of the transfer case reach the same speed, a spring-loaded shift collar automatically engages the main shaft hub to the driveline sprocket.

Many newer RWD-based vehicles have an automatic 4WD feature that switches from 2WD to 4WD when the transfer case shift control module receives wheel rotating slip information from the wheel sensors. The transfer case shift control module then engages the transfer case motor/encoder to go into 4WD. The transfer case has the typical three gear selection positions. When any of these positions is selected, the transfer case motor is locked and the transfer case stays at the gear selected. However, when the control switch is in the AUTO (or A4WD) position, the transfer case is in the adaptive mode and will cycle to 4WD when necessary.

Some pickups and SUVs are fitted with an AWD transfer case. In this system, AWD is always activated. A viscous clutch is used as a torque distribution device. Normal torque distribution is 35 percent to the front axle and 65 percent to the rear.

A take-off from electronically operated AWD systems for trucks is Chrysler's Quadra-drive. This system uses a gerotor pump to react to variations between front- and rear-axle speeds. When the front and rear driveshafts are rotating at the same speed, the gerotor pump produces no pressure and everything is normal. If a rear wheel loses traction and begins to spin, the speed difference between the front and rear axles builds hydraulic pressure in the gerotor. This pressure gradually locks the clutch pack inside the gerotor pump, transferring power to the front axle. This system also uses gerotor pumps in each axle. This provides limited-slip operation at both axles. All of the torque splitting takes place without driver intervention.

Locking Hubs

Shop Manual
Chapter 8, page 371

Many 4WD systems on trucks and utility vehicles use front-wheel locking hubs. These hubs connect the front wheels to the front-drive axles when they are in the locked position. Manual locking hubs require that a lever or knob (Figure 8-29) be turned by hand to the 2WD or 4WD position. Automatic locking hubs can be locked by shifting into 4WD and moving forward slowly. Some are unlocked by slowly backing up the vehicle. On other 4WD systems, a front-axle lock is used in place of individual locking hubs. Some late-model vehicles require no change to the hubs.

To reduce wear on the driveline of the front axle the hubs are unlocked when the vehicle is in 2WD, which allows the wheels to rotate independently of the drive axles. If the hubs remain in the 4WD or locked position when the vehicle is in 2WD, the driver will notice a drop in fuel economy and excessive noise from the front axle.

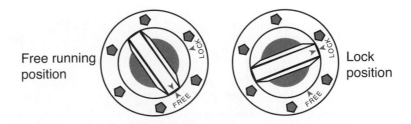

Figure 8-29 Knob positions for manual locking hubs.

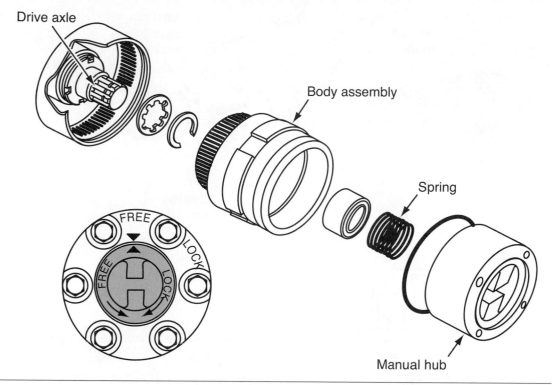

Drive axle

Body assembly

Spring

Manual hub

Figure 8-30 Action of a locking hub.

A **locking hub** is a type of clutch that engages or disengages the outer ends of the front-axle shafts from the wheel hub. A handle located in the center of manual hubs is turned to lock or unlock the hubs. This handle applies or releases spring tension on the hub's clutch. When the hub is in the locked position, the ring of the clutch is set onto the splines of the axle shaft (Figure 8-30). When the hub is in the unlocked position, spring pressure forces the clutch ring away from the axle shaft, thereby disconnecting the wheel hub and the axle.

Automatic Locks

Although automatic hubs (Figure 8-31) are more convenient for the driver, they do have a disadvantage. Many self-locking hubs are designed to unlock when the vehicle is moved in reverse. Therefore, if the vehicle is stuck and needs to back out of a trouble spot, only RWD will be available to move it. Other automatic hubs unlock immediately when 4WD is disengaged without the need to back up. On these systems, the hubs are automatically locked, regardless of the direction in which the vehicle is moving.

Locking hubs are not needed with full-time 4WD. The wheels and hubs are always engaged with the axle shafts. The interaxle differential or transfer case prevents damage and undue wear to the parts of the powertrain.

Some 4WD vehicles use a vacuum motor or mechanical linkage to move a splined sleeve to connect or disconnect the front-drive axle (Figure 8-32). With this system, locking hubs are not needed. When 2WD is selected, one axle is disconnected from the front differential. As a result, all engine torque moves to the side of the differential with the axle disconnected. This is due to normal differential action. When the vehicle is shifted into 4WD, the shift collar connects the two sections of the axle shaft.

Other axle disconnects are operated electrically. An electric motor can be used to connect and disconnect the axle (Figure 8-33). This system allows for a smooth transition from 2WD to 4WD. General Motors uses a system whereby selecting 4WD on the selector switch energizes a heating element in the axle disconnect that heats a gas, causing the plunger to operate the shift mechanism.

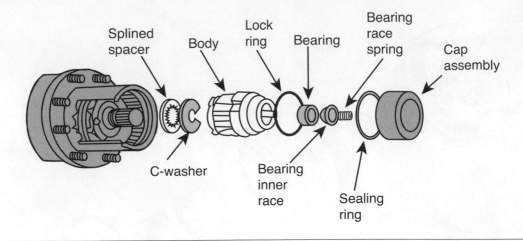

Splined spacer   Body   Lock ring   Bearing   Bearing race spring   Cap assembly

C-washer   Bearing inner race   Sealing ring

Figure 8-31 Disassembled automatic locking hub.

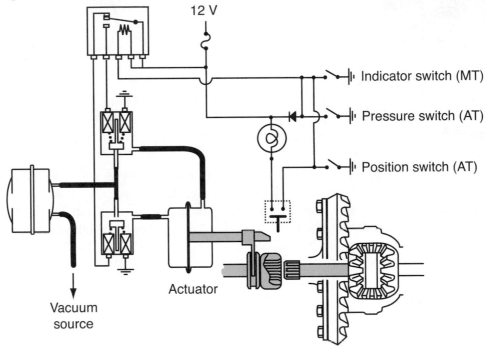

12 V

Indicator switch (MT)

Pressure switch (AT)

Position switch (AT)

Actuator

Vacuum source

Figure 8-32 Toyota's Automatic Disconnecting Differential (ADD) system.

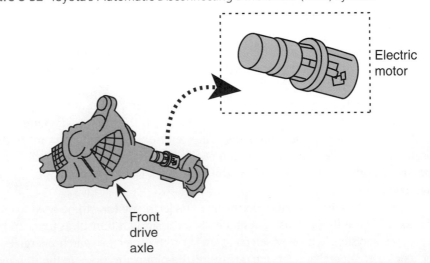

Electric motor

Front drive axle

Figure 8-33 An electric motor disconnects and connects the axles in this system.

Four-Wheel-Drive Suspensions

The suspension components necessary for 4WD vehicles are basically the same as for FWD and RWD vehicles. When the vehicle is based on a FWD setup, the addition of the rear axle requires a rear suspension similar to those found on RWD models (Figure 8-34).

When the 4WD vehicle is based on a RWD platform, provisions for the front-drive axle must be made. Basically there are two types of front-drive axles used on 4WD vehicles: the solid axle and the independent front suspension axle assembly. The solid type is basically a rear axle turned so that the pinion gear faces the center of the vehicle. The front-drive shaft is equipped with a slip yoke at the transfer case end to accommodate the movement of the axle. The ends of the axle housing are fitted with steering knuckles that allow for steering (Figure 8-35).

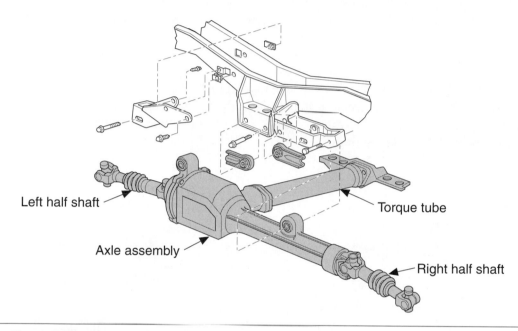

Figure 8-34 AWD rear-drive axle assembly.

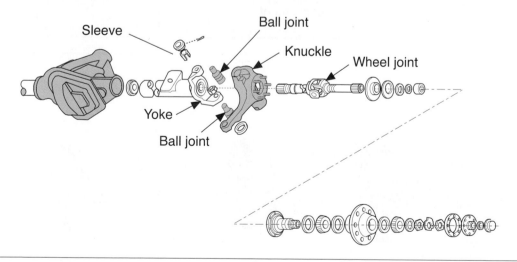

Figure 8-35 Exploded view of a front drive axle's steering knuckle and hub.

Independent Suspensions

An independent front (IDF) suspension axle (Figure 8-36) has the differential housing mounted solidly to the vehicle's frame. Short drive shafts or half shafts connect the differential to the wheels. These half shafts are suspended by springs and are able to respond independently to road surfaces. The half shafts are normally equipped with CV joints and U-joints to accommodate this movement. The ends of the shafts fit into the steering knuckles. IDF suspension improves the vehicle's handling characteristics (Figure 8-37).

To provide independent front suspension, some vehicles have one half shaft and one solid axle for the front drive axle (Figure 8-38). The half shaft is able to move independently of the solid axle, thereby giving the vehicle the ride characteristics desired from an IDF suspension.

The most common IDF suspension on 4WD vehicles was introduced by Ford Motor Company in 1980. This system utilizes two steel axle carriers and a third U-joint in the right axle shaft next to the differential. The carrier and joints allow each front wheel to move up and down independently. Instead of a one-piece tube to serve as the axle housing, the tube is split into two shorter sections and joined together at a pivot point. The axle shaft can flex near the pivot because of the centrally located third U-joint.

With IDF suspension, each wheel is able to react to the road conditions individually. Bumps are absorbed by each front wheel instead of some of the shock being transferred through the solid axle to the other wheel. Normally, these vehicles are equipped with coil springs, although some heavy-duty models use leaf springs.

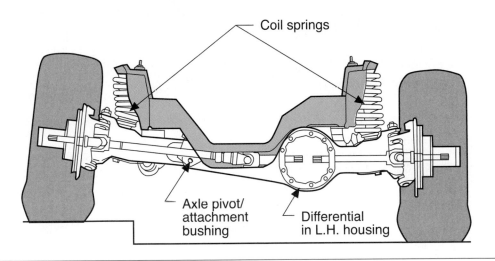

Figure 8-36 An independent suspension front drive axle assembly.

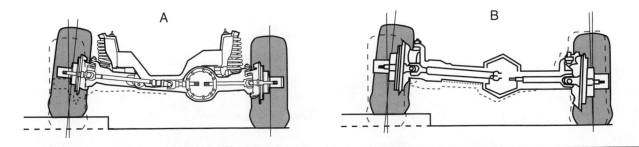

Figure 8-37 Comparison of an (A) independent front suspension and a (B) solid drive axle assembly in response to a bump.

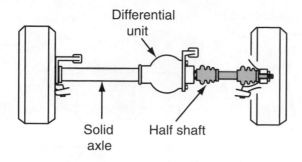

Figure 8-38 A front drive axle with one solid tube on one side and a half shaft on the other.

The wheel bearings in front- and rear-axle assemblies of 4WD vehicles are typically located so that they do not support the weight of the vehicle and are held in the hub assembly in the same way as other full-floating axle bearings.

Suspension Modifications

When a vehicle is available in either 2WD or 4WD, the suspension will differ with the drivetrain. As an example of this, let us study the suspension of a 2000 Dodge Caravan. The regular model is a 2WD with FWD. This model has an AWD option. The front suspensions of these two models are identical. The rear suspension, however, is different, mainly to support the weight and resist the torque of the rear-drive axle. The rear axle used on FWD applications is mounted to the rear leaf spring using isolator bushings at the axle mounting brackets (Figure 8-39). FWD applications also have a track bar connected to the rear axle and the frame. The rear axle used on AWD applications is mounted on a multileaf spring and does not use isolator bushings between the rear axle and the leaf springs. A track bar is not used on AWD models (Figure 8-40).

When a normally RWD vehicle is equipped with 4WD, the front suspension must be modified to support the weight of, and make room for, the drive axle. Let us do another comparison to show how different the suspensions on the same vehicle can be. A Ford Expedition is available in both 2WD and 4WD. Besides modifying the steering knuckles to accept half shafts on the 4WD model, the major change in suspension design is the use of torsion bars on 4WD models (Figure 8-41), instead of coil springs (Figure 8-42). This change requires unique mounting and supporting hardware and as a result, the suspensions of the two models are very different.

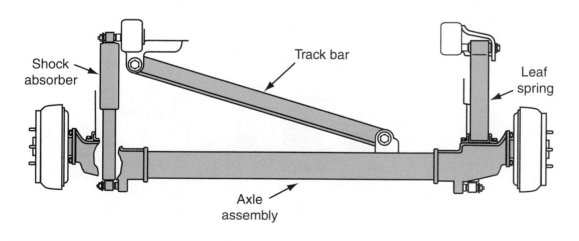

Figure 8-39 A typical rear suspension of a FWD minivan.

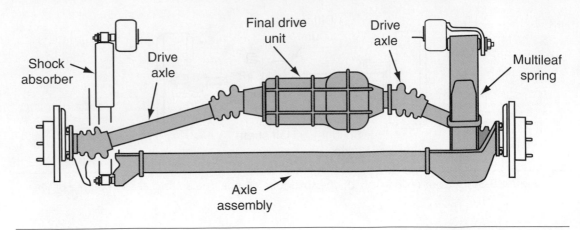

Figure 8-40 The rear suspension of the same minivan shown in Figure 8-39 when it is equipped with AWD.

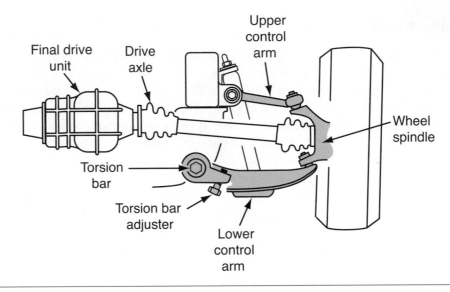

Figure 8-41 The front suspension on a 4WD Ford Expedition.

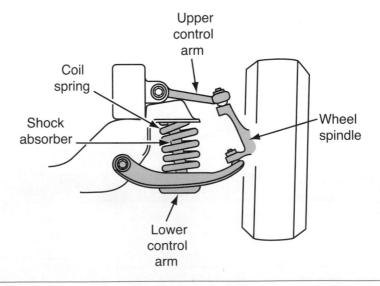

Figure 8-42 The front suspension on a 2WD Ford Expedition.

Summary

❑ With 4WD, engine power can flow to all four wheels, which can greatly increase a vehicle's traction when traveling in adverse conditions and can also improve handling as side forces generated by the turning of a vehicle or by wind gusts will have less of an effect on a vehicle that has power applied to the road on four wheels.

❑ FWD cars are modified by adding a transfer case, a rear drive shaft, and a rear axle with a differential. Some are equipped with a center differential.

❑ 4WD systems use a separate transfer case that allows the driver to choose or feature automatic controls that select the transfer of the engine's power to either two or four wheels.

❑ The transfer case is usually mounted to the side or rear of the transmission. A drive shaft is used to connect the transmission to the transfer case, or a chain or gear drive within the transfer case receives the engine's power from the transmission and transfers it to the drive shafts leading to the front- and rear-drive axles.

❑ The front wheel hubs on most 4WD vehicles must be locked or unlocked by the driver to provide for efficient operation in 2WD or 4WD. The rear wheel hubs are always engaged.

❑ Integrated full-time 4WD systems use computer controls to enhance full-time operation, adjusting the torque split depending on which wheels have traction.

❑ On-demand 4WD systems power a second axle only after the first begins to slip.

❑ Many 4WD units are equipped to allow the driver to select in and out of 4WD. The systems capable of operating in both 2WD and 4WD are called part-time 4WD systems. Full-time 4WD systems cannot be selected out of 4WD. Some units feature the option of selecting part-time or full-time capabilities.

❑ During turns, the front wheels travel a greater distance than the rear wheels. This is because the front wheels move through a wider arc than the rear wheels.

❑ Part-time 4WD systems are designed to be used only when driving off the road or on slippery surfaces. When 4WD is engaged, a part-time system locks the front and rear axles together.

❑ A transfer case is equipped with a gear or chain drive to transmit power to one or both of the drive axles. Some transfer cases are equipped with two speeds: a high range for normal driving and a low range for especially difficult terrain.

❑ Two-speed transfer cases are controlled by a shift lever that typically has four positions: 2WD HIGH, which engages only the rear axle and is used for all dry-road driving, 4WD HIGH, which engages both axles and is used at any speed on slippery surfaces, NEUTRAL, which disengages both axles, and 4WD LOW, which engages both axles and lowers the ratios of the entire driveline.

❑ The purpose of a transfer case is to transfer torque from the output of the transmission to the vehicle's front and rear axles.

❑ Some transfer cases use a spur or helical gear set to provide for the speed ranges and for engagement and disengagement of 4WD. The front axle is engaged by shifting a sliding or clutching gear into engagement with the driven gear on the drive shaft for the front wheels inside the transfer case. The sliding gears and clutches are driven through splines on the shafts.

❑ Some "shift-on-the-fly" systems use a magnetic clutch in the transfer case to bring the front drive shaft, differential, and drive axles to the same speed as the transmission. When the speeds are synchronized, an electric motor in the transfer case completes the shift.

Terms to Know

Annulus gear

Center differential

Chain drive

Driveline windup

Electromagnetic clutch

Full-time 4WD

Locking hubs

On-demand 4WD

Part-time 4WD

Pin-and-rocker joint

Planetary gear set

Round-pin

Sprocket

Sun gear

Transfer case

❏ Drive chains are often used to link the input and output shafts in a transfer case and serve as a link that does not influence gear ratios.

❏ A simple planetary gear set is made up of three gears. In the center of the gear set is the sun gear around which all other gears in the set revolve. Meshing with the sun gear are three or four planetary pinion gears whose pinions are held together by the planetary pinion carrier, or planetary carrier. The carrier holds the gears in place while allowing them to rotate around their shafts. On the outside of the planetary pinions is the ring gear that has teeth around its inside circumference that mesh with the teeth of the planetary pinions. Each planetary pinion gear is mounted to the planetary carrier by a pin or shaft.

❏ Low- and high-range speeds are available from the planetary gear set. Speed reduction provides the low range and direct drive provides the high range. Speed reduction is provided when the ring gear is held stationary. When the transfer case's shift lever is moved to low range, the planetary gear assembly slides forward on its shaft and engages with a locking plate that is bolted to the case and engages with the teeth of the ring gear to hold it stationary.

❏ Most 4WD systems on trucks and utility vehicles use front-wheel driving hubs that can be disengaged from the front axle when the vehicle is operating in the 2WD mode. When unlocked in 2WD, the front wheels still turn, but the entire front drivetrain, including the front axles, front differential, front drive shaft, and certain parts of the transfer case, stops turning.

❏ Automatic hubs lock when the driver shifts into 4WD and normally, automatically unlock when the vehicle is driven in reverse for a few feet. Automatic hubs are engaged by the rotational force of the axle shafts whenever the transfer case is in 4WD.

❏ A locking hub is a type of clutch that disengages the outer ends of the axle shafts from the wheel hub.

❏ A handle located in the center of manual hubs is turned to lock or unlock the hubs. This control handle applies or releases spring tension on the hub's clutch. When the hub is in the locked position, spring pressure causes the clutch to engage to the inner hub that is connected to the axle shaft.

❏ Locking hubs are not needed with full-time 4WD. The wheels and hubs are always engaged with the axle shafts. The interaxle differential or transfer case prevents damage to and undue wear of the parts of the powertrain.

❏ Driveline windup occurs when the drive axles are rotating at different speeds on dry pavements or when the vehicle is turning a corner.

❏ The suspension components necessary for 4WD vehicles are basically the same as for FWD and RWD vehicles.

❏ The solid front 4WD axle is basically a rear axle turned so that the pinion gear faces the center of the vehicle.

❏ An independent front (IDF) suspension axle has the differential housing mounted solidly to the vehicle's frame. Short drive shafts or half shafts connect the differential to the wheels. These half shafts are suspended by springs and are able to respond independently to road surfaces.

❏ The wheel bearings in front and rear axle assemblies of 4WD vehicles are typically located so that they do not support the weight of the vehicle and are held in the hub assembly in the same way as other full-floating axle bearings.

Review Questions

Short Answer Essays

1. What are the advantages of 4WD?

2. What are the main differences between 4WD and AWD?

3. What is the primary purpose of a transfer case?

4. Describe the operation of front locking wheel hubs.

5. What are the differences between an integrated 4WD system and an on-demand 4WD system?

6. What are the primary differences between a full-time and part-time 4WD system?

7. Name the three main driveline components that are added to a 2WD vehicle to make it a 4WD vehicle.

8. What is the primary purpose of an interaxle differential?

9. Why are chain drives used in many transfer cases?

10. Briefly explain the operation of a simple planetary gear set.

Fill-in-the-Blanks

1. A typical car has _____ wheels, _____ of which can deliver power; therefore, they could be designated as 4 × 2 vehicles. A 4WD vehicle is designated as a 4 × 4 because it has _____ wheels and power can be applied to all _____ wheels.

2. A simple planetary gear set is made up of three gears. In the center of the gear set is the _____ gear. Meshing with this gear are three or four _____ gears. These are held together by the _____ . This holds the gears in place while allowing them to rotate around their shafts. On the outside of the planetary gear set is the _____ gear. This gear has teeth around its inside circumference.

3. In a typical planetary gear set, speed reduction is provided when the _____ gear is held stationary.

4. _____ occurs when the drive axles are rotating at different speeds on dry pavements or when the vehicle is turning a corner.

5. To provide for _____ _____ suspension, some vehicles have one half shaft and one solid axle for the front-drive axle.

6. Common ways of connecting and disconnecting the front axles on a 4WD vehicle include having locking hubs, _____ motors, _____ motors, and mechanical _____ .

7. Two-speed transfer cases typically have four positions: _____ , which engages only the rear axle and is used for all dry-road driving, _____ , which engages both axles and is used at any speed on slippery surfaces, _____ , which disengages both axles, and _____ , which engages both axles and lowers the ratios of the entire driveline, and should only be used at low speeds and on very demanding terrain.

8. Some vehicles use a planetary gear set to couple the front and rear axles. This gear set functions as a _____ .

9. _____ _____ are not used on full-time 4WD systems because the axles and hubs are always driven by the transfer case.

10. An IDF suspension 4WD axle has the _____ housing mounted solidly to the vehicle's frame and uses short _____ _____ to connect the differential to the wheels.

Multiple Choice

1. When discussing the purpose of a transfer case, *Technician A* says that it transfers power to an additional drive axle. *Technician B* says that most have two speeds, which affect the overall gear ratio of the vehicle.
   Who is correct?
   A. A only
   B. B only
   C. Both A and B
   D. Neither A nor B

2. *Technician A* says that power can be locked into all four wheels by turning the locking hubs into the locked position. *Technician B* says that some transfer cases use an electric switch to engage both drive axles.
   Who is correct?
   A. A only
   B. B only
   C. Both A and B
   D. Neither A nor B

3. When discussing 4WD systems used on cars, *Technician A* says tha cars are equipped the same as a truck would be. *Technician B* says that some cars do not use a transfer case, but they are equipped with a third differential unit.
   Who is correct?
   A. A only
   B. B only
   C. Both A and B
   D. Neither A nor B

4. When discussing the gear sets of a transfer case, *Technician A* says that most use a planetary gear set to provide for the different gear selections. *Technician B* says that a chain and sprocket assembly is normally used instead of gears to connect the input shaft with the output shaft.
   Who is correct?
   A. A only
   B. B only
   C. Both A and B
   D. Neither A nor B

5. When discussing the various gear positions of a transfer case, *Technician A* says that when the transfer case is in low, the overall gear ratio is numerically increased. *Technician B* says that

when the transfer case is in the high position, the vehicle operates in an overdrive mode due to the decrease in torque multiplication.
Who is correct?

A. A only
B. B only
C. Both A and B
D. Neither A nor B

6. When discussing the different shift positions of a transfer case, *Technician A* says that 4WD HIGH engages both axles and is used at any speed on slippery surfaces. *Technician B* says that 4WD Low engages both axles and lowers the ratios of the entire driveline.
Who is correct?

A. A only
B. B only
C. Both A and B
D. Neither A nor B

7. *Technician A* says that the systems capable of operating in both 2WD and 4WD are called part-time 4WD systems. *Technician B* says that full-time 4WD systems cannot be selected out of 4WD.
Who is correct?

A. A only
B. B only
C. Both A and B
D. Neither A nor B

8. *Technician A* says that when a vehicle is turning a corner, the rear wheels travel a greater distance than the front wheels. *Technician B* says that the front wheels move through a wider arc than the rear wheels when the vehicle is making a turn.
Who is correct?

A. A only
B. B only
C. Both A and B
D. Neither A nor B

9. *Technician A* says that part-time 4WD systems are designed to be used only when driving off the road or on slippery surfaces. *Technician B* says that AWD systems are intended for off-the-road use only.
Who is correct?

A. A only
B. B only
C. Both A and B
D. Neither A nor B

10. When discussing the characteristics of full-time 4WD vehicles, *Technician A* says that when they turn a corner, the front wheels cannot rotate at the same speed as the rear wheels, causing one set of wheels to scuff along the pavement. *Technician B* says that most full-time 4WD vehicles use a viscous clutch between the front and rear axles.
Who is correct?

A. A only
B. B only
C. Both A and B
D. Neither A nor B

Advanced Four-Wheel-Drive Systems

Upon completion and review of this chapter, you should be able to:

❑ Use the correct terminology when discussing 4WD systems.

❑ Describe the different designs of 4WD systems and their applications.

❑ Compare and contrast the components of part- and full-time 4WD systems.

❑ Discuss the purpose of an interaxle differential and the design variations used by the industry.

❑ Discuss the purpose, operation, and application of a viscous coupling in 4WD systems.

❑ Explain the operation of some common AWD systems.

A BIT OF HISTORY

In 1912, the United States Army tested the viability of trucks by hosting a 1,509-mile race that took the four competing trucks off and on the road to complete the test. The winner was a 4WD truck.

Introduction

While most full-sized 4WD trucks and SUVs (Figure 9-1) are design variations of basic RWD vehicles (Figure 9-2), most passenger cars and smaller SUVs equipped with 4WD are based on FWD designs. These modified FWD systems consist of a transaxle and differential to drive the front wheels, plus some type of mechanism for connecting the transaxle to a rear driveline. In many cases this mechanism is a simple clutch or differential.

If a vehicle has full-time 4WD, a controllable differential is built into the transfer case. The purpose of this interaxle differential is to compensate for any difference in front-wheel and rear-wheel speeds and to allow the front and rear axles to operate at their own speeds.

Shop Manual
Chapter 9, page 393

Although most full-time systems cannot be selected out of 4WD, some, such as Jeep's full-time system, called Selec-Trac®, can be disengaged for better fuel economy and less drivetrain wear in dry conditions.

Figure 9-1 A 4WD sport utility vehicle going off the road.

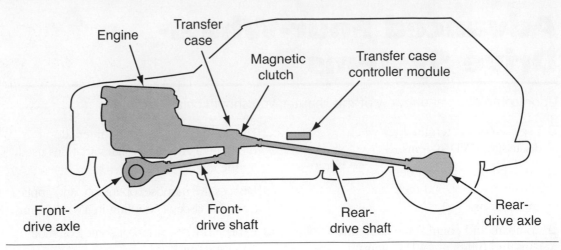

Figure 9-2 Typical electronic AWD system on an RWD-based vehicle.

A full-time system is ideal for use when driving through rapidly changing weather conditions or on surfaces that fluctuate between good and bad, such as when driving in and out of rain or on mostly clear roads spotted with patches of snow.

During turns, the front wheels travel a greater distance than the rear wheels. This is because the front wheels move through a wider arc than the rear wheels (Figure 9-3). With full-time 4WD, an open center differential allows the front wheels to travel farther or turn faster than the rear wheels without slipping.

The use of an open differential in the transfer case does have a disadvantage. If the wheels on either axle lose traction and begin to spin, the differential continues to deliver maximum torque to the axle with the minimum traction. As a result, insufficient torque to move the vehicle may be provided to the wheels that still have traction.

Limited-Slip Differentials

To overcome this problem, many full-time transfer cases are equipped with a limited-slip differential. These units may use a viscous clutch, cone clutch (Figure 9-4), or a multiple-plate clutch

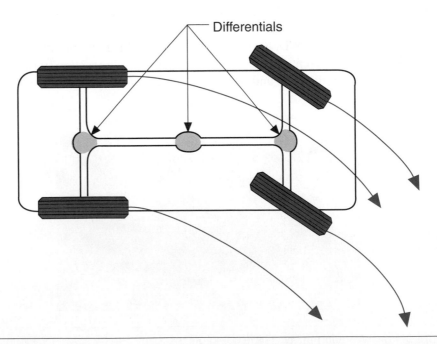

Figure 9-3 When turning a corner, the front wheels travel through a wider arc than the rear wheels.

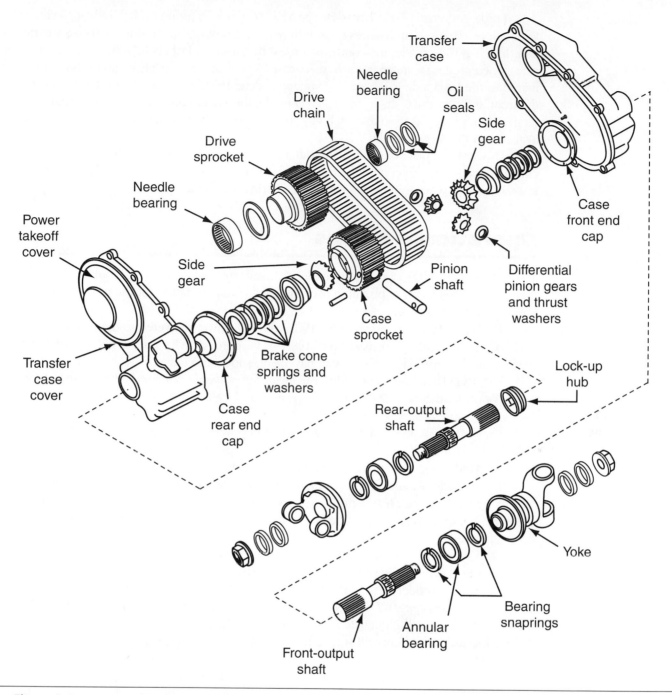

Figure 9-4 4WD transfer case equipped with a limited-slip differential. A commonly used full-time system equipped with a limited-slip differential in the transfer case was called the Quadra-Trac®. It was built by the Warner Gear Division of the Borg-Warner Corporation.

assembly to control the differential action. By using a limited-slip differential, the amount of torque delivered to the axle with the least traction can be limited. This allows more torque to be delivered to the axle with the most traction.

When differential action is not desired, the center differential (on some models) can be locked to provide torque to all four wheels. The lock position is intended for and must be used only on soft surfaces, such as dirt, mud, and snow. Typically, the differential is locked by an electromagnetic clutch. The clutch may be driver-controlled through a switch on the dash or may be computer-controlled in response to wheel speed sensors (normally the same speed sensors as used for ABS).

Ford's Automatic Four-Wheel-Drive-(A4WD) system is found on late-model Navigators and Expeditions. This system features three different modes of operation but is called a full-time system. In the A4WD mode, the computer varies the torque split between the front and rear drive axles by controlling a clutch in the transfer case. When the rear wheels begin to slip, torque transmitted to the front axle is increased. Slippage is determined by measuring the rotation speeds of the front and rear drive shafts. In the 4WD mode, the driver can select between HIGH and LOW range. A shift motor is used to switch the drive ranges in the transfer case. The LOW range is not available in the A4WD mode.

This system does not use a center differential or viscous coupling; rather, an electromagnetic clutch is duty-cycled to control the torque split. This system not only responds to wheel spin but also anticipates when slip may occur by monitoring the throttle position.

Operational Modes

The names of a 4WD's operational modes will vary by manufacturer; it would be difficult to list them all. The following explanation uses the most common names for the most commonly available operational modes.

The 4WD AUTO mode provides 4WD with full power delivered to the rear axle and to the front axle as required for increased traction. This is appropriate for normal on-road operating conditions, such as dry road surfaces, wet pavement, snow, and gravel.

The 4WD HIGH mode provides 4WD with full power to both axles. It is intended for severe winter or off-road conditions, such as deep snow and ice (where no dry or wet pavement remains uncovered), and shallow sand.

The 4WD LOW mode supplies 4WD with full power to both axles and includes a lower gear ratio for low speed. It is intended only for off-road applications that require extra power including deep sand, steep grades, and pulling a boat and trailer out of the water.

The vehicle should not be operated in 4WD HIGH and 4WD LOW on dry or merely wet pavement. Doing so will produce excessive noise, increase tire wear, and may damage driveline components. These modes are intended for use only on consistently slippery or loose surfaces.

Certain procedures must be followed when selecting the operational modes. This is important, regardless of how the system is advertised. Always follow the instructions given for the particular make and model vehicle. Failure to do so may result in system damage.

Shifting between 4WD AUTO and 4WD HIGH: When the control is moved to 4WD HIGH, the indicator light will illuminate in the instrument cluster. When the control is moved to 4WD AUTO, the indicator light will turn off. Either shift can be done at a stop or while driving at any speed.

Shifting from 4WD AUTO or 4WD HIGH to 4WD LOW: When shifting into LOW, the vehicle must first be brought to a complete stop. Then depress the brake pedal. Place the automatic transmission gearshift into neutral, or if the vehicle has a manual transmission, depress the clutch pedal. Move the 4WD control to the 4WD LOW position. If the vehicle is equipped with an electronic shift 4WD system, and the control is moved to 4WD LOW while the vehicle is moving, the system will not engage and no damage will occur to the 4WD system. The 4WD LOW indicator lamp will come on momentarily when the ignition is turned on and when 4WD LOW is selected. It will flash if the system requires service.

Shifting from 4WD LOW to 4WD AUTO or 4WD HIGH: When shifting out of LOW, the vehicle must be at a complete stop. Depress the brake pedal. Place the automatic transmission gearshift into neutral, or if the vehicle has a manual transmission, depress the clutch pedal. Move the 4WD control to the 4WD AUTO or 4WD HIGH position. The 4WD HIGH indicator lamp will come on momentarily when the ignition is turned on and when 4WD HIGH is selected. It will flash if the system requires service.

AUTHOR'S NOTE: It seems that every time there is a new model car or truck that offers 4WD, there is a new 4WD system. Each of these systems has some feature that it did not have before or one the competition does not have. It is hard to predict what will happen next. I highly recommend that you look at the owner's manual and any labeling that may be around the 4WD controls before doing anything with the vehicle, (including driving). If you do not know how the thing is supposed to work, how can you tell if it is working the way it should?

All-Wheel-Drive Systems

AWD systems do not give the driver the option of 2WD or 4WD. They always drive all wheels. AWD vehicles are usually passenger cars that are not designed for off-road operation. The AWD vehicle is designed to increase vehicle performance in poor traction situations, such as icy or snowy roads, and in emergencies. AWD gives the vehicle operator maximum control in adverse operating conditions by biasing the driving torque to the axle with driving traction. The advantage of AWD can be compared to walking on snowshoes. Snowshoes prevent the user from sinking into the snow by spreading the body weight over a large surface. AWD vehicles spread the driving force over four wheels rather than two wheels when needed.

When a vehicle travels over the road, the driving wheels transmit a tractive force to the road's surface. The ability of each tire to transmit tractive force is a result of vehicle weight pressing the tire into the road's surface and the coefficient of friction between the tire and the road. If the road's surface is dry and the tire is dry, the coefficient of friction is high and four driving wheels are not needed. If the road's surface is wet and slippery, the coefficient of friction between the tire and road is low. The tire loses its coefficient of friction on slippery road surfaces, which could result in loss of control by the operator. Unlike a 2WD vehicle, an AWD vehicle spreads the tractive effort to all four driving wheels. In addition to spreading the driving torque, the AWD vehicle biases the driving torque to the axle that has the traction only when it is needed.

All-wheel-drive (AWD) systems automatically react to normal or slippery road conditions without depending on the driver to decide when to shift into 4WD. AWD systems are not designed for off-the-road use; rather, they are designed to increase vehicle performance in poor traction situations. AWD allows for maximum control by transferring a large portion of the engine's torque to the axle with the most traction.

AWD systems do not use a conventional two-speed transfer case. Most AWD systems are based on FWD systems with a transaxle. A single-speed transfer case is added to the transaxle or an interaxle differential is added to split power between the front and rear axles (Figure 9-5). On some designs, the center differential locks automatically or the driver can manually lock it with a

Shop Manual
Chapter 9, page 396

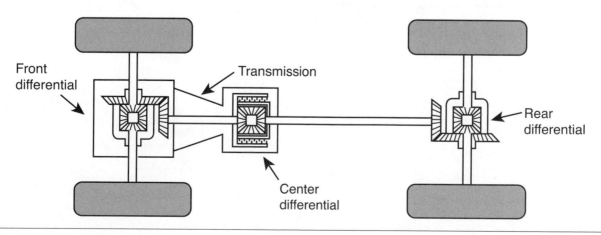

Figure 9-5 Typical AWD powertrain layout.

switch. Many AWD systems are fitted with a viscous coupling in addition to or in place of a center differential (Figure 9-6). The viscous coupling allows limited slip between the front and rear drive wheels. This provides power to all four wheels but prevents driveline windup.

Some AWD systems are called integrated systems. Wheel speed sensors send information to a computer that locks the center and axle differentials as needed to maintain traction. The transaxle and transfer case are an integrated unit. The center differential may incorporate an electronically controlled lockup clutch or viscous coupling. A low-range 4WD mode may be provided by an extra gear set in the transaxle. To maintain good traction at all times, wheel speed sensors send wheel spin information to a computer that locks the center and rear differentials as needed.

Other AWD systems are referred to as on-demand 4WD systems. In these systems, engine torque is transmitted primarily to one axle. Torque is automatically transferred to the other axle when traction is poor on the primary drive axle. Because this system responds to present road conditions, it may be operated on all road surfaces.

Many automatic AWD systems are electronically controlled and are based on a FWD drivetrain. The rear drive shaft extends from the transaxle to the rear drive axle. To transmit the power to the rear, a multiple-plate clutch is used. This clutch serves as an interaxle differential and permits a speed difference between the front and rear drive axles. Sensors monitor front and rear axle speeds, engine speed, and load on the engine and driveline. An electronic control unit receives information from the sensors and controls a solenoid that operates on a **duty cycle** (Figure 9-7) to control the fluid flow that engages the transfer clutch (Figure 9-8). The **duty solenoid** pulses, cycling on and off very rapidly, which develops a controlled slip condition. As a result, the transfer clutch operates like an interaxle differential and allows for a power split from 95 percent FWD and 5 percent RWD to 50 percent FWD and 50 percent RWD. This power split takes place so rapidly that the driver is unaware of the traction problem.

Center Differentials

When a vehicle is operating in 4WD and the front and rear axles are turning at the same speed, driveline windup can result. Windup is best described as a push-pull action. It is most evident

Shop Manual
Chapter 9, page 405

An electronic control unit receives information from the sensors and controls as a solenoid that operates on a **duty cycle** to control the fluid flow that engages the transfer clutch.

The control solenoid in the transaxle may be called a **duty solenoid**.

Duty cycle is also called a jitter cycle.

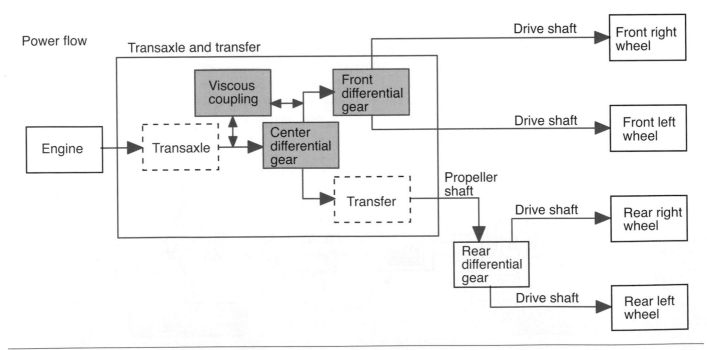

Figure 9-6 A chart showing the power flow through a viscous clutch-type center differential.

232

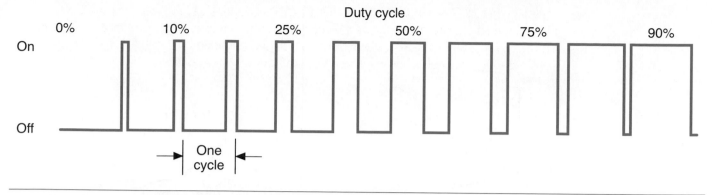

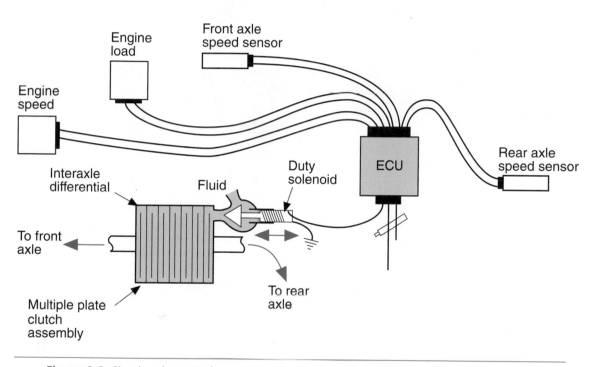

Figure 9-7 Duty cycle is the percentage of on-time per cycle. Duty cycle can be changed; however, total cycle time remains constant.

Figure 9-8 Simple schematic for an electronically controlled AWD system.

when the 4WD vehicle is operating on a dry surface and making a turn. Because the front wheels turn at a different arc than the rears, the tires must slide or drag on the pavement to complete the turn. Not only does windup cause excessive tire wear, the push-pull action can destroy a transfer case. On slippery surfaces, the tires are able to slide, preventing windup and damage. This is why part-time 4WD should only be used on slippery surfaces.

A center differential is used in full-time 4WD systems to prevent driveline windup. It can be located in the transfer case between the front-axle output shaft and the rear-axle output shaft. The center differential allows the front and rear axles to turn at different speeds. This eliminates windup and allows for improved handling during turns. The center differential is often fitted to a viscous coupling (Figure 9-9) to transfer power to the other axle. If and when the viscous coupling overheats, it will slip, and power is no longer applied to the other axle.

Some high-performance AWD cars use a vacuum system to allow the driver to lock the center and/or rear differential. This control allows the driver to select which set of wheels should receive the majority of the engine's torque.

Shop Manual
Chapter 9, page 398

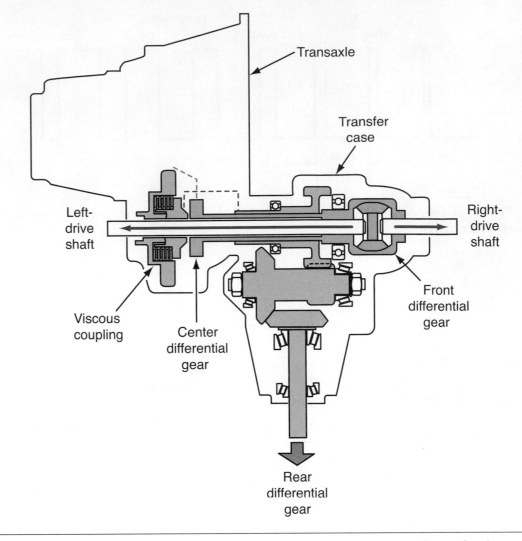

Figure 9-9 Power flow through a transfer case fitted with a viscous coupling and an interaxle differential.

The center or inter-axle differential is incorporated into the transfer case and splits power, in some proportion, to the front and rear drive axles.

Typically, on manual transmission models, the driver can lock the center differential with a switch. On the automatic transmission model, the center differential locks automatically, depending on which transmission range the driver selects and whether there is slippage between the front and rear wheels.

Electronically controlled AWD is found in several automobiles. The drivetrains on these vehicles are designed with a front transaxle and two front drive shafts (each with its CV U-joints). The rear drive shaft extends from the transaxle extension to the rear axle drive pinion and ring gear, two rear drive shafts, U-joints, and driving wheels (Figure 9-10). Remember, there must be some type of interaxle differential in full-time, AWDs.

Most electronically controlled systems respond to inputs from a variety of sensors. The control module monitors these inputs and adjusts the torque bias at the center, front, or rear differentials. Of course, the number of inputs the system monitors and what it controls determines the complexity of the system. The most complex AWD system ever developed was used on the Porsche 959, which did not wait for wheel slip before transferring torque. The system monitored the vehicle's steering angle, lateral acceleration, throttle position, yaw rate, and wheel spin in an attempt to transfer the torque where it was needed before it was absolutely necessary. Similar systems will be commonplace in the future.

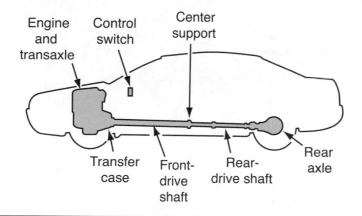

Figure 9-10 Location of the AWD components on a typical FWD car.

Viscous Couplings

The viscous clutch (Figure 9-11) is used in the driveline of vehicles to drive the axle with low tractive effort, taking the place of the interaxle differential. In existence for several years, the viscous clutch is installed to improve the mobility factor under difficult driving conditions. It is similar in action to the viscous clutch described for the cooling system fan. The viscous clutch in AWD is a self-contained unit. When it malfunctions, it is simply replaced as an assembly. The viscous clutch assembly is very compact, permitting installation within a front transaxle housing. Viscous clutches operate automatically while constantly transmitting power to the axle assembly as soon as it becomes necessary to improve driving wheel traction. This action is also known as biasing driving torque to the axle with tractive effort. The viscous clutch assembly is designed similarly to a multiple-disc clutch with alternating driven and driving plates.

AUTHOR'S NOTE: Keep in mind that an open differential sends power to the axle or wheel that has the *least* traction. Viscous coupling differentials send power to the axle or wheel that has the *most* traction.

The handling characteristics of 4WD vehicles are excellent in all driving conditions except when turning on a dry pavement. When turning a corner, the front wheels cannot rotate at the same speed as the rear wheels. This causes one set of wheels to scuff along the pavement. To overcome this tendency, many full-time 4WD and AWD vehicles use a viscous coupling between the front and rear axles.

Viscous couplings are often referred to as viscous clutches because they engage and disengage power flow.

Shop Manual
Chapter 9, page 399

When a viscous coupling provides holding ability for two parts of a planetary or bevel gearset, it is called a viscous control.

Figure 9-11 A viscous clutch.

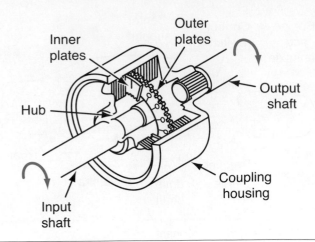

When a viscous coupling directly connects two shafts, it is called a viscous transmission.

Figure 9-12 The main components of a typical viscous coupling.

A **viscous coupling** (Figure 9-12) is basically a drum with some thick fluid in it that houses several closely fitted, thin steel discs. One set of plates is connected to the front wheels and the other to the rear. The advantage of the viscous coupling is that it splits the engine torque according to the needs of each axle.

Viscous couplings also may be used in the center and the front or rear axle differentials. They provide a holding force between two shafts as they rotate under force. Like a limited-slip differential, a viscous coupling transfers torque to the opposite drive wheel when the other wheel has less traction. A viscous clutch often takes the place of the interaxle differential. Viscous clutches operate automatically as soon as it becomes necessary to improve wheel traction.

High performance AWD vehicles use a viscous coupling in the center and rear differentials to improve cornering and maneuvering at higher speeds. The combination of a center differential viscous coupling with open front and rear differentials improves the vehicle's distribution of braking forces and is compatible with antilock braking systems.

Operation of a Viscous Coupling

A viscous coupling consists of an input shaft connected to a set of splined drive plates, an additional set of driven plates splined internally to the drum, and an output shaft splined to the drum. The drive plates and driven plates are alternately stacked and evenly spaced (Figure 9-13). The drum contains an exact amount of thick silicone oil. **Silicone oil** is used because it has high shearing resistance and volumetric change when its temperature increases. There is very little silicone fluid in a viscous coupling case. About one quarter of the case contains fluid at rest. The fluid expands rapidly to fill the case, and most couplings are designed with air bleeds to permit this to happen. There are various designs of couplings, all of which have different viscosity characteristics.

Shearing means to cut through.

Based on practical experience, vehicles operating with this clutch transmit power automatically, smoothly, and with the added benefit of the fluid being capable of dampening driveline shocks. When a difference in speed of 8 percent exists between the input shaft driven by the driving axle with tractive effort and the other axle, the clutch plates begin shearing (cutting) the special silicone fluid. The shearing action causes heat to build within the housing very rapidly, which results in the silicone fluid stiffening. The stiffening action causes a locking action between the clutch plates to take place within approximately one tenth of a second. The locking action results from the stiff silicone fluid becoming very hard for the plates to shear. The stiff silicone fluid transfers power flow from the driving to the driven plates. The driving shaft is then connected to the driven shaft through the clutch plates and stiff silicone fluid.

The viscous clutch has a self-regulating control. When the clutch assembly locks up, there is very little, if any, relative movement between the clutch plates. Because of this, silicone fluid

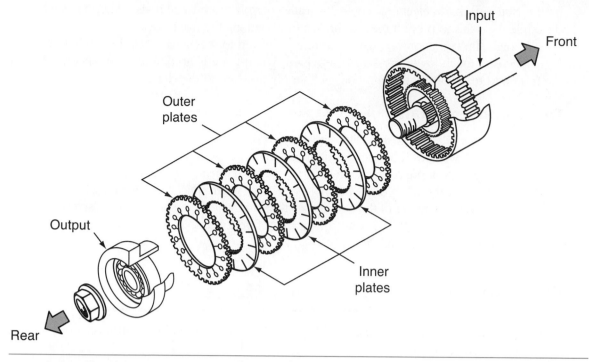

Figure 9-13 Disassembled view of a viscous coupling.

temperature drops, which reduces pressure within the clutch housing. But as speed fluctuates between the driving and driven members, heat increases, causing the silicone fluid to stiffen. Speed differences between the driving and driven members regulate the amount of slip in a viscous clutch driveline. The viscous clutch takes the place of the interaxle differential, biasing driving torque to the normally undriven axle during difficult driving conditions.

A viscous coupling is a speed sensitive limited slip. When it rotates very slowly, very little torque is transferred to the fixed side. As the speed difference increases, so does the torque transferred. A viscous clutch is most often used as, or part of, the center differential in an AWD system. With a center viscous clutch, nearly all of the power is normally transferred to one axle until more transfer is needed. Normally 90 percent of the available torque is sent to one axle and the remaining 10 percent to the other. If the system is based on a FWD vehicle and the front wheels begin to spin, the viscous clutch will transfer torque to the rear axle. If the viscous clutch is connected to a center differential, power is normally equally split between the front and rear axles.

Haldex Clutch

Some AWD vehicles are equipped with a Haldex clutch, which serves as a center differential. This clutch unit distributes the drive force variably between two axles. In a typical application, the Haldex unit mounts in front of the rear differential and receives torque from the front axle.

The Haldex unit has three main parts: the hydraulic pump driven by the slip between the axles or wheels, a wet multidisc clutch, and an electronically controlled valve. The unit is much like a hydraulic pump in which the housing and a piston are connected to one shaft and a piston actuator connected to the other.

When a front wheel slips, the input shaft to the Haldex unit spins faster then its output shaft. This causes the pump to immediately generate oil flow. The oil flow and pressure engages the multidisc clutch to send power to the rear wheels. This happens extremely quickly because an electric pump and accumulator keep the circuit primed.

The oil from the pump flows to the clutch's piston to compress the clutch pack. The oil returns to the reservoir through a controllable valve, which adjusts the oil pressure and the force on the clutch pack. An electronic control module controls the valve and also determines when to decouple the axles to prevent the rear brakes from braking the front axle.

In high slip conditions, a high pressure is delivered to the clutch pack: in tight curves or at high speeds, a much lower pressure is provided. When there is no difference in speed between the front and rear axles, the pump does not supply pressure to the clutch pack.

Quaife Differentials

Quaife Automatic Torque Biasing Differential (ATB) differentials are also used as the center differential in AWD vehicles. The Quaife differential is a gear-operated torque biasing differential with no clutch packs or preloading to transfer torque from one axle to the other, and torque transfer occurs automatically when one wheel loses traction. When quickly accelerating an AWD vehicle, the initial weight transfer to the rear causes the front wheels to spin. With a viscous clutch at the center differential, the fluid needs to heat up before it transfers more torque. With a Quaife unit, the transfer is almost immediate.

Newer 4WD Design Variations

There are many variations of advanced 4WD and AWD systems found on today's vehicles. We will take a look at some of them.

Acura and Honda's VTM-4

Late-model Honda and Acura SUVs may be equipped with a Variable Torque Management-4 Wheel Drive (VTM-4) system. This system operates the vehicle in front-wheel drive most of the time to ensure maximum fuel economy. However, the system's computer is also programmed to know when the vehicle could benefit from rear-wheel power by reading signals from the engine control system. Basically, the level of torque delivery, front to rear, is determined by the rate of acceleration and wheel slip.

Unlike most electronically controlled systems, this system does not wait until there is wheel slip to send torque to the rear wheels. It anticipates the need for all-wheel drive and engages the rear wheels whenever the vehicle is accelerating. During acceleration, the computer activates a pair of clutch packs to connect the ends of the right- and left-rear axle shafts.

The VTM-4 does not use a rear differential; rather, there is a simple ring and pinion setup with the clutches providing differential action. Each drive clutch includes an electromagnetic coil, a ball-cam device, and a set of 19 wet clutch plates, similar to those used in an automatic transmission. Nine of the clutch plates are connected to the axle shafts and 10 are connected to the ring gear. Each drive axle clutch plate is positioned between two ring gear clutch plates. When the control unit determines that torque should be transferred to the rear wheels, current is sent to the two electromagnetic coils. The magnetic field around the coils pulls on steel plates located next to each coil. Friction between those steel plates and the adjoining cam plates cause the cam plates to begin turning. As they do, the balls in the cam plates roll up curved ramps. This action creates an axial thrust against clutch-engagement plates. This thrust, in turn, compresses the clutch plate assemblies.

Depending on how much current goes to the coils, the system can vary the amount of torque distributed to each rear wheel from zero to a maximum of 55 percent. Torque sent to the rear wheels is infinitely variable and is directly proportional to the current sent to the coils by the computer.

Like other systems, the VTM-4 also kicks in when it senses wheel slip, and drivers can lock the system at the maximum rear torque setting by depressing a button on the dash. The lock mode is intended to be used in extremely slippery or stuck conditions. The lock mode only works in first, second, and reverse gears. The coils stay fully energized up to 6 mph and then gradually cut back until, at 18 mph, the system automatically shuts off.

Audi Quattro

Audi's Quattro permanent all-wheel drive system transfers from the front to the rear and side to side as needed. On cars with longitudinally mounted engines, a Torsen differential distributes power and compensates for differences in wheel rotation when cornering. A Torsen differential relies on helical gears to send torque to the different driving wheels; if the car has a transverse engine, a Haldex clutch is used.

The center differential compensates for the speed differences between the front and rear axles and distributes engine power between the front and rear wheels. The system automatically regulates the distribution of power within milliseconds. This action is based engine speed and torque, wheel spreads, and longitudinal and lateral acceleration rates.

The locking function of the center differential and the Electronic Differential Lock make sure the car can still move with only one wheel able to transfer engine power to the road.

If one of the wheels on an axle loses grip and starts spinning, power is transferred to the other wheel by the axle's differential. The electronic differential lock transfers the excess power of the spinning wheel to the other wheels with better traction.

If both wheels on an axle start spinning, the locking of the center differential sends most of the torque to the other axle.

Chrysler AWD

DaimlerChrysler offers an AWD option on their Town & Country and Caravan minivans. These vans are FWD platforms with a rear driveline added. The rear driveline assembly consists of the rear carrier, a torque tube, an overrunning clutch assembly, a vacuum operated dog clutch, and a viscous coupling. During normal operating conditions, the van operates as a FWD vehicle with 90 percent of the torque allocated to the front drive axles. As wheel slippage occurs, the viscous clutch will transmit more torque to the rear wheels. The amount of torque sent to the rear axle is determined solely by the amount of front wheel slippage.

The rear carrier houses an open differential, and half-shafts connect the differential to the rear wheels. A torque tube is used to connect the transaxle to the rear carrier. Inside the torque tube is the torque shaft that does the actual transmitting of power to the rear. The torque shaft rotates on bearings located in the torque tube. The vacuum reservoir and solenoid assembly are attached to the top of the torque tube.

The torque tube attaches to the overrunning clutch case (Figure 9-14) that is attached directly to the rear carrier. The overrunning clutch allows the rear wheels to overrun the front wheels during rapid front-wheel braking. The overrunning action allows the braking system to react the same as it would on a 2WD vehicle, with no push from the rear wheels. The overrunning clutch has a separate oil sump.

The dog clutch (Figure 9-15) provides AWD in reverse by bridging and locking out the overrunning clutch. The dog clutch has projecting ears that fit into the indents of the overrunning clutch's outer race (Figure 9-16). When pushed into this outer race, the dog clutch connects the torque shaft to the rear carrier. The dog clutch is operated by a double-acting vacuum servo. Two vacuum solenoids are controlled by the reverse lamp switch. When energized, the solenoids engage the dog clutch. A spring in the servo disengages the dog clutch if vacuum is shut off.

At the front of the torque tube is the viscous coupling, which controls the amount of torque transmitted to the rear wheels. Like most viscous couplings, this unit increases the amount of torque to the rear wheels in response to increases of front-wheel slip.

Ford's Control Trac II

This on-demand AWD system was released with Ford's small SUV, the Escape, and does not offer fully locked 4WD; rather, it sends torque to the rear wheels only when there is slippage of the front wheels. The system does not use an interaxle differential and both the front and rear differentials are open.

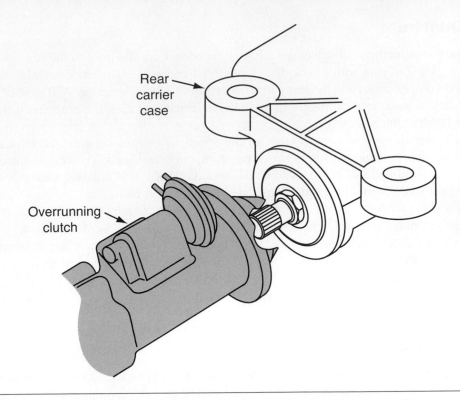

Figure 9-14 The overrunning clutch case is mounted directly to the rear carrier.

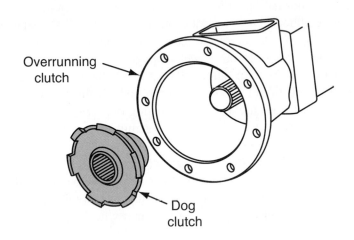

Figure 9-15 The dog clutch used to bypass the overrunning clutch.

The system has two primary modes of operation: 4 × 4 AUTO and 4 × 4 ON. In the 4 × 4 AUTO mode, all of the engine's torque is sent to the front wheels until some slip is detected. When this happens, a rotary blade coupling (Figure 9-17), which is similar to a viscous coupling, generates enough pressure to activate a multiple-disc clutch that sends up to 100 percent of the torque to the rear wheels. The rotary blade coupling relies on a three-bladed fan enclosed in a chamber filled with silicone fluid. When the front wheels slip the fan spins through the fluid, causing the fluid's temperature to rise. This rise in temperature causes the fluid to expand and, in turn, causes the pressure of the fluid to increase. The pressure of the fluid is used to activate the clutch assembly. This is different than a viscous coupling. With a viscous coupling, the fluid transmits the torque. In Ford's setup, the fluid activates the clutch and the clutch transmits the torque. The amount of torque transmitted to the rear is proportional to the speed of the fan and is infinitely variable.

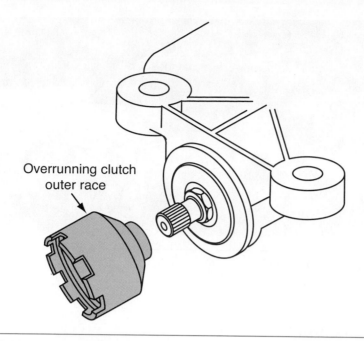

Figure 9-16 The teeth of the dog clutch lock into the indents on the overrunning clutch outer race when reverse gear is selected.

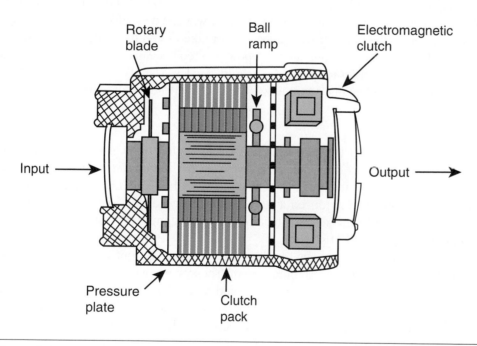

Figure 9-17 A rotary blade coupling as used in the Control Trac II system.

When the driver selects the 4 × 4 ON mode, an electromagnetic clutch is activated. As this clutch is energized, it locks a ball ramp to the input shaft and exerts pressure on the clutch pack. Now when the front wheels slip, the electromagnetic clutch and multiple clutch assembly spin via the ball ramp with the input shaft and torque is transmitted to the rear wheels. This is the only condition or operational mode in which the front and rear axles are somewhat locked together.

Figure 9-18 An AWD Subaru.

Hyundai On-Demand

When Hyundai introduced its first SUV, the vehicle had an option for on-demand AWD. This SUV is a FWD vehicle by design. To provide AWD, the vehicles were equipped with a compact unit that contained the front differential and a power takeoff for the rear axle. This unit also contained a viscous coupling and two planetary gear sets that shared a common stationary ring gear. The sun gears of the planetary sets were connected to the front wheel axle shafts. Engine torque spun one planet carrier whereas the other drove the rear power takeoff and the viscous coupling. Slippage at either or both front wheels accelerated the power takeoff relative to the rear wheels, causing the viscous coupling to transfer torque to the rear drive axles. The planetary gears allowed a maximum of 40 percent of the engine's torque to be transferred to the rear axles.

Subaru's VTD

Most AWD Subarus (Figure 9-18) send nearly all of the engine's power (90 percent) to the front wheels. This design prevents driveline windup and gives the vehicles stability on dry surfaces. When the front wheels begin to slip, more power is sent to the rear. This more-torque-to-the-front bias changed when Subaru introduced a stability control system on some models. Subaru also incorporated the AWD system into its stability control system. The result is Subaru's Variable Torque Distribution (VTD) AWD system.

The VTD system splits torque by sending 45 percent to the front axle. Traction and stability controls are part of the torque distribution system. The system uses a planetary gear-type center differential to vary the torque between the front and rear axles according to weight transfer during acceleration and deceleration as indicated by throttle position. The traction and stability controls brake the wheel or axle that is spinning to keep the vehicle heading in the correct direction.

Summary

Terms to Know
Duty cycle
Duty solenoid
Silicone oil
Viscous coupling

❏ With 4WD, engine power can flow to all four wheels, which can greatly increase a vehicle's traction when traveling in adverse conditions and can also improve handling because side forces generated by the turning of a vehicle or by wind gusts will have less of an effect on a vehicle that has power applied to the road on four wheels.

❏ FWD cars are modified by adding a transfer case, a rear drive shaft, and a rear axle with a differential. Some are equipped with a center differential.

❏ Normally AWD systems use a center differential, viscous coupling, or transfer clutch assembly to transmit engine power to the front and rear axles.

❏ Integrated full-time 4WD systems use computer controls to enhance full-time operation, adjusting the torque split depending on which wheels have traction.

❏ Most on-demand 4WD systems power a second axle only after the first begins to slip.

❏ AWD vehicles are not designed for off-road operation; rather, they are designed to increase vehicle performance in poor traction situations, such as icy or snowy roads. AWD allows for maximum control by transferring a large portion of the engine's power to the axle with the most traction. Most AWD designs use a center differential to split the power between the front and rear axles.

❏ The interaxle differential allows for different front and rear driveline shaft speeds to prevent driveline windup but may also result in a loss of traction in very slippery conditions.

❏ A viscous coupling is basically a drum with some thick fluid that houses several closely fitted, thin steel disks. One set of the discs is connected to the front wheels and the other to the rear.

❏ An open differential sends power to the axle or wheel that has the *least* traction. Viscous coupling differentials send power to the axle or wheel that has the *most* traction.

❏ In a typical viscous coupling, one of the two shafts that have external splines meshes with the viscous coupling housing's internal splines that also mesh with the viscous coupling plates. The other shaft rotates on seals in the housing. The plates rotate in a reservoir of silicone oil.

❏ The viscous coupling plates rotate at different speeds and can shear the silicone fluid with ease.

Review Questions

Short Answer Essays

1. How does a viscous clutch assembly work? Why is it used in AWD systems?

2. What are the main differences between 4WD and AWD?

3. In basic construction, what is the major difference between full-sized 4WD trucks and SUVs and smaller 4WD trucks and SUVs?

4. Some 4WD systems offer a low range. What is its purpose?

5. What types of limited-slip units are commonly used in center differentials?

6. What is a major advantage of an AWD system when compared to a conventional 4WD system?

7. Briefly explain how a viscous coupling works.

8. What is the primary purpose of an interaxle differential?

9. Some consider the system used in a Porsche 959 as the most complex AWD system ever developed, why?

10. Briefly explain how a Haldex clutch works.

Fill-in-the-Blanks

1. Normally, AWD systems use a _____ _____ ,

   _____ _____ , or _____

   _____ assembly to transmit engine power to the front and rear axles.

2. A viscous coupling is a _____ _____ limited slip.

3. The Quaife differential is a _____-_____ torque biasing differential.

4. The interaxle differential prevents _____ _____ but may also result is a loss of traction in very slippery conditions.

5. A _____ differential allows for different front and rear axle speeds.

6. Silicone oil is used in a viscous coupling because its _____ is not affected very much by temperature and it has high _____ resistance and _____ changes at high temperatures.

7. When the plates of a viscous coupling rotate at different speeds, the plates _____ the fluid.

8. When a viscous coupling provides holding ability for two parts of a plannetary or bevel gear set, it is called a _____ _____.

9. An open differential sends power to the axle or wheel that has the _____ traction. Viscous coupling differentials send power to the axle or wheel that has the _____ traction.

10. Center and rear differentials may be fitted with a limited-slip differential that uses either a _____ clutch or a _____ coupling.

Multiple Choice

1. *Technician A* says that some AWD systems have a center differential. *Technician B* says that some AWD vehicles have a viscous clutch. Who is correct?
A. A only
B. B only
C. Both A and B
D. Neither A nor B

2. When discussing Hyundai's on-demand 4WD system, *Technician A* says that these vehicles are equipped with a compact unit that contains the front differential, a power takeoff for the rear axle, a viscous coupling, two electromagnetic clutches, and two planetary gear sets. *Technician B* says that the common ring gear of the planetary sets is held stationary and the sun gears are connected to the front-wheel axle shafts. Who is correct?
A. A only
B. B only
C. Both A and B
D. Neither A nor B

3. When discussing on-demand 4WD systems, *Technician A* says that these systems are automatically controlled. *Technician B* says that these systems drive the front and rear axles equally until some slip is detected.

Who is correct?
A. A only
B. B only
C. Both A and B
D. Neither A nor B

4. *Technician A* says that during turns the front wheels travel a greater distance than the rear wheels. *Technician B* says that when a vehicle is operating in 4WD and the front and rear axles are turning at the same speed, driveline windup can result. Who is correct?
A. A only
B. B only
C. Both A and B
D. Neither A nor B

5. When discussing an Acura MDX with VTM-4, *Technician A* says that the system operates the vehicle in RWD most of the time to ensure maximum fuel economy. *Technician B* says that unlike most electronically controlled systems, this system does not wait until there is wheel slip to send torque to the other axle. Who is correct?
A. A only
B. B only
C. Both A and B
D. Neither A nor B

6. *Technician A* says that an open differential sends power to the axle or wheel that has the least slip. *Technician B* says that viscous coupling differentials send power to the axle or wheel that has the most traction. Who is correct?
   A. A only
   B. B only
   C. Both A and B
   D. Neither A nor B

7. When discussing Audi's Quattro all-wheel drive system: *Technician A* says that the system transfers power from the front to the rear and side to side as needed. *Technician B* says that Quattro equipped cars with a transverse engine use a Haldex clutch. Who is correct?
   A. A only
   B. B only
   C. Both A and B
   D. Neither A nor B

8. When discussing AWD systems, *Technician A* says that they normally use a center differential, viscous coupling, or transfer clutch assembly to transmit and split engine power to the front and rear axles. *Technician B* says that nearly all AWD vehicles are designed for off-road operation and extremely poor road conditions. Who is correct?
   A. A only
   B. B only
   C. Both A and B
   D. Neither A nor B

9. When discussing DaimlerChrysler's AWD option on their minivans, *Technician A* says that the amount of torque sent to the rear axle is determined solely by the amount of front wheel slippage. *Technician B* says that the system equally splits engine power to the front and rear axles when the brakes are applied quickly and hard. Who is correct?
   A. A only
   B. B only
   C. Both A and B
   D. Neither A nor B

10. When discussing viscous clutches, *Technician A* says that the driver activates the viscous clutch through a selector switch to improve driving wheel traction. *Technician B* says that the gears inside the clutch assembly rotate in the viscous fluid. Who is correct?
    A. A only
    B. B only
    C. Both A and B
    D. Neither A nor B

Drivetrain Electrical and Electronic Systems

Upon completion and review of this chapter, you should be able to:

❏ Understand and explain the basic principles of electricity.

❏ Understand and explain the basic difference between electricity and electronics.

❏ Understand and define the terms voltage, current, and resistance.

❏ Name the various electrical components and their uses in electrical circuits.

❏ Understand and describe the purpose and operation of a clutch safety switch.

❏ Understand and describe the purpose and operation of a reverse lamp switch.

❏ Understand and describe the purpose and operation of an upshift light and a high gear switch.

❏ Understand and describe the location and operation of ABS speed sensor circuits.

❏ Understand and describe the purpose and operation of a shift blocking circuit.

Introduction

Although a manual transmission is not typically electrically operated or controlled, a few accessories of the car are controlled by the transmission. The driveline is also fitted with sensors that give vital information to computers that control other systems of the automobile. There are a few transmissions that have their shifting controlled or limited by electronics. To understand the operation of the accessories, sensors, and controls, you must have a good understanding of electricity and electronics.

There is often confusion concerning the terms electrical and electronic. In this book, electrical and electrical systems will refer to wiring and electrical parts. Electronics will mean computers and other black box type–items used to control engine and vehicle systems.

A basic understanding of electrical principles is important for proper diagnosis of any system that is monitored, controlled, or operated by electricity. Although the subject is normally covered in a separate course, a quick overview of electricity and its principles is presented here.

Basic Electricity

All things are made up of atoms. An atom is the smallest particle of something. Atoms are very small and cannot be seen with the eye. This may be the reason many technicians struggle to understand electricity. The basics of electricity focus on atoms. Understanding the structure of the atom is the first step to understanding how electricity works. The following principles describe atoms, which are the building blocks of all materials.

❏ In the center of every atom is the nucleus.
❏ The nucleus contains positively-charged particles called protons and particles called neutrons that have no charge.
❏ Negatively-charged particles called electrons orbit around every nucleus.
❏ Every type of atom has a different number of protons and electrons, but each atom has an equal number of protons and electrons. Therefore, the total electrical charge of an atom is zero, or neutral.

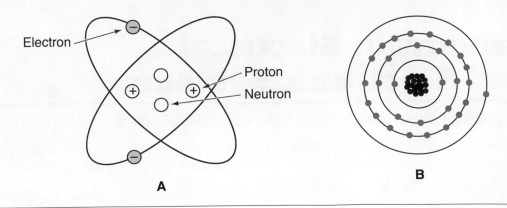

Figure 10-1 Basic structure of (A) a simple atom and of (B) a copper atom.

In all atoms, the electrons are arranged in different orbits, called shells. Each shell contains a specific or certain number of electrons. For example, a copper atom (Figure 10-1) has 29 electrons and 29 protons. The 29 electrons are arranged in shells. The outer shell has only one electron. This outer shell needs 32 electrons to be completely full. This means that the one electron in the outer shell is loosely tied to the atom and can be easily removed.

The looseness or tightness of the electrons in orbit around the nucleus of an atom explains the behavior of electricity. Electricity is caused by the flow of electrons from one atom to another. The release of energy as one electron leaves the orbit of one atom and jumps into the orbit of another is **electricity** (Figure 10-2). The key behind creating electricity is to give a reason for the electrons to move.

There is a natural attraction of electrons to protons. Electrons have a negative charge and are attracted to something with a positive charge. When an electron leaves the orbit of an atom, the atom then has a positive charge. An electron moves from one atom to another because the atom next to it appears to be more positive than the one it is orbiting. An electrical power source provides for a more positive charge and in order to allow for a continuous flow of electricity, it supplies free electrons. In order to have a continuous flow of electricity, three things must be present: an excess of electrons in one place, a lack of electrons in another place, and a path between the two places.

Two power or energy sources are used in an automobile's electrical system. These are based on a chemical reaction and on magnetism. A car's battery is a source of chemical energy. A chemical reaction in the battery provides for an excess of electrons in one place and a lack of electrons in another place. Batteries have two terminals, a positive and a negative. Basically, the negative terminal is the outlet for the electrons, and the positive terminal is the inlet for the electrons to get to the protons. The chemical reaction in a battery causes a lack of electrons at the positive (+) terminal and an excess at the negative (–) terminal. This creates an electrical imbalance, causing the electrons to flow through the path provided by a wire. A simple example of this process is shown in the battery and light arrangement in Figure 10-3.

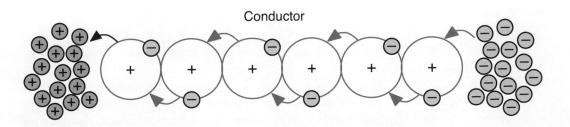

Figure 10-2 Electricity is the flow of electrons from one atom to another.

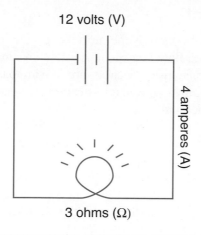

12 volts (V)

4 amperes (A)

3 ohms (Ω)

Figure 10-3 A simple light circuit.

The chemical process in the battery continues to provide electrons until the chemicals become weak. At that time, either the battery has run out of electrons or all of the protons are matched with an electron. When this happens, there is no longer a reason for the electrons to want to move to the positive side of the battery. It no longer looks more positive. Fortunately, the vehicle's charging system restores the battery's supply of electrons. This allows the chemical reaction in the battery to continue almost indefinitely.

Electricity and magnetism are interrelated. One can be used to produce the other. Moving a wire (a conductor) through an already existing magnetic field can produce electricity. This process of producing electricity through magnetism is called *induction*. In a generator, a coil of wire is moved through a magnetic field. In an alternator, a magnetic field is moved through a coil of wire. In both cases, electricity is produced. The amount of electricity that is produced depends on a number of factors, including the strength of the magnetic field, the number of wires that pass through the field, and the speed at which the wire moves through the magnetic field.

Measuring Electricity

Electrical **current** is a term used to describe the movement or flow of electricity. The greater the number of electrons flowing past a given point in a given amount of time, the more current the circuit has. This current, like the flow of water or any other substance, can be measured. **Voltage** is electrical pressure (Figure 10-4). Voltage is the force developed by the attraction of the electrons to the protons. The more positive one side of the circuit is, the more voltage is present in the circuit. Voltage does not flow; rather, it is the pressure that causes current flow. When any substance flows, it meets resistance. The resistance to electrical flow can be measured.

Voltage is the force that causes electrons to move.

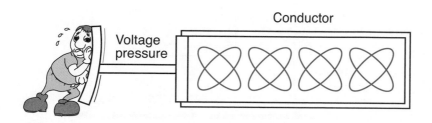

Conductor

Voltage pressure

Figure 10-4 Voltage is the force that causes electrons to move.

Electrical Flow (Current)

The **ampere** is the unit for measuring electric current. One ampere means that 6.25 billion electrons are flowing past a given point in one second.

In **direct current (DC)**, the electrons flow in one direction only; in **alternating current (AC)**, the electrons change direction at a fixed rate.

The unit for measuring electrical current is the **ampere**, usually called an amp. The instrument used to measure electrical current flow in a circuit is called an **ammeter**.

In the flow of electricity, millions of electrons are moving past any given point at the speed of light. The electrical charge of any one electron is extremely small. It takes millions of electrons to make a charge that can be measured.

There are two types of electrical flow, or current: **direct current (DC)** and **alternating current (AC)**. In direct current, the electrons flow in one direction only. The example of the battery and light shown earlier is based upon direct current. In alternating current, the electrons change direction at a fixed rate. Most automobile circuits operate on DC current, while the current in homes and buildings is AC. Generators and alternators produce AC volts, but the AC voltage is converted to DC before it is released into the rest of the electrical system.

Resistance

In every atom, the electrons resist being moved out of their shell. The amount of resistance depends on the type of atom. As explained earlier, in some atoms (such as those in copper) there is very little resistance to electron flow because the outer electron is loosely held. In other substances there is more resistance to flow because the outer electrons are tightly held.

The resistance to current flow produces heat. This heat can be measured to determine the amount of resistance. A unit of measured resistance is called an **ohm**. Resistance can be measured by an instrument called an **ohmmeter**.

Pressure

One **volt** is the amount of pressure (force) required to move 1 amp of current through a resistance of 1 ohm. Voltage is measured by an instrument called a **voltmeter**.

In electrical flow, some force is needed to move the electrons between atoms. This force is the pressure that exists between a positive and less positive point within an electrical circuit. This force, also called electromotive force (EMF), is measured in units called **volts**. One volt is the amount of pressure (force) required to move 1 ampere of current through a resistance of 1 ohm. Voltage is measured by an instrument called a **voltmeter**.

Circuits

When electrons are able to flow along a path (wire) between two points, an electrical circuit is formed. An electrical circuit is considered complete when there is a path that connects the positive and negative terminals of the electrical power source. Somewhere in the circuit there must be a load or resistance to control the amount of current in the circuit. Most automotive electrical circuits use the chassis as the path to the negative side of the battery (Figure 10-5). Electrical components have a lead that connects them to the chassis. These are called the chassis ground connections. In a complete circuit, the flow of electricity can be controlled and applied to do

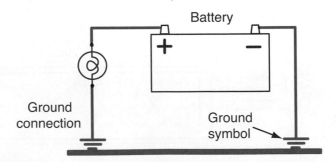

Figure 10-5 A simple light circuit using the vehicle as the negative conductor for the circuit.

useful work, such as light a headlamp or turn over a starter motor. Components that use electrical power put a load on the circuit and consume electrical energy.

Although electrons in a circuit actually flow from negative to positive (the electron theory of current flow), automotive electrical diagrams continue to assume that current flow is from positive to negative (the conventional theory of current flow).

The amount of current that flows in a circuit is determined by the resistance in that circuit. As resistance goes up, the current goes down. The energy used by a load is measured in volts. Amperage stays constant in a circuit but the voltage is dropped as it powers a load. Measuring voltage drop determines the amount of energy consumed by the load.

Ohm's Law

To understand the relationship between current, voltage, and resistance in a circuit, you need to know **Ohm's law**, the basic law of electricity. This law states that it takes one volt of electrical pressure to push one ampere of electrical current through one ohm of resistance. As such, the law provides a mathematical formula for determining the amount of current, voltage, or resistance in a circuit when two of these are known. The basic formula is: Voltage = Current multiplied by Resistance.

Although the basic premise of this formula is calculating unknown values in an electrical circuit (Figure 10-6), it also helps to define the behaviors of electrical circuits. A knowledge of these behaviors is important to an automotive technician.

If voltage does not change, but there is a change in the resistance of the circuit, the current will change. If resistance increases, current decreases. If resistance decreases, current will increase. If voltage changes, so must the current or resistance. If the resistance stays the same and current decreases, so will voltage. Likewise, if current increases, so will the voltage.

AUTHOR'S NOTE: The electrical property you will most often measure is voltage. Understanding Ohm's law and how conditions will affect voltage is your key to diagnosis. Whenever you get a voltage measurement that is not what you expected or what the specification calls for, think of resistance! For example, if you measure the voltage at both ends of a wire, you would expect the readings to be the same. However, if the readings aren't alike, you know there is some resistance between those two points. The higher the difference, the higher the resistance!

Electrical Circuits

A complete electrical circuit exists when electrons flow along a path between two points. In a complete circuit, resistance must be low enough to allow the available voltage to push electrons between the two points. Most automotive circuits contain five basic parts.

Ohm's law, the basic law of electricity states that it takes one volt of electrical pressure to push one ampere of electrical current through one ohm of resistance. The basic formula is: Voltage = Current multiplied by Resistance.

Voltage (E) = Current (I) times Resistance (R), therefore

$$E = I \times R.$$

Current (I) = Voltage (E) divided by Resistance (R), therefore

$$I = E/R.$$

Resistance (R) = Voltage (E) divided by Current (I), therefore

$$R = E/I.$$

Figure 10-6 Ohm's law.

1. **Power sources**, such as the battery or alternator, that provide the energy needed to create electron flow

2. **Conductors**, such as copper wires, that provide a path for current flow

3. **Loads**, which are devices that use electricity to perform work, such as light bulbs, electric motors, or resistors

4. **Controllers**, such as switches or relays, that direct the flow of electrons

5. **Protection devices**, such as fuses, circuit breakers, and fusible links.

Shop Manual
Chapter 10, page 433

A complete circuit must have a complete path from the power source to the load and back to the source. With the many circuits on an automobile, this would require hundreds of wires connected to both sides of the battery. To avoid this, automobiles are equipped with power distribution centers or fuse blocks that distribute battery voltage to various circuits. The positive side of the battery is connected to the fuse block, and power is distributed from there (Figure 10-7).

As a common return circuit, auto manufacturers use a wiring style that involves using the vehicle's metal frame (Figure 10-8) as part of the return circuit. The load is often grounded directly to the metal frame, which then acts as the return wire in the circuit. Current passes from the battery, through the load, and into the frame. The frame is connected to the negative terminal of the battery through the battery's ground wire. This completes the circuit.

An electrical component, such as an AC generator, is often mounted directly to the engine block, transmission case, or frame. This direct mounting effectively grounds the component without the use of a separate ground wire. In other cases, however, a separate ground wire must be run from the component to the frame or another metal part to ensure a sound return path. The increased use of plastics and other nonmetallic materials in body panels and engine parts has made electrical grounding more difficult. To assure good grounding back to the battery, some manufacturers now use a network of common grounding terminals and wires.

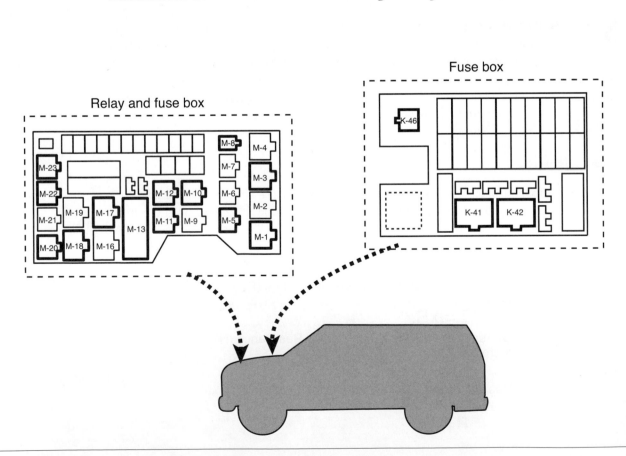

Figure 10-7 Location of fuse and relay boxes on a typical 4WD vehicle.

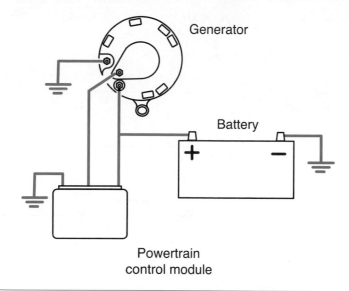

Figure 10-8 To eliminate the need to run separate return wires, many accessories and automotive components are grounded to the vehicle's chassis.

Circuit Components

Automotive electrical circuits contain a number of different types of electrical devices. The more common components are outlined in the following sections.

Resistors are used to limit current flow (and thereby voltage) in circuits where full current flow and voltage are not needed. Resistors are devices specially constructed to introduce a measured amount of electrical resistance into a circuit (Figure 10-9). In addition, some other components use resistance to produce heat and even light. An electric window defroster is a specialized type of resistor that produces heat. Electric lights are resistors that get so hot they produce light.

Resistors in common use in automotive circuits are of three types: fixed value, stepped or tapped, and variable.

Shop Manual
Chapter 10, page 434

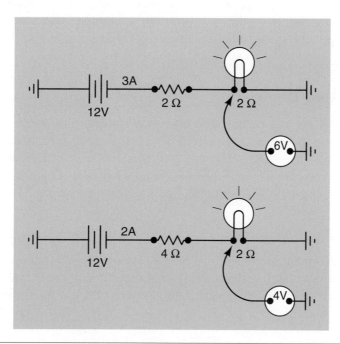

Figure 10-9 The effect of resistors in electrical circuits.

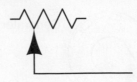

Figure 10-10 A rheostat.

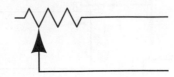

Figure 10-11 A potentiometer.

Fixed value resistors are designed to have only one rating, which should not change. These resistors are used to control voltage, such as in an automotive ignition system.

Tapped or stepped resistors are designed to have two or more fixed values, available by connecting wires to the several taps of the resistor. Heater motor resistor packs, which provide for different fan speeds, are an example of this type of resistor.

Variable resistors are designed to have a range of resistances available through two or more taps and a control. Two examples of this type of resistor are rheostats and potentiometers. **Rheostats** have two connections (Figure 10-10), one to the fixed end of a resistor and one to a sliding contact with the resistor. Turning the control moves the sliding contact away from or toward the fixed end tap, increasing or decreasing the resistance. **Potentiometers** have three connections (Figure 10-11), one at each end of the resistance and one connected to a sliding contact with the resistor; turning the control moves the sliding contact away from one end of the resistance, but toward the other end.

Another type of variable resistor is the thermistor. This resistor is designed to change in values as its temperature changes. Although most resistors are carefully constructed to maintain their rating within a few ohms through a range of temperatures, the thermistor is designed to change its rating. Thermistors are used to provide compensating voltage in components or to determine temperature. As a temperature sender, the thermistor is connected to a voltmeter calibrated in degrees. As the temperature rises or falls, the resistance also changes. This changes the reading on the meter.

Circuit Protective Devices

When overloads or shorts in a circuit cause too much current to flow, the wiring in the circuit heats up, the insulation melts, and a fire can result unless the circuit has some kind of protective device. Fuses, fuse links, maxi-fuses, and circuit breakers are designed to provide protection from high current. These protection devices open the circuit when high current is present. As a result, the circuit no longer works, but the wiring and the components are saved from damage.

Switches

Electrical circuits are usually controlled by a switch of some type. Switches do two things. They turn the circuit on or off and they direct the flow of current in a circuit. Switches can be under the control of the driver or can be self-operating through a condition of the circuit, the vehicle, or the environment.

Rheostats are variable resistors with two connections that are used to change current flow through a circuit.

Potentiometers are variable resistors with three connections that are typically used to change voltage.

Shop Manual
Chapter 10, page 448

Shop Manual
Chapter 10, page 450

Contacts in a switch can be of several types, each named for the job they do or the sequence in which they work. A hinged-pawl switch is the simplest type of switch. It either makes or breaks the current in a single conductor or circuit. It is a single-pole, single-throw (SPST) switch. The throw refers to the number of output circuits, and the pole refers to the number of input circuits made by the switch.

Another type of SPST switch is the momentary contact switch. The spring-loaded contact on this switch keeps it from making the circuit except when pressure is being applied to the button. A horn switch is of this type. Because the spring holds the contacts open, the switch has a further designation: normally open. In the case in which the contacts are held closed except when the button is pressed, the switch is designated normally closed.

Single-pole, double-throw switches have one wire in and two wires out. This type of switch allows the driver to select between two circuits, such as high-beam or low-beam headlights.

Switches can be designed with a great number of poles and throws. The transmission neutral start switch may have two poles and six throws and is referred to as a multiple-pole, multiple-throw (MPMT) switch. It contains two movable wipers that move in unison across two sets of terminals. The dotted line shows that the wipers are mechanically linked, or ganged. The switch closes a circuit to the starter in either P (park) or N (neutral) and to the back-up lights in R (reverse) (Figure 10-12).

Most switches are combinations of hinged-pawl and push-pull switches, with different numbers of poles and throws. Some special switches are required, however, to satisfy the circuits of modern automobiles. A mercury switch is sometimes used to detect motion in a component, such as the one used in the engine compartment to turn on the compartment light.

A temperature-sensitive switch usually contains a bimetallic element heated either electrically or by some component in which the switch is used as a sensor. When engine coolant is below or at normal operating temperature, the engine coolant temperature sensor is in its normally open condition. If the coolant exceeds the temperature limit, the bimetallic element bends the two contacts together, and the switch to the indicator or the instrument panel is closed. Other applications for heat-sensitive switches are time delay switches and flashers.

Relays

A **relay** (Figure 10-13) is an electric switch that allows a small amount of current to control a much larger one. It consists of a control circuit. When the control circuit switch is open, no current flows to the coil, so the windings are deenergized. When the switch is closed, the coil is energized, turning the soft iron core into an electromagnet and drawing the armature down. This closes the power circuit contacts, connecting power to the load circuit. When the control switch is opened, the current stops flowing in the coil, the electromagnet disappears, and the armature is released, which breaks the power circuit contacts.

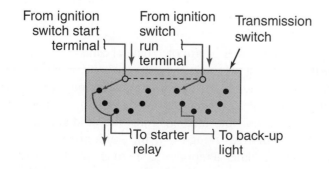

Figure 10-12 A multiple-pole, multiple-throw neutral start safety switch.

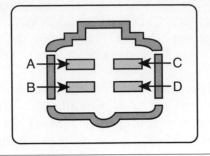

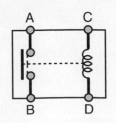

Figure 10-13 A typical relay.

Shop Manual
Chapter 10, page 455

Solenoids

Solenoids (Figure 10-14) are also electromagnets with movable cores used to translate electrical current flow into mechanical movement. The movement of the core causes something else, such as a lever, to move. They also can close electrical contacts, acting as a relay at the same time.

Electromagnetism Basics

Electricity and magnetism are related. One can be used to create the other. Current flowing through a wire creates a magnetic field around the wire. Moving a wire through a magnetic field creates current flow in the wire.

Many automotive components, such as alternators, ignition coils, starter solenoids, and magnetic pulse generators operate using principles of electromagnetism.

Although almost everyone has seen magnets at work, a simple review of magnetic principles is in order to ensure a clear understanding of electromagnetism.

A substance is said to be a magnet if it has the property of magnetism—the ability to attract such substances as iron, steel, nickel, or cobalt. These are called magnetic materials.

A magnet has two points of maximum attraction, one at each end of the magnet. These points are called poles, with one being designated the north pole and the other the south pole. When two magnets are brought together, opposite poles attract, whereas similar poles repel.

A magnetic field, called a **field of flux**, exists around every magnet (Figure 10-15). The field consists of invisible lines along which a magnetic force acts. These lines emerge from the north pole and enter the south pole, returning to the north pole through the magnet itself. All lines of force leave the magnet at right angles to the magnet. None of the lines cross each other. All lines are complete.

Magnets can occur naturally in the form of a mineral called magnetite. Artificial magnets can also be made by inserting a bar of magnetic material inside a coil of insulated wire and passing a heavy direct current through the coil. This principle is very important in understanding certain automotive electrical components. Another way of creating a magnet is by stroking the magnetic material with a bar magnet. Both methods force the randomly arranged molecules of the magnetic material to align themselves along north and south poles.

Artificial magnets can be either temporary or permanent. Temporary magnets are usually made of soft iron. They are easy to magnetize but quickly lose their magnetism when the

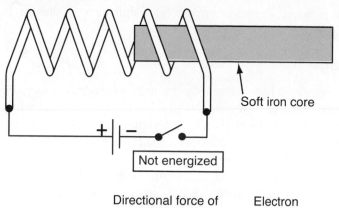

Soft iron core

+ | − ⊸ ⊸

Not energized

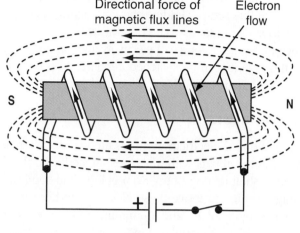

Directional force of
magnetic flux lines

Electron
flow

S

N

+ | − ⊸ ⊸

Note: Core will move to center of coil regardless
of directional force of magnetic flux lines.

Energized

Figure 10-14 A solenoid.

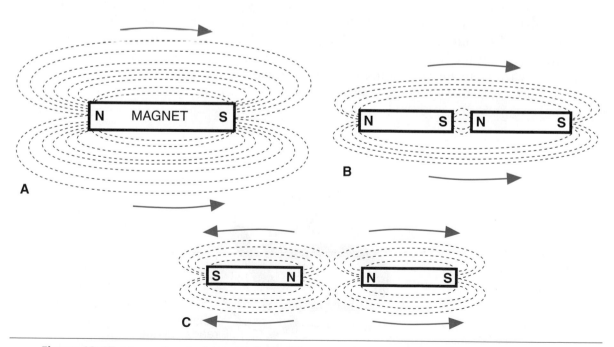

N MAGNET S

A

N S N S

B

S N N S

C

Figure 10-15 Magnetic principles: (A) field of flux around a magnet, (B) unlike poles attract each other, and (C) like poles repel.

magnetizing force is removed. Permanent magnets are difficult to magnetize, but once magnetized they retain this property for very long periods.

Induced Voltage

Now that we have explained how current can be used to generate a magnetic field, it is time to examine the opposite effect of how magnetic fields can produce electricity. Consider a straight piece of conducting wire with the terminals of a voltmeter attached to both ends. If the wire is moved across a magnetic field, the voltmeter registers a small voltage reading (Figure 10-16). A voltage has been induced in the wire.

It is important to remember that the conducting wire must cut across the flux lines to induce a voltage. Moving the wire parallel to the lines of flux does not induce voltage. The wire need not be the moving component in this setup. Holding the conducting wire still and moving the magnetic field at right angles to it also induces voltage in the wire.

The wire or conductor becomes a source of electricity and has a polarity or distinct positive and negative end. However, this polarity can be switched depending on the relative direction of movement between the wire and magnetic field. This is why an alternator produces alternating current.

Conductors and Insulators

Controlling and routing the flow of electricity requires the use of materials known as conductors and insulators. Conductors are materials with a low resistance to the flow of current. If the number of electrons in the outer shell or ring of an atom is less than four, the force holding them in place is weak. The voltage needed to move these electrons and create current flow is relatively small. Most metals, such as copper, silver, and aluminum are excellent conductors.

Copper wire is by far the most popular conductor used in automotive electrical systems. Wire wound inside of electrical units, such as ignition coils and generators, usually has a very thin baked-on insulating coating. External wiring is often covered with a plastic-type insulating material that is highly resistant to environmental factors like heat, vibration, and moisture. Where flexibility is required, the copper wire will be made of a large number of very small strands of wire woven together.

When the number of electrons in the outer ring is greater than four, the force holding them in orbit is very strong and very high voltages are needed to move them. These materials are known as **insulators**. They resist the flow of current. Thermal plastics are the most common electrical insulators used today. They can resist heat, moisture, and corrosion without breaking down.

When there are four electrons in the outer ring, the material is a semiconductor. A semiconductor is neither a conductor nor an insulator.

Insulators are materials like thermal plastics that do not let electrons flow easily through them.

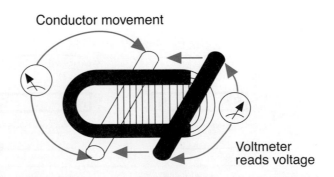

Figure 10-16 Moving a conductor through a magnetic field induces a voltage.

Basics of Electronics

Computerized engine controls and other features of today's cars would not be possible if it were not for electronics. For purposes of clarity, let us define electronics as the technology of controlling electricity. Electronics has become a special technology beyond electricity. Transistors, diodes, semiconductors, integrated circuits, and solid-state devices are all considered to be part of electronics rather than just electrical devices. But keep in mind that all the basic laws of electricity apply to electronic controls.

Semiconductors

A **semiconductor** is a material or device that can function as either a conductor or an insulator, depending on how its structure is arranged. Semiconductor materials have less resistance than an insulator, but more resistance than a conductor. Some common semiconductor materials include silicon (Si) and germanium (Ge).

In semiconductor applications, materials have a crystal structure. This means that their atoms do not lose and gain electrons as the atoms in conductors do. Instead, the atoms in these semiconductor materials share outer electrons with each other. In this type of atomic structure, the electrons are tightly held and the element is stable.

Because the electrons are not free, crystals cannot conduct current. These materials are called electrically inert materials. A small amount of trace element must be added for these to function as semiconductors. The addition of these traces, called impurities, allows the material to function as a semiconductor. The type of impurity added determines what type of semiconductor will be produced.

The **diode** is the simplest semiconductor device. A diode allows current to flow in one direction, but not in the opposite direction. Therefore, it can function as a switch, acting as either a conductor or insulator, depending on the direction of current flow.

A variation of the diode is the zener diode. This device functions like a standard diode until a certain voltage is reached. When the voltage level reaches this point, the zener diode allows current to flow in the reverse direction. Zener diodes are often used in electronic voltage regulators.

A **transistor** is an electronic device produced by joining three sections of semiconductor materials. Like the diode, it is very useful as a switching device, functioning as either a conductor or an insulator.

One transistor or diode is limited in its ability to do complex tasks. However, when many semiconductors are combined into a circuit, they can perform complex functions.

An **integrated circuit** is simply a large number of diodes, transistors, and other electronic components, such as resistors and capacitors, all mounted on a single piece of semiconductor material. This type of circuit has a tremendous size advantage. It is extremely small. Circuitry that used to take up entire rooms can now fit into a pocket. The principles of semiconductor operation remain the same in integrated circuits—only the size has changed.

The increasingly small size of integrated circuits is very important to automobiles. This means that electronics is no longer confined to simple tasks, such as rectifying alternator current. Enough transistors, diodes, and other solid-state components can be installed in a car to make logic decisions and issue commands to other areas of the engine. This is the foundation of computerized-control systems.

The computer (Figure 10-17) has taken over many of the tasks in cars and trucks that were formerly performed by vacuum, electromechanical, or mechanical devices. When properly programmed, they can carry out explicit instructions with blinding speed and almost flawless consistency.

The computer is called the PCM, ECM, central processor, or microprocessor.

A typical electronic control system is made up of sensors, actuators, and related wiring that is tied to a computer.

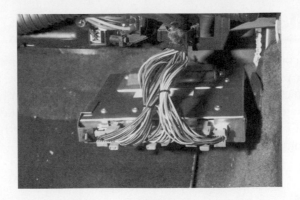

Figure 10-17 A typical automotive computer.

Until recent years, the only electrical part connected to the transmission was the back-up switch. This was normally mounted so the external shift linkage opened and closed the circuit.

Shop Manual
Chapter 10, page 453

Sensors

All sensors perform the same basic function. They detect a mechanical condition (movement or position), chemical state, or temperature condition and change it into an electrical signal that can be used by the computer to make decisions. The computer makes decisions based on information it receives from sensors. Each sensor used in a particular system has a specific job to do. Together these sensors provide enough information to help the computer form a complete picture of vehicle operation. Even though there are a variety of different sensor designs, they all fall under one of two operating categories: reference voltage sensors or voltage generating sensors.

Reference voltage (Vref) sensors provide input to the computer by modifying or controlling a constant, predetermined voltage signal (Figure 10-18). This signal, which can have a reference value from 5 to 9 volts, is generated and sent out to many sensors by a reference voltage regulator located inside the processor. The term processor is used to describe the actual metal box that houses the computer and its related components. Because the computer knows that a certain

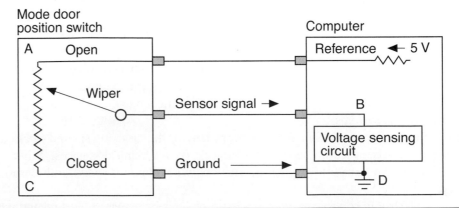

Figure 10-18 A potentiometer alters the reference voltage in response to the movement of the wiper and sends a signal back to the voltage sensing circuit of the computer.

voltage value has been sent out, it can indirectly interpret things like motion, temperature, and component position, based on what comes back. For example, consider the operation of the throttle position sensor (TP sensor). During acceleration (from idle to wide-open throttle), the computer monitors throttle plate movement based on the changing reference voltage signal returned by the TP sensor. (The TP sensor is a type of variable resistor known as a rotary potentiometer that changes circuit resistance based on throttle shaft rotation.) As TP sensor resistance varies, the computer is programmed to respond in a specific manner (for example, increase fuel delivery or alter spark timing) to each corresponding voltage change.

Most sensors presently in use are variable resistors (potentiometers). They modify a voltage to or from the computer, indicating a constantly changing status that can be calculated, compensated for, and modified. That is, most sensors simply control a voltage signal from the computer. When varying internal resistance of the sensor allows more or less voltage to ground, the computer senses a voltage change on a monitored signal line. The monitored signal line may be the output signal from the computer to the sensor (one- and two-wire sensors), or the computer may use a separate return line from the sensor to monitor voltage changes (three-wire sensors).

While most sensors are variable reference voltage, there is another category of sensor—the **voltage generating devices** (Figure 10-19). These sensors include components like the Hall-effect switch, oxygen sensor (zirconium dioxide), and knock sensor (piezoelectric), that are capable of producing their own input voltage signals. This varying voltage signal, when received by the computer, enables the computer to monitor and adjust for changes in the computerized engine control system.

In addition to variable resistors, two other commonly used reference voltage sensors are switches and thermistors. Switches provide the necessary voltage information to the computer so that vehicles can maintain the proper performance and driveability. Thermistors are special types of resistors that convert temperature into a voltage. Regardless of the type of sensors used in electronic control systems, the computer is incapable of functioning properly without input signal voltage from sensors.

Communication Signals

Most input sensors are designed to produce a voltage signal that varies within a given range (from high to low, including all points in between). A signal of this type is called an **analog** signal. Unfortunately, the computer does not understand analog signals. It can only read a **digital** binary signal, which is a signal that has only two values—on or off (Figure 10-20).

To overcome this communication problem, all analog voltage signals are converted to a digital format by a device known as an analog-to-digital converter (A/D converter). The A/D converter is located in a section of the processor that receives the input signal. Some sensors like

Analog is a type of signal that carries a constant or changing voltage.

Digital is a type of signal that is caused by switching the circuit on and off.

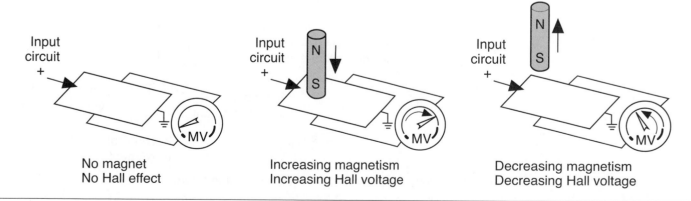

Input circuit +
No magnet
No Hall effect

Input circuit +
Increasing magnetism
Increasing Hall voltage

Input circuit +
Decreasing magnetism
Decreasing Hall voltage

Figure 10-19 Hall-effect principles of voltage induction.

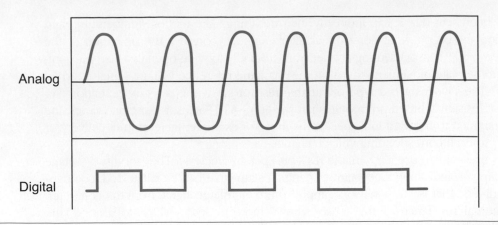

Figure 10-20 Analog signals are constantly variable, whereas digital signals are either on or off, or high or low.

Automatic transmission equipped vehicles have a neutral safety switch that only allows the starter to work when park or neutral is selected.

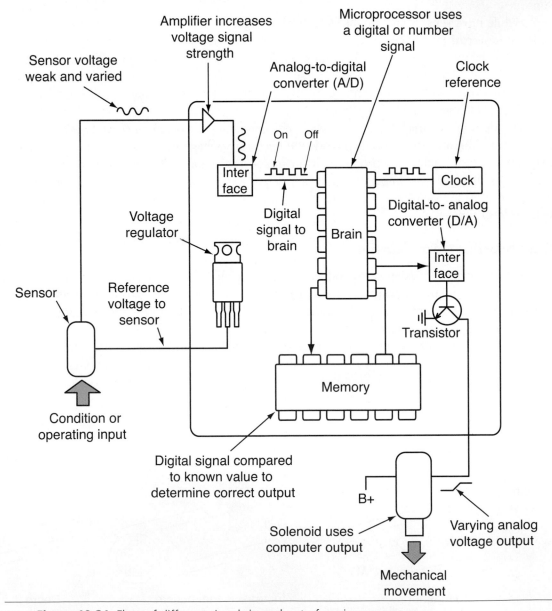

Amplifier increases voltage signal strength

Microprocessor uses a digital or number signal

Sensor voltage weak and varied

Analog-to-digital converter (A/D)

Clock reference

On   Off

Inter face

Clock

Voltage regulator

Digital signal to brain

Digital-to-analog converter (D/A)

Brain

Inter face

Sensor

Reference voltage to sensor

Transistor

Condition or operating input

Memory

Digital signal compared to known value to determine correct output

B+

Solenoid uses computer output

Varying analog voltage output

Mechanical movement

Figure 10-21 Flow of different signals in and out of a microprocessor.

the Hall-effect switch produce a digital or square wave signal that can go directly to the computer as input (Figure 10-21).

A computer's memory holds the programs and other data, such as vehicle calibrations, which the microprocessor refers to in performing calculations. To the computer, the program is a set of instructions or procedures that it must follow. Included in the program is information that tells the microprocessor when to retrieve input (based on temperature, time, etc.), how to process the input, and what to do with it once it has been processed.

Actuators

After the computer has assimilated and processed the information, it sends output signals to control devices called actuators. These actuators, which are solenoids, switches, relays, or motors, physically act on or carry out a decision the computer has made.

Actuators are electromechanical devices that convert an electrical current into mechanical action. This mechanical action can then be used to open and close valves, control vacuum to other components, or open and close switches. When the microcomputer receives an input signal indicating a change in one or more of the operating conditions, the microcomputer determines the best strategy for handling the conditions. The microcomputer then controls a set of actuators to achieve a desired effect or strategy goal. In order for the computer to control an actuator, it must rely on a component called an **output driver**.

Output drivers are also located in the processor (along with the input conditioners, microprocessor, and memory) and operate by the digital commands issued by the microcomputer. Basically, the output driver is nothing more than an electronic on and off switch that the computer uses to control the ground circuit of a specific actuator.

The **output driver** is basically an electronic on and off switch that the computer uses to control the ground circuit of a specific actuator.

Shop Manual
Chapter 10, page 450

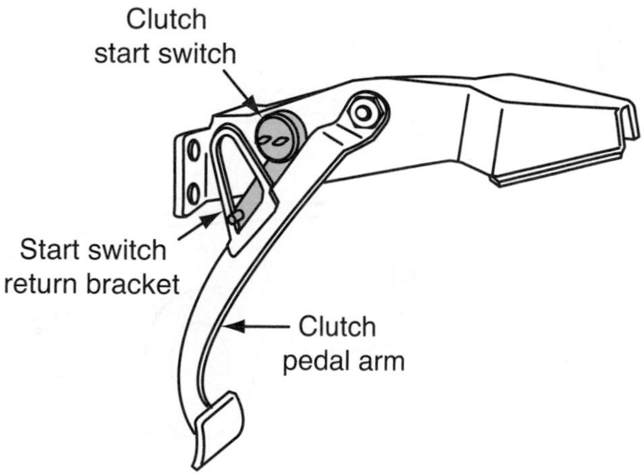

Clutch Safety Switch

The **clutch safety switch** (Figure 10-22) is connected into the starting circuit. The switch prevents the starting of the engine unless the clutch pedal is fully depressed. The switch is normally open when the clutch pedal is released. When the clutch pedal is depressed, the switch closes and completes the circuit between the ignition switch and the starter solenoid. Sometimes a neutral switch is also used, and this switch bypasses the clutch switch when the transmission is shifted into neutral.

Figure 10-22 The clutch switch responds to the movement of the clutch pedal.

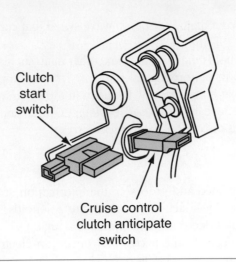

Figure 10-23 Various clutch switches used with cruise control systems.

Disengaging the clutch also reduces the current draw of the starter when the engine is being started in cold weather. By disengaging the clutch, the starter motor only rotates the engine. When the clutch is engaged, the starter motor must also turn the input shaft of the transmission even when the transmission is in neutral. This can provide quite a load on the starter, especially when it is cold and the transmission oil is thick.

Clutch switches are also used in cruise control circuits (Figure 10-23). When the driver depresses the clutch pedal, power for the cruise control system is shut off. This prevents high throttle operation when there is no load on the engine.

Reverse Lamp Switch

Shop Manual
Chapter 10, page 452

All vehicles sold in the United States after 1971 are required to have backup (reverse) lights. Backup lights illuminate the area behind the vehicle and warn other drivers and pedestrians that the vehicle is moving in reverse (Figure 10-24). Typically power for the lamps is supplied through

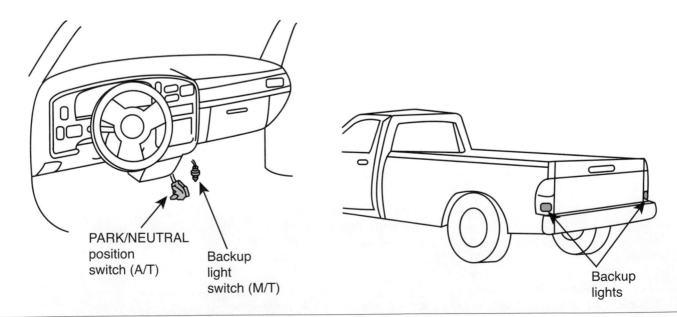

Figure 10-24 Typical backup light system.

the ignition switch when it is in the on position. When the driver shifts the transmission into reverse, the contacts of the backup light switch are closed; this completes the light circuit. Most manual transmissions are equipped with a separate switch located on the transmission (Figure 10-25) but can be mounted to the shift linkage away from the transmission. If the switch is mounted in the transmission, the shifting fork will close the switch and complete the circuit whenever the transmission is shifted into reverse gear. If the switch is mounted on the linkage, the switch is closed directly by the linkage.

Some late-model Corvettes have a redundant reverse switch to ensure reliability of the keyless entry and starting system. The car must be in reverse gear to start and to lock after it is shut off.

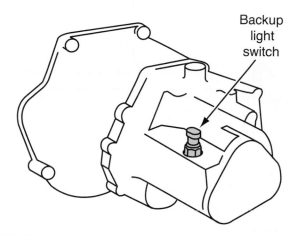

Figure 10-25 Typical location of a backup light switch in a transaxle.

High Gear Switch

Some early emission-controlled cars are equipped with a transmission-controlled spark (TCS) switch. This emission control switch closes the circuit only when the transmission is shifted into high gear. Closing the switch allows ignition vacuum advance to operate, thereby preventing vacuum advance during all other gears.

Some manual transmission equipped vehicles with cruise control systems also use a high gear switch to prevent engagement of cruise control when the transmission is in a lower gear. When high gear is selected, cruise control is enabled.

Upshift Lamp Circuit

Upshift and shift lamps inform the driver when to shift into the next gear in order to maximize fuel economy. These lights are controlled by the PCM, which activates the light according to engine speed, engine load, and vehicle speed (Figure 10-26). Basically these light circuits operate like a vacuum gauge. When engine load is low, engine vacuum is high. And when engine load is high, vacuum is low. The shift light will come on whenever there is high vacuum. The shift lamp is lit at those engine speeds and loads in which engine vacuum is high and the transmission is in a forward gear. The shift light stays on until the transmission is shifted or the engine's operating conditions change. This circuit works in all forward gears except high gear, in which a high gear switch disables the circuit.

Engine load is determined by the throttle position sensor and the manifold absolute pressure (MAP) sensor.

ABS Speed Sensor Circuits

Wheel speed sensors are used in antilock brake systems to measure the speed of the wheels. The tip of the sensor is located on the steering knuckle near a toothed ring or rotor. The toothed ring is

Shop Manual
Chapter 10, page 453

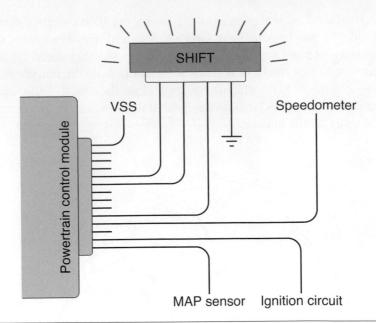

Figure 10-26 Simplified SHIFT light circuit.

typically attached to the drive axle shaft and rotates at the same speed as the wheel. As the ring spins, a voltage is induced in the sensor. The strength and frequency of this voltage varies with the wheel's speed. In some antilock brake systems, the wheel speed sensor is mounted at each wheel. The toothed ring is part of the outer CV joint or axle assembly (Figure 10-27). In other systems, the toothed ring is mounted next to the differential ring gear (Figure 10-28). In these cases, one sensor monitors the speed of the entire axle assembly instead of the individual wheels. Some transmissions are fitted with a sensor (Figure 10-29), which is typically used to monitor vehicle speed.

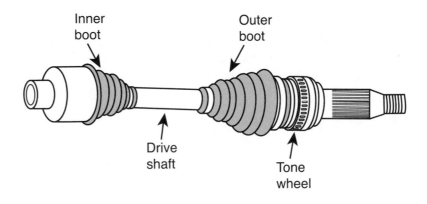

Figure 10-27 Location of speed sensor toothed ring (tone wheel) on a half shaft.

Shop Manual
Chapter 10, page 460

Shift Blocking

Some six-speed transmissions have a feature called shift blocking or *skip shift*. This prevents the driver from shifting into second or third gears from first gear, when the coolant temperature is above 50 degrees C (122 degrees F), the speed of the vehicle is between 12 and 22 mph (20 and 29 km/h), and the throttle is opened 35 percent or less. Shift blocking occurs to ensure good fuel economy and keeps the vehicle in compliance with federal fuel economy standards. These trans-

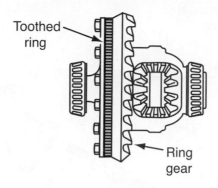

Figure 10-28 Toothed ring (excitor ring) mounted to a differential's ring gear.

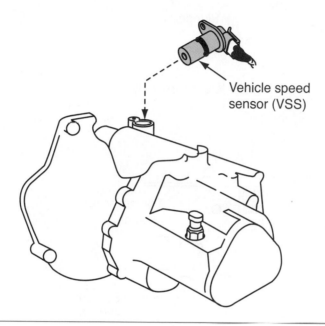

Figure 10-29 Typical location of a VSS mounted in a transaxle.

missions are equipped with reverse lockout (Figure 10-30), as are some others. This feature prevents the engagement of reverse whenever the vehicle is moving forward.

Shift blocking is controlled by the PCM (Figure 10-31). A solenoid is used to block off the shift pattern from first gear to second or third. The driver moves the gearshift from its up position to a lower position, as if shifting into second, and fourth gear is selected. The solenoid does not impede downshifting.

Electrical Clutches

Shop Manual
Chapter 10, page 457

Some older vehicles were equipped with a clutchless manual transmission. These offered the driver control of the transmission without needing to depress the clutch during gear changes. These transmissions promised to offer the advantages of an automatic transmission with those of a manual transmission. In short, they were designed to work the clutch according to needs, instead of the driver's foot. Often the clutch is misused and performance, fuel economy, and durability suffer because a driver does not use the clutch wisely. One of the most commonly found clutchless transmissions was the Volkswagen "Automatic Stickshift." Although these systems were

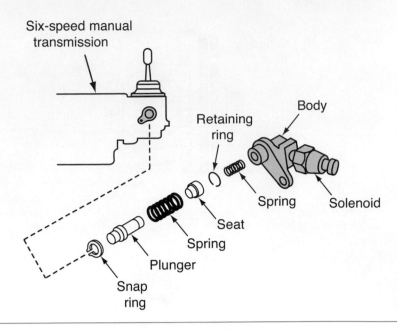

Figure 10-30 Location of reverse lockout assembly.

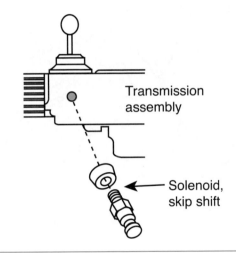

Figure 10-31 Location of skip shift solenoid.

clever and convenient, they were dropped because they were very complicated, expensive to produce, and gear shifting felt very abrupt. With the advances made in the electronics world, these systems are once again under development.

The clutch is released by a small electric motor that is controlled by a computer. The computer responds to sensors that send information about engine speed, transmission speed, throttle position, gearshift position, and driver intent (Figure 10-32). A gearshift lever switch responds only to fore and aft movements and is load sensitive. When the driver moves the shifter to change gears, the control module actuates the hydraulic clutch system. If the driver quickly and forcibly moves the shifter, the clutch will be disengaged and engaged quickly. If the gearshift is moved slowly, the clutch action will also be slower. A switch allowing the driver to select the harshness and quickness of the shifts allows the computer to control the clutch according to the intent of the driver, as well as to the movement of the shifter.

When the vehicle is stopped and in gear, the clutch is controlled by throttle movement. The clutch remains disengaged until the driver applies pressure on the throttle; then the clutch engages to match the movement of the throttle as well as the driver's intent.

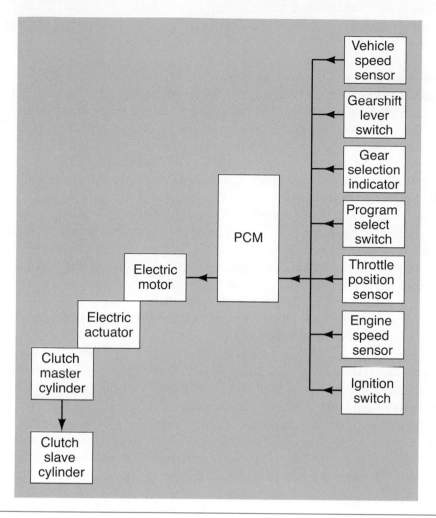

Figure 10-32 Simplified electric clutch circuit.

Self-Shifting Manual Transmissions

Self-shifting manual transmissions are currently being developed for passenger vehicles and use on Formula One racecars. These manual transmissions work like typical manual transmissions except they have electronic or hydraulic actuators to shift the gears and to work the clutch. Some designs allow the driver to shift gears using buttons or paddles on the steering wheel, while others are fully automatic.

These transmissions are quite different from regular automatic transmissions, as they do not use the expensive equipment that makes an automatic transmission work. They don't have finely machined valves in elaborate fluid channels, planetary gear sets, or torque converters. All of these add cost to the production of a transmission.

Self-shifting manual transmissions have fewer parts than an automatic. They are basic manual transmissions with a motor connected to the shift forks and a clutch actuator. A computer controls both of these. The computer is programmed to shift the transmission at the correct time, in the correct sequence, and to activate the clutch and allow for precise shifting. These features will allow the clutch assembly to last a long time, because the number one thing that causes premature clutch failure is the driver. With the computer working the clutch, the driver will not affect the operation or durability of the clutch.

In operation, the driver requests a gear change by using the shifting mechanism (shift lever, push buttons, etc.). There is no gearshift linkage or cable; instead, a sensor at the shifter sends a signal to the controller. The controller, in turn, commands the actuators to open/close the clutch and disengage/engage the gear sequence with very fast response times. Engine torque is

Self-shifting
manual
transmissions work
like typical manual
transmissions except
they have electronic
or hydraulic
actuators to shift the
gears and to work
the clutch.

controlled during the shift by either directly controlling the throttle or ignition/fuel injection system to provide smooth shifts.

Volkswagen/Audi Direct Shift Gearbox (DSG). The Direct Shift Gearbox (DSG) is a six-speed manual transaxle. Through the use of an integrated twin multiplate clutch, two gears can be engaged at the same time. When the car is moving, one gear is engaged. When the next gearshift point is approached, the next appropriate gear is preselected, but its clutch is kept disengaged. The shifting process opens the clutch of the activated gear and closes the other clutch at the same time. The gear change takes place under load so that a permanent flow of power is maintained.

The transaxle is based on two clutches, two input shafts, and a Mechatronic module that serves as the brains for the transaxle.

The clutches are hydraulically controlled. Clutch #1 is fixed to the odd numbered gears plus reverse and clutch #2 controls the even gears. The arrangement hints at being two parallel three-speed transmissions built into one housing. This is how two gears can be engaged at the same time. As an example of this action, when the car is being driving in third gear, fourth gear is already engaged but is not yet active. As soon as the ideal shift point is reached, the clutch for third gear opens when the other closes, activating fourth gear. The opening and closing of the clutches happen at the same time, producing a smooth and quick shift. The entire shift process is completed in a few hundredths of a second.

The most complex component of the DSG is the mechatronic control module. The module is located in the transaxle and contains actuators and a valve chest with 12 sensors that convert physical or chemical information into electrical information. A heat exchanger is bolted to the housing to keep the computer at desired temperatures.

The module determines and manages data for controlling the clutches, input and output shafts, cooling, gear selection, pressures, as well as various malfunction security levels. Five modulation valves, five shift valves, and numerous other valves are used to control the transaxle. A connector allows the transfer of data from the control unit to the vehicle's multiplex system and sends information from the vehicle and engine to the DSG computer.

BMW Sequential M Gearbox (SMG). Similar to the DSG, BMW's Sequential M Gearbox (SMG) is a self-shifting manual transmission. The system is based on a regular six-speed transmission with numerous solenoids. Shifting characteristics are based on inputs from a variety of sensors (Figure 10-33), driver selected programs, and the programmed logic in the control module. Basically, the control module interrupts the engine's power for just milliseconds and causes gear changes electrohydraulically when it opens and closes the clutch.

Engaging reverse gear is done in the same way as a manual transmission, with the gearshift lever. When the car is parked while engaged in a gear, it cannot roll.

An additional function is called the *climbing assistant*. This permits pulling away on forward slopes without rolling back. It can be used both in the sequential and automatic mode and for forward and reverse travel. All the driver needs to do is to press the brake pedal with the car stationary and to pull the rocker switch on the steering wheel for a short period of time. When the brake is released again, the car is ready to drive away within two seconds without first rolling away in an uncontrolled manner.

Another feature, the *acceleration assistant*, provides maximum acceleration from a stop. To activate this program, the driver selects the S6 driving program, pushes selector lever forward, and keeps it in that position when the car is stationary. The driver now fully depresses the accelerator pedal and the optimum engine speed for maximum acceleration is automatically set. When the driver then releases the selector lever, the car accelerates as quickly as possible with a minimum amount of wheel slip.

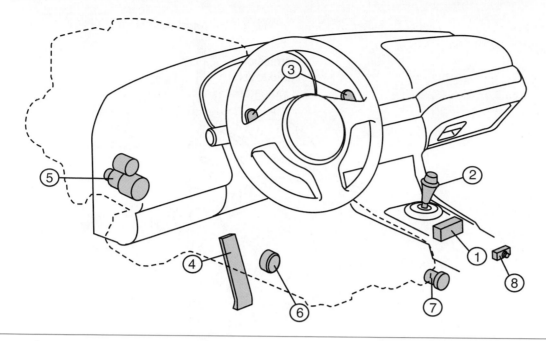

Figure 10-33 SMG system: 1—Drivelogic control module, 2—gearshift, 3—paddles, 4—accelerator input, 5—hydraulic unit. 6—clutch posititition sensor, 7,—input shaft speed sensor and transmission oil temperature sensor, 8—gear selector position sensor.

Other Electronic Systems

As mentioned in the previous chapter, some 4WD vehicles use an electric switch to control the transfer case. Some models use an electronic control unit to control the engagement and disengagement of 4WD. These systems use a clutch pedal position switch, an electric shift motor, a shift position sensor, and a speed sensor mounted to the transfer case, or speed sensors at each wheel. A few systems use a vacuum motor in place of the electric shift motor. These vacuum units are controlled by solenoids. The selector switch tells the computer when the clutch pedal has been fully depressed to allow the transfer case to be shifted from 4WD HIGH to 4WD LOW. The shift position sensor informs the computer as to what gear range the transfer case is in. The speed sensor tells the control unit when the vehicle is stopped, allowing shifts from HIGH to LOW to occur.

The rear axle differential, on some model vehicles, is equipped with an electromagnetic clutch for lockup of the differential. This clutch is the same as used in 4WD systems to lock the center differential and may be controlled by a switch or by a traction control computer. When one rear wheel loses traction and speeds up, the computer energizes the clutch, which locks up the differential and provides equal power to both rear wheels.

The electromagnetic clutch consists of an actuator coil, an armature, and a stack of steel clutch plates. When the clutch is energized, a magnetic field is produced to compress the clutches; this locks the differential.

Some models are equipped with a low transmission fluid level indicator (Figure 10-34). This type of warning light is normally a simple circuit consisting of a sensor or sending unit, wires, and the lamp. The sending unit provides a path for ground when the fluid is low, thereby lighting the warning lamp. When the fluid level is high enough to keep the switch open, the lamp remains off.

Shop Manual
Chapter 10, page 458

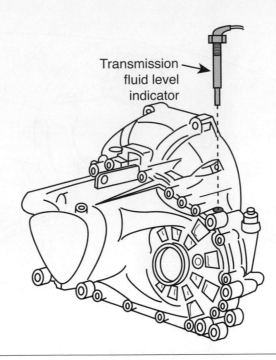

Transmission fluid level indicator

Figure 10-34 Typical transmission fluid level indicator.

Summary

❑ A basic understanding of electrical principles is important for effective electrical diagnosis and for the proper operation and interpretation of diagnostic tools.

❑ The release of energy as one electron leaves the orbit of one atom and jumps into the orbit of another is electricity.

❑ Two power or energy sources are used in an automobile's electrical system; these are based on a chemical reaction and on magnetism.

❑ A car's battery is a source of chemical energy. A chemical reaction in the battery provides for an excess of electrons in one place and a lack of electrons in another place.

❑ The flow of electricity is called current and is measured in amperes. There are two types of electrical flow: direct current (DC) and alternating current (AC).

❑ Resistance to current flow produces heat. The amount of resistance is measured in ohms.

❑ In a complete electrical circuit, the flow of electricity is controlled and applied to perform tasks, such as lighting headlights and turning over the starter motor. Circuit testers are used to identify shorted and open circuits.

❑ Voltage is electrical pressure and is measured in volts.

❑ For electrical flow to occur, there must be an excess of electrons in one place, a lack of electrons in another, and a path between the two places.

❑ Ohm's law states that it takes one volt of electrical pressure to push one ampere of electrical current through one ohm of resistance.

❑ If voltage does not change, but there is a change in the resistance of the circuit, the current will change. If resistance increases, current decreases. If resistance decreases, current increases. If voltage changes, so must the current or resistance. If the resistance stays the same and the current decreases, so will the voltage. Likewise, if the current increases, so will the voltage.

❑ Voltage is electrical pressure and is measured in volts.

❑ For electrical flow to occur, there must be an excess of electrons in one place, a lack of electrons in another, and a path between the two places.

❑ Ohm's law states that it takes one volt of electrical pressure to push one ampere of electrical current through one ohm of resistance.

❑ If voltage does not change, but there is a change in the resistance of the circuit, the current will change. If resistance increases, current decreases. If resistance decreases, current increases. If voltage changes, so must the current or resistance. If the resistance stays the same and the current decreases, so will the voltage. Likewise, if the current increases, so will the voltage.

❑ The mathematical relationship between current, resistance, and voltage is expressed in Ohm's law, $E = IR$, where voltage is measured in volts, current in amperes, and resistance in ohms.

❑ Resistors in common use in automotive circuits are of three types: fixed value, stepped or tapped, and variable.

❑ Fixed value resistors are designed to have only one rating, which should not change. These resistors are used to control voltage, such as in an automotive ignition system.

❑ Rheostats have two connections, one to the fixed end of a resistor and one to a sliding contact with the resistor.

❑ Potentiometers have three connections, one at each end of the resistance and one connected to a sliding contact with the resistor. Turning the control moves the sliding contact away from one end of the resistance, but toward the other end.

❑ Electrical schematics are diagrams with electrical symbols that show the parts and how electrical current flows through the vehicle's electrical circuits. They are used in troubleshooting.

❑ The strength of an electromagnet depends on the number of current-carrying conductors and what is in the core of the coil. Inducing a voltage requires a magnetic field producing lines of force, conductors that can be moved, and movement between the conductors and the magnetic field so that the lines of force are cut.

❑ Fuses, fuse links, maxi-fuses, and circuit breakers protect circuits against overloads. Switches control on and off and direct current flow in a circuit. A relay is an electric switch. A solenoid is an electromagnet that translates current flow into mechanical movement. Resistors limit current flow.

❑ A semiconductor is a material or device that can function as either a conductor or an insulator, depending on how its structure is arranged.

❑ An integrated circuit is simply a large number of diodes, transistors, and other electronic components, such as resistors and capacitors, all mounted on a single piece of semiconductor material.

❑ The diode allows current to flow in one direction but not in the opposite direction.

❑ Transistors are used as switching devices.

❑ Computers are electronic decision-making centers. Input devices called sensors feed information to the computer. The computer processes this information and sends signals to controlling devices. A typical electronic control system is made up of sensors, actuators, a microcomputer, and related wiring.

❑ Most input sensors are variable resistance or reference types, switches, and thermistors.

Terms to Know (continued)

Reference voltage (V_{ref}) sensors

Relay

Rheostats

Semiconductor

Self-shifting manual transmission

Solenoids

Transistor

Voltage

Voltage generating devices

Voltmeter

Volts

❏ All sensors detect a mechanical condition, chemical state, or temperature condition and change it into an electrical signal that can be used by the computer to make decisions.

❏ All sensors are either reference voltage sensors or voltage generating sensors.

❏ Most input sensors are designed to produce a voltage signal that varies within a given range called an analog signal.

❏ A computer does not understand analog signals. It can only read a digital binary signal, which is a signal that has only two values—on or off.

❏ After the computer has assimilated and processed the information, it sends output signals to control devices called actuators, which are solenoids, switches, relays, or motors that physically act on or carry out a decision the computer has made.

❏ The clutch start switch prevents starting of the engine unless the clutch pedal is fully depressed.

❏ Most manual transmissions are equipped with a clutch safety switch located on the transmission, but they can be mounted to the shift linkage away from the transmission.

❏ Upshift and shift lamps inform the driver when to shift into the next gear in order to maximize fuel economy. These lights are controlled by the PCM, which activates the light according to engine speed, engine load, and vehicle speed.

❏ Wheel speed sensors are used in antilock brake systems to measure the speed of the wheels. The tip of the sensor is located on the steering knuckle near a toothed ring or rotor. The toothed ring is typically attached to the drive axle shaft and rotates at the same speed as the wheel.

❏ Some transmissions are fitted with a sensor, which is typically used to monitor vehicle speed.

❏ Shift blocking prevents the driver from shifting into second or third gears from first gear, when the coolant temperature is above 50 degrees C (122 degrees F), the speed of the vehicle is between 12 and 22 mph (20 and 29 km/h), and the throttle is opened 35 percent or less.

❏ Shift blocking is controlled by the PCM. A solenoid is used to block off the shift pattern from first gear to second or third.

❏ Self-shifting manual transmissions are not based on automatic transmissions.

❏ Some transmissions are equipped with a low transmission fluid level indicator.

Review Questions

Short Answer Essays

1. Name the two energy sources used in automobile electrical systems.

2. For electrical flow to occur, what must be present?

3. Define electricity.

4. What is the difference between voltage and current?

5. What must be present in a circuit in order to cause electricity to flow?

6. State Ohm's law.

7. Describe the differences between a rheostat and a potentiometer.

8. What is the difference between a fixed resistor and a variable resistor?

9. What types of sensors are typically used in an automotive computer system?

10. What is the purpose of shift blocking?

Fill-in-the-Blanks

1. The two power or energy sources used in an automobile's electrical system are based on a _____ _____ and on _____ .

2. Current is measured in _____; electrical voltage is measured in _____; and electrical resistance is measured in _____ .

3. Backup lights are normally controlled by a switch located in the _____ or controlled by the movement of the _____ .

4. _____ , _____ , _____ _____ , and _____ _____ are used to protect circuits against current overloads.

5. _____ control on and off and direct current flow in a circuit.

6. The center of an atom is known as the _____ .

7. Resistance is measured in _____ .

8. A _____ is a material or device that can function as either a conductor or an insulator, depending on how its structure is arranged.

9. A computerized circuit depends on two types of signals: _____ and _____ .

10. Wheel speed sensors are used in antilock brake systems to measure the speed of the wheels. The toothed ring of the sensor is typically attached to the _____ _____ , _____ , or is pressed onto the _____ of the differential.

Multiple Choice

1. *Technician A* says that magnetism is a source of electrical energy in an automobile. *Technician B* says that chemical reaction is a source of electrical energy in an automobile. Who is correct?
   A. A only
   B. B only
   C. Both A and B
   D. Neither A nor B

2. When discussing the behavior of electricity, *Technician A* says that if voltage does not change, but there is a change in the resistance of the circuit, the current will change. *Technician B* says that if resistance increases, current decreases. Who is correct?
   A. A only
   B. B only
   C. Both A and B
   D. Neither A nor B

3. *Technician A* says that rheostats have three connections, one at each end of the resistance and one connected to a sliding contact with the resistor. *Technician B* says that potentiometers have two connections, one at the power end of the resistance and one connected to a sliding contact with the resistor. Who is correct?
   A. A only
   B. B only
   C. Both A and B
   D. Neither A nor B

4. *Technician A* says that electrical resistance is the pressure that causes current to flow in a circuit. *Technician B* says that if there is zero resistance in a circuit, a maximum amount of current will flow in the circuit. Who is correct?
   A. A only
   B. B only
   C. Both A and B
   D. Neither A nor B

5. *Technician A* says that a diode allows current to flow in one direction but not in the opposite direction. *Technician B* says that transistors are used as switching devices. Who is correct?
   A. A only
   B. B only
   C. Both A and B
   D. Neither A nor B

6. *Technician A* says that the flow of electricity is called current. *Technician B* says that the flow of electricity is measured in amperes. Who is correct?
   A. A only
   B. B only
   C. Both A and B
   D. Neither A nor B

7. *Technician A* says that if the resistance in a circuit changes, so must the voltage or current. *Technician B* says that if voltage changes, so must the current or resistance. Who is correct?
   A. A only
   B. B only
   C. Both A and B
   D. Neither A nor B

8. *Technician A* says that upshift and shift lamps are controlled by vehicle speed. *Technician B* says that upshift and shift lamps are controlled by throttle position. Who is correct?
   A. A only
   B. B only
   C. Both A and B
   D. Neither A nor B

9. When discussing batteries, *Technician A* says that a chemical reaction in the battery provides for an excess of electrons in one part of the battery. *Technician B* says that a chemical reaction in the battery provides for a lack of electrons in one part of the battery. Who is correct?
   A. A only
   B. B only
   C. Both A and B
   D. Neither A nor B

10. *Technician A* says that there are two types of electrical flow: direct current (DC) and alternating current (AC). *Technician B* says that DC is used to operate most automotive electrical circuits. Who is correct?
   A. A only
   B. B only
   C. Both A and B
   D. Neither A nor B

Appendix

Driveline Related Associations

ATRA	Automatic Transmission Rebuilders Association	Oxnard, CA
ETI	Equipment and Tool Institute	Research Triangle Park, NC
MAP	Motorist Assurance Program	Washington, DC
	Manual Transmission Technical Group	New York, NY
SEMA	Specialty Equipment Market Association	Diamond Bar, CA

Glossary
Glosario

Abrasion   Wearing or rubbing away of a part.
Abrasión   Desgaste o rozamiento de una pieza.

Acceleration   An increase in velocity or speed.
Aceleración   Aumento de velocidad o celeridad.

Accident   Something that happens unintentionally and is a consequence of doing something else.
Accidente   Suceso imprevisto que ocurre como consecuencia de otro.

Actuator   A device that carries out the instructions of a computer. Normally these devices are solenoids or motors.
Actuador   Dispositivo que cumple las instrucciones de una computadora. Generalmente estos dispositivos son solenoides o motores.

Adhesives   Chemicals used to hold gaskets in place during the assembly of an engine. They also aid the gasket in maintaining a tight seal by filling in the small irregularities on the surfaces and by preventing the gasket from shifting due to engine vibration.
Adhesivos   Productos químicos que sirven para sujetar las guarniciones en su lugar durante el montaje de un motor. También ayudan a la guarnición a mantener una junta de estanqueidad hermética al rellenar las pequeñas irregularidades en las superficies, y al impedir que la guarnición se desplace debido a la vibración del motor.

Alignment   An adjustment to a line or to bring into a line.
Alineación   Ajuste a una línea o poner en línea.

All-wheel drive (AWD)   Commonly refers to a full-time 4WD system.
Tracción en las cuatro ruedas   Usualmente se refiere a un sistema en el cual la fuerza motriz se reparte a tiempo completo entre las cuatro ruedas.

Alternating current (AC)   A type of current that changes its direction as it flows through a circuit.
Corriente alterna (AC)   Tipo de corriente que cambia de dirección al fluir por un circuito.

Ammeter   A meter used to measure electrical current.
Amperímetro   Aparato de medida para una corriente eléctrica.

Ampere   The unit of measure for electrical current.
Amperio   Unidad de medida para una corriente eléctrica.

Analog   A type of signal that carries a changing or constant voltage.
Análogo   Un tipo de señal que lleva una voltaje que cambia o que es constante.

Antifriction bearing   A bearing designed to reduce friction. This type of bearing normally uses ball or roller inserts to reduce the friction.
Cojinete de antifricción   Cojinete diseñado para disminuir la fricción. Normalmente a este tipo de cojinete se le insertan bolas o rodillos para disminuir la fricción.

Arbor press   A small hand-operated shop press used when only a light force is required against a bearing, shaft, or other part.

Prensa para calar   Pequeña prensa de taller accionada a mano y utilizada solamente cuando se requiere una fuerza ligera contra un cojinete, un árbol u otra pieza.

Asbestos   A material that was commonly used as a gasket material in places where temperatures are great. This material is being used less frequently today because of health hazards that are inherent to the material.
Asbesto   Material que usualmente se utilizaba como material de guarnición en lugares donde las temperaturas eran muy elevadas. Hoy en día se utiliza dicho material con menos frecuencia debido a los riesgos para la salud inherentes al mismo.

Atom   The smallest particle of an element in which all the chemical characteristics of the element are present.
Átomo   Partícula más pequeña en la cual están presentes todos los atributos.

Automatic locking/unlocking hubs   Front wheel hubs that can engage or disengage themselves from the axles automatically.
Cubos automáticos de cierre/descerrar   Cubos ubicados en las ruedas delanteras que automáticamente pueden engranarse o desengranarse de los ejes.

Automatic transmission   A transmission in which gear or ratio changes are self-activated, eliminating the necessity of hand-shifting gears.
Transmisión automática   Sistema en el cual los cambios de velocidades o de relación se activan automáticamente, eliminando así la necesidad de emplear la palanca del cambio de velocidades.

Axial   Parallel to a shaft or bearing bore.
Axial   Paralelo a un árbol o calibre de cojinete.

Axis   The center line of a rotating part, a symmetrical part, or a circular bore.
Pivote   Línea central de una pieza giratoria, una pieza simétrica, o un calibre circular.

Axle   The shaft or shafts of a machine on which the wheels are mounted.
Eje   Árbol o árboles de una máquina sobre el cual se montan las ruedas.

Axle carrier assembly   A cast-iron framework that can be removed from the rear axle housing for service and adjustment of the parts.
Conjunto portador del eje   Armazón de hierro fundido que se puede remover del puente trasero para la reparación y el ajuste de las piezas.

Axle housing   Designed in the removable carrier or integral carrier types to house the drive pinion, ring gear, differential, and axle shaft assemblies.
Puente trasero   Diseñado en los tipos de portador desmontables o enterizos para alojar los conjuntos del piñón de mando, de la corona, del diferencial y del árbol motor.

Axle ratio   The ratio between the rotational speed (rpm) of the drive shaft and that of the driven wheel; gear reduction through the differential, determined by dividing the number of teeth on the ring gear by the number of teeth on the drive pinion.
Relación del eje   Relación entre las revoluciones por minuto (rpm) del árbol de mando y las de la velocidad de la rueda accionada; la des-

multiplicación de engranajes por medio del diferencial, determinada al dividir el número de dientes en la corona por el número de dientes en el piñón de mando.

Axle shaft A shaft on which the road wheels are mounted.

Árbol motor Árbol sobre el cual se montan las ruedas de la carretera.

Axle shaft end thrust A force exerted on the end of an axle shaft that is most pronounced when the vehicle turns corners and curves.

Empuje longitudinal del árbol motor Fuerza ejercida sobre el extremo de un árbol motor. Dicha fuerza es más marcada cuando el vehículo dobla una esquina o una curva.

Axle shaft tubes These tubes are attached to the axle housing center section to surround the axle shaft and bearings.

Tubos del árbol motor Estos tubos se fijan a la sección central del puente trasero para rodear el árbol motor y los cojinetes.

Backlash The amount of clearance or play between two meshed gears.

Contragolpe Cantidad del espacio libre u holgura entre dos engranajes.

Balance Having equal weight distribution. The term is usually used to describe the weight distribution around the circumference and between the front and back sides of a wheel and tire assembly. Uneven weight distribution causes vibrations when the wheel is spun because the center of gravity of the wheel and tire assembly does not line up with the center line of the axle. A wheel and tire are balanced by adding weights to the wheel rim opposite the heavy section.

Equilibrio Que tiene igual distribución de peso. Usualmente este término se emplea para describir la distribución de peso alrededor de la circunferencia y entre las partes delantera y trasera de un conjunto de rueda y llanta. La distribución desigual de peso puede causar vibraciones cuando se gira la rueda porque el centro de gravedad del conjunto de rueda y llanta no está alineado con la línea central del eje. Una rueda y una llanta están en equilibrio cuando se le añade más peso a la llanta opuesta a la sección pesada.

Balanced resistance A situation in which two objects, such as axle shafts, present the same resistance to driving rotation.

Resistencia equilibrada Situación en la que dos objetos, tales como los árboles motores, ofrecen la misma resistencia al mando de rotación.

Ball bearing An antifriction bearing consisting of a inner and outer race with hardened steel balls that roll between the two races. This bearing is typically used to support the load of a shaft.

Cojinete de bolas Cojinete antifricción que consiste de un anillo interno y externo con bolas de acero templadas que giran entre los dos anillos, y que apoya la carga del árbol.

Ball joint A suspension component that attaches the control arm to the steering knuckle and serves as its lower pivot point. The ball joint gets its name from its ball-and-socket design. It allows both up and down motion as well as rotation. In a MacPherson strut FWD suspension system, the two lower ball joints are nonload carrying.

Junta esférica Componente de la suspensión que fija el brazo de mando al muñón de dirección, y sirve como punto de pivote inferior para el muñón de dirección. Así se le llama a la junta esférica por su diseño de rótula. Permite tanto el movimiento de ascenso y descenso, como el de rotación. En un sistema de suspensión de tracción delantera montante MacPherson, las dos juntas esféricas inferiores no tienen capacidad de carga.

Bearing The supporting part that reduces friction between a stationary and rotating part or between two moving parts.

Cojinete Pieza de soporte que disminuye la fricción entre una pieza fija y una pieza giratoria o entre dos piezas móviles.

Bearing cage A spacer that keeps the balls or rollers in a bearing in proper position between the inner and outer races.

Jaula de cojinete Espaciador que mantiene las bolas o los rodillos en posición correcta en el cojinete, entre los anillos interior y exterior.

Bearing caps In the differential, caps held in place by bolts or nuts that, in turn, hold bearings in place.

Casquillos de cojinete En el diferencial, los casquillos sujetados por tornillos o tuercas que a su vez sujetan los cojinetes en su lugar.

Bearing cone The inner race, rollers, and cage assembly of a tapered-roller bearing. Cones and cups must always be replaced in matched sets.

Cono de cojinete El anillo interior, los rodillos y el conjunto de jaula de un cojinete de rodillos cónicos. Se deben, reemplazar siempre los conos y las rótulas en conjuntos que hagan juego.

Bearing cup The outer race of a tapered-roller bearing or ball bearing.

Rótula de cojinete Anillo exterior de un cojinete de rodillos cónicos o cojinete de bolas.

Bearing race The surface on which the rollers or balls of a bearing rotate. The outer race is the same thing as the cup, and the inner race is the one closest to the axle shaft.

Anillo de cojinete Superficie sobre la cual giran los rodillos o las bolas de un cojinete. El anillo exterior es lo mismo que la rótula, y el anillo interior es el que se encuentra más cerca del árbol motor.

Belleville spring A tempered spring steel cone-shaped plate used to increase the mechanical force in a pressure plate assembly.

Muelle de Belleville Lámina de muelle de acero templado en forma cónica utilizada para ayudar a la fuerza mecánica en un conjunto de placa de presión.

Bell housing A housing that fits over the clutch components and connects the engine and the transmission. A common term for a clutch housing.

Alojamiento de campana Alojamiento que se monta sobre los componentes del embrague y que conecta el motor y la transmisión. Término común para el alojamiento del embrague.

Bellows Rubber protective covers with accordion-like pleats used to contain lubricants and exclude contaminating dirt or water.

Fuelles Cubiertas protectoras de caucho con pliegues en forma de acordeón, utilizadas para contener los lubricantes y evitar la contaminación por polvo o agua.

Bevel spur gear A gear that has teeth with a straight center line cut on a cone.

Engranaje recto cónico Engranaje que tiene dientes con un corte recto de línea central cortado en forma cónica.

Bolt torque The turning effort required to offset resistance as the bolt is being tightened.

Par de torsión del perno Esfuerzo giratorio requerido para neutralizar la resistencia al apretarse el perno.

Boots See bellows.

Botas Véase fuelles (bellows).

Brake horsepower (bhp) Power delivered by the engine and available for driving the vehicle; bhp = torque × rpm/5,252.

Potencia en caballos indicada al freno (bhp) Energía descargada por el motor y disponible para accionar el vehículo; bhp = par de torsión × rpm/5.252.

Brinnelling Rough lines worn across a bearing race or shaft due to impact loading, vibration, or inadequate lubrication.

Acción de Brinnell Líneas toscas a través de un anillo de cojinete o un árbol, causadas por la carga de un impacto, vibración o engrase inadecuado.

Bronze An alloy of copper and tin.

Bronce Aleación de cobre y estaño.

Burnish To smooth or polish by the use of a sliding tool under pressure.

Bruñir Suavizar o pulir con una herramienta deslizante bajo presión.

Burr A feather edge of metal left on a part being cut with a file or other cutting tool.

Rebaba Bisel de metal que queda en una pieza que ha sido cortada con una lima u otra herramienta de corte.

Bushing A cylindrical lining used as a bearing assembly and made of steel, brass, bronze, nylon, or plastic.

Buje Forro en forma cilíndrica utilizado como un conjunto de cojinete, hecho de acero, latón, bronce, nylon o plástico.

C-clip A C-shaped clip used to retain the drive axles in some rear axle assemblies.

Grapa-C Grapa en forma de C utilizada para retener los ejes de mando en algunos conjuntos del eje trasero.

Cage A spacer used to keep the balls or rollers in proper relation to one another. In a CV joint, the cage is an open metal framework that surrounds the balls to hold them in position.

Jaula Espaciador utilizado para sujetar las bolas o los rodillos en relación correcta entre sí. En una junta de velocidad constante, la jaula es un armazón abierto de metal que rodea las bolas para que se mantengan en su posición.

Camber A suspension alignment term used to define the amount that the centerline of a wheel is tilted inward or outward from the true vertical plane of the wheel. If the top of the wheel is tilted inward, the camber is negative. If the top of the wheel is tilted outward, the camber is positive.

Combadura Un término de alineamiento de suspensión que denota cuán inclinada está, hacia adentro o hacia afuera, la línea central de una rueda del verdadero plano vertical de la rueda. Si la parte superior de la rueda está inclinada hacia adentro, la combadura es negativa. Si la parte superior de la rueda está inclinada hacia afuera, la combadura es positiva.

Canceling angles Opposing operating angles of two U-joints cancel the vibrations developed by the individual U-joint.

Angulos de supresión Los ángulos de funcionamiento opuestos de dos juntas universales cancelan las vibraciones producidas por la junta universal individual.

Cardan Universal joint A nonconstant velocity U-joint consisting of two yokes with their forked ends joined by a cross. The driven yoke changes speed twice in 360 degrees of rotation.

Junta de cardán Junta universal de velocidad no constante que consiste de dos horquillas con sus extremos unidos por una cruz. El yugo accionado cambia de velocidad dos veces en una rotación de 360°.

Carrier An object that bears, cradles, moves, or transports some other object or objects.

Portador Objeto que apoya, acojina, mueve o transporta otro u otros objetos.

Case-harden To harden the surface of steel. The carburizing method used on low-carbon steel or other alloys to make the case or outer layer of the metal harder than its core.

Cementar Endurecer la superficie del acero. El método carburizante empleado en el acero de bajo contenido en carbón u otras aleaciones para hacer que la caja o capa exterior del metal sea más dura que el núcleo.

Castellate Formed to resemble a castle battlement, as in a castellated nut.

Entallar Formado para que se asemeje a la almena de un castillo, como una tuerca de corona.

Castellated nut A nut with six raised portions or notches through which a cotter pin can be inserted to secure the nut.

Tuerca de corona Tuerca con seis partes elevadas o muescas a través de las cuales se puede insertar un pasador de chaveta para sujetar la tuerca.

Caustic A material has the ability to destroy or eat through something. Caustic materials are considered extremely corrosive.

Cáustico Material que tiene la habilidad de destruir o corroer algo. A los materiales cáusticos se les considera extremadamente corrosivos.

Center-support bearing Ball-type bearing mounted on a vehicle cross member to support the drive shaft and provide better installation angle to the rear axle.

Silleta de suspensión central Cojinete de tipo bola montado sobre la traviesa de un vehículo para apoyar el árbol de mando y proveer un mejor ángulo de montaje al eje trasero.

Center section The middle of a live axle housing containing the drive pinion, ring gear, and differential assembly.

Sección central Centro del puente trasero integral que contiene el piñón de mando, la corona y el conjunto del diferencial.

Centering joint A ball socket joint placed between two Cardan U-joints to ensure that the assembly rotates on center.

Junta centradora Junta de rótula montada entre dos juntas de cardán para asegurar que el montaje gire en el centro.

Centrifugal clutch A clutch that uses centrifugal force to apply a higher force against the friction disc as the clutch spins faster.

Embrague centrífugo Embrague que utiliza fuerza centrífuga para aplicar mayor fuerza contra el disco de fricción cuando el embrague gira con más rapidez.

Centrifugal force The force acting on a rotating body that tends to move it outward and away from the center of rotation. The force increases as rotational speed increases.

Fuerza centrífuga Fuerza que acciona un cuerpo giratorio y que tiende a moverlo hacia afuera y más lejos del centro de rotación. La fuerza aumenta cuando aumenta la velocidad de rotación.

Chamfer A bevel or taper at the edge of a hole or a gear tooth.

Chaflán Bisel o cono al borde de un agujero o diente de engranaje.

Chamfer face A beveled surface on a shaft or part that allows for easier assembly. The ends of FWD drive shafts are often chamfered to make installation of the CV joints easier.

Superficie achaflanada Superficie biselada en un árbol o pieza que facilita el montaje. Los extremos de los árboles de mando de tracción delantera se achaflanan para facilitar la instalación de las juntas CV.

Chase To straighten up or repair damaged threads.

Roscar Enderezar o reparar roscas averiadas.

Chasing To clean threads with a tap.

Filetear Limpiar las roscas con un macho de roscar.

Chassis The vehicle frame, suspension, and running gear. It includes the control arms, struts, springs, trailing arms, sway bars, shocks, steering knuckles, and frame. The drive shafts, CV joints, and transaxle are not part of the chassis or suspension.

Chasís E sel armazón del vehículo, la suspensión y el tren de ruedas. En vehículos de tracción delantera, incluye los brazos de mando, los montantes, los muelles, los brazos traseros, las barras de oscilación lateral, los amortiguadores, los muñones de dirección y el armazón. Los árboles de mando, las juntas de velocidad constante y el transeje no forman parte del chasís o de la suspensión.

Circlip A split steel snap ring that fits into a groove to hold various parts in place. Circlips are often used on the ends of FWD drive shafts to retain the CV joints.

Grapa circular Anillo de resorte hendido, en acero, que se inserta en una ranura para sujetar varias piezas en su lugar. Con frecuencia se utilizan las grapas circulares en los extremos de árboles de mando de tracción delantera para retener las juntas de velocidad constante.

Clashing Grinding sound heard when gear and shaft speeds are not the same during a gearshift operation.

Entrechoque Rechinamiento que se escucha cuando las velocidades del engranaje y del árbol no son iguales durante el cambio de velocidades.

Clearance The space allowed between two parts, such as between a journal and a bearing.

Espacio libre Espacio permitido entre dos piezas, por ejemplo, entre un muñón y un cojinete.

Close ratio A relative term for describing the gear ratios in a transmission. If the gears are numerically close, they are said to be close ratio. This design gives quicker acceleration at the expense of initial acceleration and fuel economy.

Relación próxima Término relativo que describe las relaciones de los engranajes en una transmisión. Si existe una proximidad numérica entre los engranajes, se dice que su relación es próxima. Este diseño permite una aceleración más rápida a costa de la aceleración inicial y del rendimiento de combustible.

Cluster assembly A manual transmission-related term applied to a group of gears of different sizes machined from one steel casting.

Conjunto desplazable Término relacionado a la transmisión manual que se aplica a un grupo de engranajes de diferentes tamaños hechos a máquina, de una pieza fundida en acero.

Cluster gear A common term for the counter gear assembly.

Engranajes desplazables Término común que denomina el conjunto de contraengranaje.

Clutch A device for connecting and disconnecting the engine from the transmission or for a similar purpose in other units.

Embrague Dispositivo que sirve para engranar y desengranar el motor de la transmisión o para un propósito parecido en otras unidades.

Clutch control cable A cable assembly with a flexible outer housing anchored at the upper and lower ends. Moving back and forth inside the flexible housing is a braided wire cable that transfers clutch pedal movement to the clutch release lever.

Cable de mando del embrague Conjunto de cable con un alojamiento exterior flexible sujetado a los extremos superior e inferior. Un cable trenzado de alambre que transfiere el movimiento del pedal del embrague a la palanca de desembrague se mueve de atrás para adelante dentro del alojamiento.

Clutch cover A term used by some manufacturers to describe a pressure plate.

Tapa del embrague Término empleado por algunos fabricantes para describir una placa de presión.

Clutch (friction) disc The friction material part of the clutch assembly that fits between the flywheel and pressure plate.

Disco de embrague (fricción) La pieza material de fricción del conjunto de embrague que encaja entre el volante y la placa de presión.

Clutch fork In the clutch, a Y-shaped member into which the throw-out bearing is assembled.

Horquilla de embrague En el embrague, la pieza en forma de Y sobre la cual se monta el cojinete de desembrague.

Clutch housing A large aluminum or iron casting that surrounds the clutch assembly. Located between the engine and transmission, it is sometimes referred to as bell housing.

Alojamiento del embrague Pieza grande fundida en hierro o en aluminio que rodea al conjunto del embrague. Ubicado entre el motor y la transmisión, a veces se llama alojamiento de campana.

Clutch linkage A combination of shafts, levers, or cables that transmits clutch pedal motion to the clutch assembly.

Articulación de embrague Combinación de árboles, palancas o cables que transmite el movimiento del pedal del embrague al conjunto del embrague.

Clutch packs A series of clutch discs and plates installed alternately in a housing to act as a driving or driven unit.

Paquetes del embrague Serie de discos y placas del embrague instalados por turno en un alojamiento para que funcionen como una unidad de accionamiento o accionada.

Clutch pedal A pedal in the driver's compartment that operates the clutch.

Pedal del embrague Pedal que hace funcionar el embrague; ubicado en el compartimiento del conductor.

Clutch pedal free-play The amount the pedal can move without applying pressure on the pressure plate.

Juego libre del pedal del embrague Amplitud de movimiento del pedal sin que éste aplique presión sobre la placa de presión.

Clutch pushrod A solid or hollow rod that transfers linear motion between movable parts; that is, the clutch release bearing and release plate.

Varilla de empuje del embrague Varilla sólida o hueca que transfiere un movimiento lineal entre piezas móviles; es decir, el cojinete de desembrague del embrague y la placa de desembrague.

Clutch safety switch See neutral start switch.

Interruptor de seguridad del embrague Véase interruptor de arranque neutro.

Clutch shaft Sometimes known as the transmission input shaft or main drive pinion. The clutch driven disc drives this shaft.

Árbol de embrague A veces llamado árbol impulsor de la transmisión o piñón principal de mando. El disco accionado del embrague acciona este árbol.

Clutch slippage Engine speed increases but increased torque is not transferred through to the driving wheels because of clutch slippage.

Deslizamiento del embrague La velocidad del motor aumenta, pero el par de torsión no se transmite a las ruedas motrices a causa del deslizamiento del embrague.

Coefficient of friction The ratio of the force resisting motion between two surfaces in contact to the force holding the two surfaces in contact.

Coeficiente de fricción Relación de la fuerza que resiste el movimiento entre dos superficies en contacto a la fuerza que mantiene el contacto de las dos superficies.

Coil spring A heavy wirelike steel coil used to support the vehicle weight while allowing for suspension motions. On FWD cars, the front coil springs are mounted around the MacPherson struts. On the rear suspension, they may be mounted to the rear axle, to trailing arms, or around rear struts.

Muelle helicoidal Espiral grueso de acero parecido al alambre, que sirve para apoyar el peso del vehículo mientras permite el movimiento de suspensión. En vehículos de tracción delantera los muelles helicoidales delanteros se suspenden alrededor de los montantes MacPherson. En la suspensión trasera pueden suspenderse del eje trasero, de los brazos traseros, o alrededor de los montantes traseros.

Coil preload springs Coil springs are made of tempered steel rods formed into a spiral that resist compression; located in the pressure plate assembly.

Muelles helicoidales de carga previa Los muelles helicoidales se fabrican de varillas de acero templado configuradas en forma de espiral que resisten la compresión; ubicados en el conjunto de placa de presión.

Coil spring clutch A clutch using coil springs to hold the pressure plate against the friction disc.

Embrague de muelle helicoidal Embrague que emplea muelles helicoidales para mantener la placa de presión contra el disco de fricción.

Companion flange A mounting flange that fixedly attaches a drive shaft to another drivetrain component.

Brida acompañante Una brida de montaje que fija un árbol de mando a otro componente del tren de mando.

Compound A mixture of two or more ingredients.

Compuesto Mezcla de dos o más ingredientes.

Concentric Two or more circles having a common center.

Concéntrico Dos o más círculos que tienen un centro común.

Concentric clutch A combination of a slave cylinder and release bearing that is mounted in the transmission's housing.

Embrague concéntrico La combinación de un cilindro secundario y un cojinete de desembrague montados en la caja de la transmisión.

Conductor A material in which electricity flows easily.

Conductor Una materia por la cual fluye fácilmente la electricidad.

Cone clutch The driving and driven parts conically shaped to connect and disconnect power flow. A clutch made from two cones, one fitting inside the other. Friction between the cones forces them to rotate together.

Embrague cónico Piezas de accionamiento y accionadas, en forma cónica, empleadas para conectar y desconectar el regulador de fuerza. Un embrague hecho de dos conos, uno que se inserta dentro del otro. La fricción entre los conos les hace girar juntos.

Constant mesh A manual transmission design that permits the gears to be constantly enmeshed regardless of vehicle operating circumstances.

Engrane constante Diseño de transmisión manual que permite que los engranajes permanezcan siempre engranados a pesar de las condiciones del funcionamiento del vehículo.

Constant mesh transmission A transmission in which the gears are engaged at all times and shifts are made by sliding collars, clutches, or other means to connect the gears to the output shaft.

Transmisión de engrane constante Transmisión en la que los engranajes están siempre engranados, y los cambios se llevan a cabo a través de chavetas deslizantes, embragues u otros medios para conectar los engranajes al árbol de rendimiento.

Constant velocity joint (also called CV joint) A flexible coupling between two shafts that permits each shaft to maintain the same driving or driven speed regardless of operating angle, allowing for a smooth transfer of power. The constant velocity joint consists of an inner and outer housing with balls in between or a tripod and yoke assembly.

Junta de velocidad constante (llamada también junta CV) Unión flexible entre dos árboles que permite que cada árbol mantenga la misma velocidad de accionamiento o accionada, a pesar del ángulo de funcionamiento, y permite que la transferencia de fuerza sea suave. La junta de velocidad constante consiste de un alojamiento interior y exterior entre el cual se insertan bolas, o un conjunto de trípode y yugo.

CVT (Continuously Variable Transmission) A transmission with no fixed forward gear ratios, typically uses pulleys and a belt rather than gears.

Mecanismo de cambio de velocidades constante y variable (CVT) Una transmisión sin engranajes delanteras fijas; generalmente se usan polispastos en vez de engranajes.

Contraction A reduction in mass or dimension; the opposite of expansion.

Contracción Disminución en masa o dimensión; lo opuesto de expansión.

Control arm A suspension component that links the vehicle frame to the steering knuckle or axle housing and acts as a hinge to allow up and down wheel motions. The front control arms are attached to the frame with bushings and bolts and are connected to the steering knuckles with ball joints. The rear control arms attach to the frame with bushings and bolts and are welded or bolted to the rear axle or wheel hubs.

Brazo de mando Componente de suspensión que une el armazón del vehículo al muñón de dirección o al puente trasero, y funciona como una bisagra para permitir el movimiento de ascenso y descenso de la rueda. Los brazos de mando delanteros se fijan al armazón con bujes y pernos y se conectan a los muñones de dirección con juntas esféricas. Se fijan los brazos de mando traseros al armazón con bujes y pernos, y se sueldan o se empernan al eje trasero o a los cubos de la rueda.

Control cable An assembly with a flexible outer housing anchored at the upper and lower ends. Moving back and forth inside the housing is a braided stainless steel wire cable that transfers pedal movement to the release lever.

Cable de mando Conjunto con un alojamiento exterior flexible sujetado a los extremos superiores e inferiores. Un cable trenzado de acero inoxidable transfiere el movimimiento del pedal a la palanca de desembrague, y se mueve hacia atrás y hacia adelante dentro del alojamiento.

Controller A device that switches an electrical circuit on and off or that changes the operation of the circuit.

Controlador Un dispositivo que enciende o apaga un circuito eléctrico o que cambia la operación del circuito.

Corrode To eat away gradually as if by gnawing, especially by chemical action.

Corroerse Carcomer gradualmente como al roer, especialmente debido a una acción química.

Corrosion Chemical action, usually by an acid, that eats away (decomposes) a metal.

Corrosión Acción química, normalmente producida por un ácido, que carcome (descompone) un metal.

Corrosivity A statement defining how likely it is that a substance will destroy or eat away at other substances.

Corrosividad Enunciado que define la posibilidad de que una sustancia destruya o carcoma otras sustancias.

Cotter pin A type of fastener made from soft steel in the form of a split pin that can be inserted in a drilled hole. The split ends are spread to lock the pin in position.

Pasador de chaveta   Tipo de aparato de fijación hecho de acero recocido en forma de pasador hendido que puede insertarse en un agujero barrenado. Se separan los extremos hendidos para fijar la chaveta en la posición correcta.

Counterclockwise rotation   Rotating in the opposite direction to the hands on a clock.

Rotación a la izquierda   Rotación en el sentido inverso a la dirección de las agujas del reloj.

Counter gear assembly   A cluster of gears designed on one casting with short shafts supported by antifriction bearings. Closely related to the cluster assembly.

Mecanismo contador   Grupo de engranajes diseñados en una sola fundición con árboles cortos apoyados por cojinetes de antifricción. Estrechamente relacionado al conjunto desplazable.

Countershaft   An intermediate shaft that receives motion from a main shaft and transmits it to a working part; sometimes called a lay shaft.

Árbol de retorno   Árbol intermedio que recibe movimiento de un árbol primario y lo transmite a una pieza móvil; llamado también árbol secundario.

Coupling   A connecting means for transferring movement from one part to another; may be mechanical, hydraulic, or electrical.

Acoplamiento   Método de conexión para transferir movimiento de una pieza a otra; puede ser mecánico, hidráulico o eléctrico.

Coupling yoke   A part of the double Cardan U-joint that connects the two U-joint assemblies.

Yugo de acoplamiento   Pieza de la junta doble de cardán que conecta los dos conjuntos de junta universal.

Cover plate   A stamped steel cover bolted over the service access to the manual transmission.

Cubreplaca   Cubreplaca de acero estampada, empernada sobre el acceso de reparación a la transmisión manual.

Critical speed   The rotational speed at which an object begins to vibrate as it turns. This is mostly caused by centrifugal forces.

Velocidad crítica   Velocidad de rotación a la cual un objeto comienza a vibrar mientras gira. Esto se debe en gran parte a fuerzas centrífugas.

Cross member   A steel part of the frame structure that transverses the vehicle body to connect the longitudinal frame rails. Cross members can be welded into place or removed from the vehicle.

Traviesa   Pieza de acero de la estructura del armazón que atraviesa la carrocería para conectar las barras longitudinales del armazón. Las traviesas se pueden soldar o remover del vehículo.

Crush sleeve   A commonly used term for the collapsible spacer in a differential assembly.

Manguito de quiebra   Término utilizado comunamente para denominar el espaciador desmontable en un conjunto de diferencial.

Current   The flow of electricity through a circuit.

Corriente   El flujo de la electricidad por un circuito.

Cushioning springs   A common name for a clutch disc's wave springs.

Muelles de acojinamiento   Nombre común para los muelles ondulares del disco del embrague.

CV joints   Constant velocity joints that allow the angle of the axle shafts to change with no loss in rotational speed.

Juntas CV   Juntas de velocidad constante que permiten que el ángulo de los árboles motores cambie sin que disminuya la velocidad de rotación.

Dead axle   An axle that only supports the vehicle and does not transmit power.

Eje portante   Eje que sirve sólo para apoyar el vehículo, y que no transmite fuerza motriz.

Deceleration   The rate of decrease in speed.

Deceleración   Disminución de la velocidad.

Deflection   Bending or movement away from normal due to loading.

Desviación   Flexión o movimiento fuera de lo normal debido a la carga.

Degree   A unit of measurement equal to 1/360th of a circle.

Grado   Unidad de medida equivalente a cada una de las 360 partes de un círculo.

Density   Compactness; relative mass of matter in a given volume.

Densidad   Compacidad; masa relativa de materia en un volumen dado.

Detent   A small depression in a shaft, rail, or rod into which a pawl or ball drops when the shaft, rail, or rod is moved. This provides a locking effect.

Retén   Pequeña depresión en un árbol, una barra o una varilla sobre el/la cual cae un trinquete o una bola cuando se mueve el árbol, la barra o la varilla. Esto provee un efecto de blocaje.

Detent mechanism   A shifting control designed to hold the manual transmission in the gear range selected.

Mecanismo de detención   Control de cambio de velocidades diseñado para mantener la transmisión manual dentro del límite del engranaje elegido.

Diagnosis   A systematic study of a machine or machine parts to determine the cause of improper performance or failure.

Diagnosis   Estudio sistemático de una máquina o de piezas de una máquina para establecer la causa del mal funcionamiento o falla.

Dial indicator   A measuring instrument with the readings indicated on a dial rather than on a thimble as on a micrometer.

Indicador de cuadrante   Instrumento de medida que muestra las lecturas en un cuadrante en vez de en un tambor como en el caso de un micrómetro.

Diaphragm spring   A circular disc shaped like a cone with spring tension that allows it to flex forward or backward. Often referred to as a Belleville spring.

Muelle de diafragma   Disco circular en forma de cono, con tensión en el muelle que le permite moverse hacia adelante o hacia atrás. Conocido también como muelle de Belleville.

Diaphragm spring clutch   A clutch in which a diaphragm spring, rather than a coil spring, applies pressure against the friction disc.

Embrague de muelle de diafragma   Embrague en el que un muelle de diafragma, en vez de un muelle helicoidal, ejerce presión contra el disco de fricción.

Differential   A mechanism between drive axles that permits one wheel to run at a different speed than the other while turning.

Diferencial   Mecanismo entre los ejes de mando que permite que una rueda gire a una velocidad diferente que la otra.

Differential action   An operational situation in which one driving wheel rotates at a slower speed than the opposite driving wheel.

Acción del diferencial   Situación de funcionamiento donde una rueda motriz gira más despacio que la rueda motriz opuesta.

Differential case   The metal unit that encases the differential side gears and pinion gears and to which the ring gear is attached.

Caja del diferencial   Unidad metálica que reviste los engranajes laterales del diferencial y los engranajes de piñón, y sobre la cual se monta la corona.

Differential drive gear   A large circular helical gear driven by the transaxle pinion gear and shaft that drives the differential assembly.

Engranaje del diferencial   Engranaje helicoidal circular grande accionado por el engranaje de piñón del transeje y el árbol, que acciona el conjunto del diferencial.

Differential housing Cast-iron assembly that houses the differential unit and the drive axles. This is also called the rear axle housing.

Alojamiento del diferencial Conjunto de hierro fundido que aloja la unidad del diferencial y los ejes de mando. Llamado también puente trasero.

Differential pinion gears Small beveled gears located on the differential pinion shaft.

Engranajes de piñón del diferencial Pequeños engranajes biselados ubicados en el árbol de piñón del diferencial.

Differential pinion shaft A short shaft locked to the differential case that supports the differential pinion gears.

Árbol de piñón del diferencial Árbol corto fijado a la caja del diferencial. Este árbol apoya los engranajes de piñón del diferencial.

Differential ring gear A large circular hypoid-type gear enmeshed with the hypoid drive pinion gear.

Corona del diferencial Engranaje hipoide circular grande engranado con el engranaje de piñón hipoide.

Differential side gears The gears inside the differential case that are internally splined to the axle shafts and are driven by the differential pinion gears.

Engranajes laterales del diferencial Engranajes dentro de la caja del diferencial que son ranurados internamente a los árboles motores, y accionados por los engranajes de piñón del diferencial.

Digital A type of signal that is caused by switching the circuit on and off.

Digital Tipo de señal causada al apagar y encender el circuito.

Diode A semiconductor that allows current to flow in one direction only.

Diodo Semiconductor que permite fluir la corriente solamente en una dirección.

Direct current (DC) A type of current that always flows in one direction, from a point of higher potential to a point of lower potential.

Corriente directa (DC) Tipo de corriente que siempre fluye en una dirección, desde el punto de energía potencial más alto hacia el punto de energía potencial más bajo.

Direct drive One turn of the input driving member compared to one complete turn of the driven member, such as when there is direct engagement between the engine and drive shaft in which the engine crankshaft and the drive shaft turn at the same rpm.

Toma directa Una vuelta de la pieza de accionamiento comparada a una vuelta completa de la pieza accionada, como por ejemplo, cuando hay un engrane directo entre el motor y el árbol de mando donde el cigüeñal del motor y el árbol de mando giran a las mismas rpm.

Disengage When the operator moves the clutch pedal toward the floor to disconnect the driven clutch disc from the driving flywheel and pressure plate assembly.

Desengranar Cuando el conductor hunde el pedal del embrague para desconectar el disco de embrague accionado del volante motor y del conjunto de la placa de presión.

Distortion A warpage or change in form from the original shape.

Deformación Abarquillamiento o cambio en la forma original de la configuración original.

Dog tooth A series of gear teeth that are part of the dog clutching action in a transmission synchronizer operation.

Diente de sierra Serie de dientes de engranaje que forman parte de la acción del embrague de garras durante una sincronización de la transmisión.

Double Cardan Universal joint A near constant velocity U-joint that consists of two Cardan Universal joints connected by a coupling yoke.

Junta doble de cardán Junta universal de velocidad casi constante compuesta de dos juntas de cardán conectadas por un yugo de acoplamiento.

Double-offset constant velocity joint Another name for the type of plunging, inner CV joint found on many GM, Ford, and Japanese FWD cars.

Junta de velocidad constante de desviación doble Otro nombre para el tipo de junta de velocidad constante interior de pistón tubular, instalada en muchos automóviles de tracción delantera japoneses, de la GM y de la Ford.

Dowel A metal pin attached to one object that, when inserted into a hole in another object, ensures proper alignment.

Pasador Chaveta metálica fijada a un objeto que, cuando se inserta dentro de un agujero en otro objeto, asegura una alineación correcta.

Dowel pin A pin inserted in matching holes in two parts to maintain those parts in fixed relation one to another.

Espiga de madera Chaveta insertada en agujeros parejos en dos piezas para mantener dichas piezas en una relación fija entre sí.

Downshift To shift a transmission into a lower gear.

Cambio de alta a baja velocidad Cambiar la transmisión a un engranaje de menos velocidad.

Driveline The Universal joints, drive shaft, and other parts connecting the transmission with the driving axles.

Línea de transmisión Las juntas universales, el árbol de mando y otras piezas que conectan la transmisión a los ejes motores.

Driveline torque Relates to rear wheel driveline and is the transfer of torque between the transmission and the driving axle assembly.

Par de torsión de la línea de transmisión Relacionado a la línea de transmisión de las ruedas traseras y es la transferencia del par de torsión entre la transmisión y el conjunto del eje motor.

Driveline wrap-up A condition in which axles, gears, U-joints, and other components can bind or fail if the 4WD mode is used on pavement where 2WD is more suitable.

Falla de la línea de transmisión Condición que ocurre cuando los ejes, los engranajes, las juntas universales, y otros componentes se traban o fallan si se emplea la tracción a las cuatro ruedas sobre un pavimento donde la tracción a las dos ruedas es más adecuada.

Drive pinion The gear that takes its power directly from the drive shaft or transmission and drives the ring gear.

Piñón de mando Engranaje que obtiene su fuerza motriz directamente del árbol de mando o de la transmisión, y que acciona la corona.

Drive pinion flange A rim used to connect the rear of the drive shaft to the rear axle drive pinion.

Brida de piñón de mando Corona utilizada para conectar la parte trasera del árbol de mando al piñón de mando del eje trasero.

Drive pinion gear One of the two main driving gears located within the transaxle or rear driving axle housing. Together the two gears multiply engine torque.

Engranaje del piñón de mando Uno de los dos mecanismos de accionamiento principales ubicados dentro del transeje o el puente trasero. Los dos engranajes actúan juntos para multiplicar el par de torsión del motor.

Drive shaft An assembly of one or two U-joints connected to a shaft or tube; used to transmit power from the transmission to the differential. Also called the propeller shaft.

Árbol de mando Conjunto de una o dos juntas universales conectadas a un árbol o tubo; utilizado para transmitir la fuerza motriz desde la transmisión hasta el diferencial. Llamado también árbol transmisor.

Drive shaft installation angle The angle the drive shaft is mounted off the true horizontal line measured in degrees.

Ángulo de montaje del árbol de mando El ángulo medido en grados al que se monta el árbol de mando fuera de la línea horizontal verdadera.

Driven disc The part of the clutch assembly that receives driving motion from the flywheel and pressure plate assemblies.

Disco accionado Pieza del conjunto del embrague que recibe su fuerza motriz del conjunto del volante y del conjunto de la placa de presión.

Driven gear The gear meshed directly with the driving gear to provide torque multiplication, reduction, or a change of direction.

Engranaje accionado Engranaje engranado directamente al mecanismo de accionamiento para proveer multiplicación o reducción de par de torsión, reducción o un cambio de dirección.

Drivetrain The components that transmit power from the engine to the drive wheels. The drivetrain consists of the clutch or torque converter, transmission, driveshaft, rear axle and differential, or transaxle and half-shafts.

Tren de mando Los componentes que transmiten potencia del motor a las ruedas motrices. El tren de mando consiste en el embrague o convertidor de par de torsión, la transmisión, el arbol de mando, el eje trasero y diferencial, o el transeje y los semiejes.

Driving axle A term related collectively to the rear driving axle assembly where the drive pinion, ring gear, and differential assembly are located within the driving axle housing.

Eje motor Término relacionado colectivamente al conjunto del eje motor trasero donde el piñón de mando, la corona, y el conjunto del diferencial. Están ubicados dentro del puente trasero.

Drop forging A piece of steel shaped between dies when hot.

Estampado Pieza de acero conformada entre troqueles mientras está caliente.

Dry-disc clutch A clutch in which the friction faces of the friction disc are dry, as opposed to a wet-disc clutch, which runs submerged in oil. The conventional type of automobile clutch.

Embrague de disco seco Embrague en el que las placas de fricción del disco de fricción están secas, lo opuesto de un disco mojado, que funciona sumergido en aceite. Tipo convencional de embrague de automóviles.

Dry friction The friction between two dry solids.

Fricción seca Fricción entre dos sólidos secos.

Dual-mass flywheel A flywheel composed of two plates connected by a damper assembly. This flywheel is designed to reduce the engine vibrations that are transmitted through the transmission.

Volante de doble masa Un volante compuesto de dos placas unidas por una asamblea amortiguadora. Este volante está diseñado para disminuir las vibraciones del motor que se transmiten por la transmisión.

Dummy shaft A shaft, shorter than the countershaft, used during disassembly and reassembly in place of the countershaft.

Árbol falso Árbol más corto que el árbol de retorno empleado durante el desmontaje y el remonte en vez del árbol de retorno.

Dynamic In motion.

Dinámico En movimiento.

Dynamic balance The balance of an object when it is in motion; for example, the dynamic balance of a rotating drive shaft.

Equilibrio dinámico Equilibrio de un objeto cuando está en movimiento; por ejemplo, el equilibrio dinámico de un árbol de mando giratorio.

Eccentric One circle within another circle wherein both circles do not have the same center, or a circle mounted off center. On FWD cars, front-end camber adjustments are accomplished by turning an eccentric cam bolt that mounts the strut to the steering knuckle.

Excéntrico Un círculo dentro de otro donde los dos círculos no comparten el mismo centro, o un círculo colocado fuera del centro. En automóviles de tracción delantera se realizan ajustes a la combadura del tren delantero girando un perno de leva excéntrica que levanta el montante al muñón de dirección.

Efficiency The ratio between the power of an effect and the power expended to produce the effect; the ratio between an actual result and the theoretically possible result.

Rendimiento La relación entre la fuerza de un efecto y la fuerza rendida para producir tal efecto; la relación entre un resultado verdadero y un resultado teóricamente posible.

Elastomer Any rubber-like plastic or synthetic material used to make bellows, bushings, and seals.

Elastómero Cualquier material plástico o sintético parecido al caucho, que se utiliza para fabricar fuelles, bujes y juntas de estanqueidad.

Electrical short An alternative path for the flow of electricity.

Corto eléctrico Camino alternativo para el flujo de electricidad.

Electromagnet A magnet formed by electrical flow through a conductor.

Electroimán Imán que resulta de la corriente eléctrica mediante un conductor.

Element A substance with only one type of atom.

Elemento Sustancia que contiene un sólo tipo de átomo.

Ellipse A compressed form of a circle.

Elipse Forma comprimida de un círculo.

End play The amount of axial or end-to-end movement in a shaft due to clearance in the bearings.

Holgadura Amplitud de movimiento axial o movimiento de extremo a extremo en un árbol debido al espacio libre en los cojinetes.

Engage When the vehicle operator moves the clutch pedal up from the floor, this engages the driving flywheel and pressure plate to rotate and drive the driven disc.

Engranar Cuando el conductor del vehículo suelta el pedal del embrague, el volante y la placa de presión se engranan para girar y accionar el disco accionado.

Engagement chatter A shaking, shuddering action that takes place as the driven disc makes contact with the driving members. Chatter is caused by a rapid grip and slip action.

Vibración de acoplamiento Movimiento de agitación y estremecimiento que ocurre cuando el disco accionado entra en contacto con las piezas de accionamiento. La causa de esta vibración es un movimiento rápido de garra y de deslizamiento.

Engine The source of a power for most vehicles. It converts burned fuel energy into mechanical force.

Motor Fuente de potencia para la mayoría de los vehículos. Convierte el combustible quemado en fuerza mecánica.

Engine torque A turning or twisting action developed by the engine measured in foot-pounds or kilogram meters.

Esfuerzo de rotación del motor Movimiento de giro o torcedura que produce el motor, medido en libras-pies o en kilográmetros.

Equilibrium Exists when the applied forces on an object are balanced and there is no overall resultant force.

Equilibrio Existe cuando las fuerzas aplicadas sobre un objecto son iguales y no resulta otra fuerza.

Evaporate Atoms or molecules break free from the body of the liquid to become gas particles.

Evaporar Los atomos o las moléculas se desprenden del liquido y ílegan a ser particulas de gas.

Expansion An increase in the size of a mass due to the movemnet of atoms and molecules as heat moves into a mass.

Expansión Aumento de tamaño de una masa a causa del movimiento de átomos y moléculas durante el movimiento del calor hacia el interior de la masa.

Extension housing An aluminum or iron casting of various lengths that encloses the transmission output shaft and supporting bearings.

Alojamiento de extensión Pieza fundida en aluminio o en hierro de longitudes variadas que encubre el árbol de rendimiento de la transmisión y los cojinetes de soporte.

External cone clutch The external surface of one part has a tapered surface to mate with an internally tapered surface to form a cone clutch.

Embrague cónico externo La superficie externa de una pieza tiene una superficie cónica para hacer juego con una superficie internamente cónica, y así formar un embrague de cono.

External gear A gear with teeth across the outside surface.

Engranaje externo Engranaje que tiene dientes a través de la superficie exterior.

Externally tabbed clutch plates Clutch plates designed with tabs around the outside periphery to fit into grooves in a housing or drum.

Placas de embrague con orejetas externas Se diseñan placas de embrague con orejetas alrededor de la periferia exterior para que puedan ajustarse a las acanaladuras de un alojamiento o de un tambor.

Extreme pressure lubricant A special lubricant for use in hypoid gear differentials; needed because of the heavy wiping loads imposed on the gear teeth.

Lubrificante para presión extrema Lubrificante especial que se utiliza en diferenciales de engranaje hipoide; necesario a causa del intenso esfuerzo al que están sometidos los dientes de los engranajes.

Face The front surface of an object.

Frente Superficie frontal de un objeto.

Fatigue The buildup of natural stress forces in a metal part that eventually causes it to break. Stress results from bending and loading the material.

Fatiga Acumulación de tensiones naturales en una pieza metálica que finalmente ocasiona una ruptura. La tensión es una consecuencia de la flexión y de la carga a las cuales está expuesto el material.

Feeler gauge A metal strip or blade finished accurately with regard to thickness, and used for measuring the clearance between two parts; such gauges ordinarily come in a set of different blades graduated in thickness by increments of 0.001 inch.

Calibrador de espesores Lámina metálica o cuchilla acabada con precisión de acuerdo al espesor que se utiliza para medir el espacio libre entre dos piezas. Dichos calibradores normalmente están disponibles en juegos de cuchillas con diferente graduacion según el espesor, en incrementos de 0,001 pulgadas.

Fiber composites A mixture of metallic threads along with a resin form a composite offering weight and cost reduction, long-term durability, and fatigue life. Fiberglass is a fiber composite.

Compuestos de fibra La mezcla de hilos metálicos y resina forman un compuesto que ofrece una disminución de peso y costo, mayor durabilidad y resistencia a la fatiga. La fibra de vidrio es un compuesto de fibra.

Final drive The final set of reduction gears the engine's power passes through on its way to the drive wheels.

Transmisión final Último juego de reductores por el cual pasa la fuerza del motor en camino a las ruedas motrices.

Final drive gears The main driving gears located in the axle area of the transaxle housing.

Engranajes de la transmisión final Mecanismos principales de accionamiento ubicados en la región del eje del alojamiento del transeje.

Final drive ratio The ratio between the drive pinion and ring gear.

Relación de la transmisión final Relación entre el piñón de mando y la corona.

First gear A small diameter driving helical- or spur-type gear located on the cluster gear assembly. First gear provides torque multiplication to get the vehicle moving.

Engranaje de primera velocidad Mecanismo de mando helicoide o recto de diámetro pequeño ubicado en el conjunto del tren desplazable. El engranaje de primera velocidad inicia la multiplicación de par de torsión para impulsar el vehículo.

Fit The contact between two machined surfaces.

Conexión Contacto entre dos superficies maquinadas.

Fixed-type constant velocity joint A joint that cannot telescope or plunge to compensate for suspension travel. Fixed joints are always found on the outer ends of the drive shafts of FWD cars. A fixed joint may be of either Rzeppa or tripod type.

Junta de velocidad constante de tipo fijo Junta que no puede extenderse o hundirse para compensar el movimiento de la suspensión. Siempre se encuentran juntas fijas en los extremos exteriores de los árboles de mando de vehículos de tracción delantera. Una junta fija puede ser de tipo Rzeppa o trípode.

Flammability A statement of how well the substance supports combustion.

Inflamabilidad Enunciado que formula cuán bien una sustancia soporta la combustión.

Flange A projecting rim or collar on an object for keeping it in place.

Brida Corona proyectada o collar en un objeto que lo mantiene en su lugar.

Flange yoke The part of the rear U-joint attached to the drive pinion.

Yugo de brida Pieza de la junta universal trasera fijada al piñón de mando.

Flexplate A lightweight flywheel used only on engines equipped with an automatic transmission. The flexplate is equipped with a starter ring gear around its outside diameter and also serves as the attachment point for the torque converter.

Placa flexible Volante liviano empleado solamente en motores equipados con transmisión automática. La placa flexible está equipada con una corona de arranque alrededor de su diámetro exterior y sirve también de punto de fijación para el convertidor del par motor.

Fluid coupling A device in the powertrain consisting of two rotating members; transmits power from the engine, through a fluid, to the transmission.

Acoplamiento fluido Mecanismo en el tren transmisor de potencia que consiste de dos piezas giratorias; transmite la fuerza desde el motor, por medio de un fluido, hasta la transmisión.

Fluid drive A drive in which there is no mechanical connection between the input and output shafts, and power is transmitted by moving oil.

Transmisión hidráulica Transmisión en la que no existe conexión mecánica alguna entre el árbol impulsor y el árbol de rendimiento. La fuerza se transmite a través del aceite motor.

Flywheel A heavy metal wheel that is attached to the crankshaft and rotates with it; helps smooth out the power surges from the engine power strokes; also serves as part of the clutch and engine-cranking system.

Volante Rueda pesada de metal que se fija al cigüeñal y que gira con él; ayuda a neutralizar las sacudidas de fuerza de las carreras motrices del motor; sirve también como parte del sistema de embrague y de arranque del motor.

Flywheel ring gear A gear, fitted around the flywheel, that is engaged by teeth on the starting motor drive to crank the engine.

Corona del volante Engranaje ajustado alrededor del volante, engranado por dientes en el mando del motor de arranque para hacer arrancar el motor.

Foot-pound (or ft.-lb.) This is a measure of the amount of energy or work required to lift 1 pound a distance of 1 foot.

Libra-pie Medida de la cantidad de energía o fuerza que se requiere para levantar una libra a una distancia de un pie.

Force Any push or pull exerted on an object; measured in pounds and ounces, or in newtons (N) in the metric system.

Fuerza Cualquier empuje o tirón que se ejerce sobre un objeto; medido en libras y onzas, o en newtons (N) en el sistema métrico.

Forward coast side The side of the ring gear tooth the drive pinion contacts when the vehicle is decelerating.

Cara de cabotaje delantera Cara del diente de la corona con el cual el piñón de mando entra en contacto mientras el vehículo desacelera.

Forward drive side The side of the ring gear tooth that the drive pinion contacts when accelerating or on the drive.

Cara de mando delantero Cara del diente de la corona con el cual el piñón de mando entra en contacto mientras acelera o está en marcha.

Four-wheel drive (4WD) On a vehicle, driving axles are found at both front and rear, so that all four wheels can be driven. 4WD is the standard abbreviation for four-wheel drive.

Tracción a las cuatro ruedas En un vehículo, los ejes motores se encuentran ubicados en las partes delantera y trasera, para que las cuatro ruedas se puedan accionar. 4WD es la abreviatura común para tracción a las cuatro ruedas.

Four wheel high A transfer case shift position in which both front and rear drive shafts receive power and rotate at the speed of the transmission output shaft.

Alto de cuatro ruedas Posición en la caja de cambios donde los árboles de mando delantero y trasero reciben fuerza, y giran a la misma velocidad que el árbol de rendimiento de la transmisión.

Frame The main understructure of the vehicle to which everything else is attached. Most FWD cars have only a subframe for the front suspension and drivetrain. The body serves as the frame for the rear suspension.

Armazón Chasís principal del vehículo al cual se fijan todas las demás piezas. La mayoría de los vehículos de tracción delantera sólo tienen un chasís que soporta la suspensión delantera y el tren de mando. La carrocería sirve como armazón para la suspensión trasera.

Free-running gears Gears that rotate independently on their shaft.

Engranajes de funcionamiento libre Engranajes que giran de manera independiente en su árbol.

Free-wheeling clutch A mechanical device that will engage the driving member to impart motion to a driven member in one direction but not the other. Also known as an "overrunning clutch."

Embrague de marcha en rueda libre Mecanismo que se engranará a la pieza de accionamiento para impulsar movimiento a una pieza accionada en una dirección pero no en la otra. Conocido también como "embrague de giro libre".

Friction The resistance to motion between two bodies in contact with each other.

Fricción Resistencia al movimiento entre dos cuerpos en contacto el uno con el otro.

Friction bearing A bearing in which there is sliding contact between the moving surfaces. Sleeve bearings, such as those used in connecting rods, are friction bearings.

Cojinete de fricción Cojinete en el cual existe un contacto deslizante entre las superficies en movimiento. Los cojinetes de manguito, como los que se utilizan en las bielas, son cojinetes de fricción.

Friction disc In the clutch, a flat disc, faced on both sides with friction material and splined to the clutch shaft. It is positioned between the clutch pressure plate and the engine flywheel. Also called the clutch disc or driven disc.

Disco de fricción Disco plano del embrague, revestido en las dos caras con material de fricción y ranurado al árbol de embrague. Está ubicado entre la placa de presión del embrague y el volante de la máquina. Llamado también disco de embrague o disco accionado.

Friction facings A hard-molded or woven asbestos or paper material that is riveted or bonded to the clutch driven disc.

Revestimiento de fricción Material de moldeado duro, de asbesto tejido o de papel que se remacha o adhiere al disco accionado del embrague.

Front bearing retainer An iron or aluminum circular casting fastened to the front of a transmission housing to retain the front transmission bearing assembly.

Retenedor del cojinete de rueda delantero Pieza circular fundida en hierro o en aluminio fijada a la parte delantera de un alojamiento de la transmisión para sujetar el conjunto del cojinete de transmisión delantero.

Front differential/axle assembly Like a conventional rear axle but having steerable wheels.

Conjunto de diferencial/eje delantero Igual que el eje trasero convencional, pero con ruedas orientables.

Front-wheel drive (FWD) The vehicle has all drivetrain components located at the front.

Tracción delantera Todos los componentes del tren de mando en el vehículo se encuentran en la parte delantera.

Fulcrum The support that provides a pivoting point for a lever.

Fulcro Soporte que le provee punto de apoyo a una palanca.

Fulcrum ring A circular ring over which the pressure plate diaphragm spring pivots.

Anillo de fulcro Anillo circular sobre el cual gira el muelle del diafragma de la placa de presión.

Full-floating rear axle An axle that only transmits driving force to the rear wheels. The weight of the vehicle (including payload) is supported by the axle housing.

Eje trasero enteramente flotante Eje que solamente transmite la fuerza motriz a las ruedas traseras. El puente trasero soporta el peso del vehículo (incluyendo la carga útil).

Full-time 4WD   Systems that use a center differential that accommodates speed differences between the two axles; necessary for on-highway operation.

Tracción a las cuatro ruedas a tiempo completo   Sistemas que utilizan un diferencial central que acomoda diferentes velocidades entre los dos ejes; necesaria para el funcionamiento en la carretera.

Fully synchronized transmission   A transmission in which all of its forward gears are equipped with a synchronizer assembly. In a manual transmission, the synchronizer assembly operates to improve the shift quality in all forward gears.

Transmisión enteramente sincronizada   Transmisión en la que todos los engranajes delanteros han sido equipados con un conjunto sincronizador. En una transmisión manual el conjunto sincronizador funciona para mejorar la calidad del desplazamiento en todos los engranajes delanteros.

Galling   Wear caused by metal-to-metal contact in the absence of adequate lubrication. Metal is transferred from one surface to the other, leaving behind a pitted or scaled appearance.

Corrosión por rozamiento   Desgaste causado por el contacto de un metal con otro metal debido a la ausencia de lubrificación adecuada. Se transfiere el metal de una superficie a la otra, lo cual deja un aspecto corroído o raspado.

Gasket   A layer of material, usually made of cork, paper, plastic, composition, or metal, or a combination of these, placed between two parts to make a tight seal.

Guarnición   Capa de un material, normalmente hecho de corcho, papel, plástico, pasta o metal, o una combinación de éstos, ubicada entre dos piezas para crear una junta de estanqueidad hermética.

Gasket cement   A liquid adhesive material or sealer used to install gaskets.

Cemento de guarnición   Material líquido adhesivo, o de juntura, utilizado para instalar guarniciones.

Gear   A wheel with external or internal teeth that serves to transmit or change motion.

Engranaje   Rueda con dientes externos o internos que sirve para transmitir o cambiar el movimiento.

Gear lubricant   A type of grease or oil blended especially to lubricate gears.

Lubrificante de engranaje   Tipo de grasa o aceite mezclado especialmente para lubrificar engranajes.

Gear ratio   The number of revolutions of a driving gear required to turn a driven gear through one complete revolution. For a pair of gears, the ratio is found by dividing the number of teeth on the driven gear by the number of teeth on the driving gear.

Relación de engranajes   Número de revoluciones de un mecanismo de accionamiento requeridas para hacer girar un engranaje accionando una revolución completa. Para un par de engranajes se obtiene la relación al dividir el número de dientes en el engranaje accionado por el número de dientes en el mecanismo de accionamiento.

Gear reduction   A situation in which, when a small gear drives a large gear, there is an output speed reduction and a torque increase that results in a gear reduction.

Reducción de engranajes   Una situación en que cuando un engranaje pequeño acciona un engranaje grande. Se produce una reducción de velocidad de rendimiento y un aumento de par de torsión que resulta en una reducción de engranajes.

Gear whine   A high-pitched sound developed by some types of meshing gears.

Silbido del engranaje   Sonido agudo producido por algunos tipos de engranajes.

Gearboxes   A slang term for transmissions.

Cajas de engranajes   Coloquialismo que significa transmisiones.

Gearshift   A linkage-type mechanism by which the gears in an automobile transmission are engaged and disengaged.

Cambio de velocidades   Mecanismo de tipo empalme a través del cual se engranan y se desengranan los engranajes en la transmisión de un vehículo.

Graphite   Very fine carbon dust with a slippery texture used as a lubricant.

Grafito   Polvo muy fino de carbón con una textura resbaladiza que se utiliza como lubrificante.

Grind   To finish or polish a surface by means of an abrasive wheel.

Esmerilar   Acabar o pulir una superficie con una rueda abrasiva.

Half shaft   Either of the two drive shafts that connect the transaxle to the wheel hubs in FWD cars. Half shafts have CV joints attached to each end to allow for suspension motions and steering. The shafts may be of solid or tubular steel and may be of different lengths.

Semieje   Cualquiera de los dos árboles de mando que conecta el transeje a los cubos de rueda en automóviles de tracción delantera. Los semiejes tienen juntas de velocidad constante fijadas a cada extremo para permitir el movimiento de suspensión y la dirección. Los semiejes pueden ser de acero sólido o tubular y pueden variar sus longitudes.

Harshness   A bumpy ride caused by a stiff suspension. Can be cured by installing softer springs or shock absorbers.

Aspereza   Viaje de muchas sacudidas ocacionadas por una suspensión rígida. Puede remediarse con la instalación de muelles más flexbiles o amortiguadores.

Heat   A form of energy caused by the movement of atoms and molecules.

Calor   Forma de energía que se origina con el movimiento de los átomos y de las moléculas.

Heat sink   A piece of material that absorbs heat to prevent the heat from settling on another component.

Fuente fría   Pieza hecha de un material que absorbe el calor para impedir que el calor penetre otro componente.

Heat treatment   Heating, followed by fast cooling, to harden metal.

Tratamiento térmico   Calentamiento seguido del enfriamiento rápido, para endurecer el metal.

Heel   The outside, larger half of the gear tooth.

Talón   Mitad exterior más grande del diente de engranaje.

Helical   Shaped like a coil spring or a screw thread.

Helicoidal   Formado como a un muelle helicoidal o un filete de tornillo.

Helical gear   Gears with the teeth cut at an angle to the axis of the gear.

Engranaje helicoidal   Engranajes que tienen los dientes cortados a un ángulo al pivote del engranaje.

Herringbone gear   A pair of helical gears designed to operate together. The angle of the pair of gears forms a V.

Engranaje bihelocoidal   Par de engranajes helicoidales diseñados para funcionar juntos. El ángulo al par de engranajes forma una V.

High gears   Third, fourth, and fifth gears in a typical transmission.

Engranajes de alta multiplicación   Engranajes de tercera, cuarta y quinta velocidad en una transmisión típica.

Horsepower   A measure of mechanical power, or the rate at which work is done. One horsepower equals 33,000 ft.-lb. (foot-pounds) of work per minute. It is the power necessary to raise 33,000 pounds a distance of 1 foot in 1 minute.

Potencia en caballos Medida de fuerza mecánica o velocidad a la que se realiza el trabajo. Un caballo de fuerza es equivalente a 33.000 libras-pies de trabajo por minuto. Es el esfuerzo necesario para levantar 33.000 libras a la distancia de un pie en un minuto.

Hotchkiss drive A type of rear suspension in which leaf springs absorb the rear axle housing torque.

Transmisión Hotchkiss Tipo de suspensión trasera en la cual unos muelles de láminas absorben el par de torsión del puente trasero.

Hub The center part of a wheel to which the wheel is attached.

Cubo Parte central de una rueda, a la cual se fija la rueda.

Hydraulic clutch A clutch that is actuated by hydraulic pressure; used in cars and trucks when the engine is some distance from the driver's compartment where it would be difficult to use mechanical linkages.

Embrague hidráulico Embrague accionado por presión hidráulica; utilizado en automóviles y camiones cuando el motor está lejos del compartimiento del conductor para dificultar la utilización de bielas motrices mecánicas.

Hydraulic fluid reservoir A part of a master cylinder assembly that holds reserve fluid.

Despósito de fluido hidráulico Parte del conjunto del cilindro primario que contiene el fluido de reserva.

Hydraulic press A piece of shop equipment that develops a heavy force by use of a hydraulic piston-and-jack assembly.

Prensa hidráulica Pieza del equipo de taller que desarrolla fuerza pesada por medio de un conjunto de gato de pistón hidráulico.

Hydraulic pressure Pressure exerted through the medium of a liquid.

Presión hidráulica Presión ejercida a través de un líquido.

Hydrocarbon A substance composed of hydrogen and carbon molecules.

Hidrocarburo Sustancia compuesta de moléculas de carbono e hidrógeno.

Hypoid gear A gear that is similar in appearance to a spiral bevel gear, but the teeth are cut so that the gears match in a position where the shaft center lines do not meet; cut in a spiral form to allow the pinion to be set below the center line of the ring gear so that the car floor can be lower.

Engranaje hipoide Engranaje parecido a un engranaje cónico con dentado espiral, pero en el cual los dientes se cortan para que los engranajes se engranen en una posición donde las líneas centrales del árbol no se crucen; cortado en forma de espiral para permitir que el piñón sea colocado debajo de la línea central de la corona, y que así el piso del vehículo sea más bajo.

Hypoid gear lubricant An extreme pressure lubricant designed for the severe operation of hypoid gears.

Lubrificante del engranaje hipoide Lubrificante de extrema presión diseñado para el funcionamiento riguroso de los engranajes hipoides.

ID Inside diameter.

ID Diámetro interior.

Idle Engine speed at which the accelerator pedal is fully released and there is no load on the engine.

Marcha mínima Velocidad del motor cuando el pedal del acelerador se suelta completamente y no hay ninguna carga en el motor.

Idler gear A gear that rides between a set of gears and does not affect the ratio but does change the direction of the output gear.

Engranaje de marcha mínima Engranaje que pasea entre un grupo de engranajes y no afecta la proporción, pero cambia la dirección del engranaje de salida.

Ignitability A statement of how easily a substance can catch fire.

Inflamabilidad Afirmación sobre la facilidad con la cual puede encenderse una sustancia.

Impermeable Materials that do not adsorb fluids.

Impermeable Materiales que no absorben fluídos.

Inboard constant velocity joint The inner CV joint, or the one closest to the transaxle. The inboard joint is usually a plunging-type joint that telescopes to compensate for suspension motions.

Junta de velocidad constante del interior Junta de velocidad constante interior, o la que está más cerca del transeje. La junta del interior es normalmente una junta de tipo sumergible que se extiende para compensar el movimiento de la suspensión.

Inclinometer A device designed with a spirit level and graduated scale to measure the inclination of a driveline assembly. The inclinometer connects to the drive shaft magnetically.

Inclinómetro Instrumento diseñado con nivel de burbuja de aire y escala graduada para medir la inclinación de un conjunto de la línea de transmisión. El inclinómetro se conecta magnéticamente al árbol de mando.

Increments A series of regular additions from small to large.

Incrementos Serie de aumentos regulares de lo pequeño a lo grande.

Independent rear suspension (IRS) The vehicle's rear wheels move up and down independently of each other.

Suspensión trasera independiente Las ruedas traseras del vehículo realizan un movimiento de ascenso y descenso de manera independiente la una de la otra.

Independent suspension A suspension system that allows one wheel to move up and down without affecting the opposite wheel. Provides superior handling and a smoother ride.

Suspensión independiente Sistema de suspensión que permite que una rueda realice un movimiento de ascenso y descenso sin afectar la rueda opuesta. Permite un manejo excelente y un viaje mucho más cómodo.

Index To orient two parts by marking them. During reassembly, the parts are arranged so the index marks are next to each other. Used to preserve the orientation between balanced parts.

Alinear Orientar dos piezas marcándolas. Durante el remonte se arreglan las piezas de modo que las indicaciones de alineación queden juntas. Utilizado para mantener la orientación entre piezas equilibradas.

Inertia The resistance to a change in motion or direction.

Inercia No hay tolerancia entre los cambios de movimiento o de dirección.

Inner bearing race The inner part of a bearing assembly on which the rolling elements, ball or roller, rotate.

Anillo de cojinete interior Parte interior de un conjunto de cojinete sobre la cual giran los elementos rodantes, o sea, las bolas o los rodillos.

Input shaft The shaft carrying the driving gear by which the power is applied, as to the transmission.

Árbol impulsor Árbol que soporta el mecanismo de accionamiento a través del cual se aplica la fuerza motriz, como por ejemplo, a la transmisión.

Inserts One of several terms that could apply to the shift plates found in a synchronizer assembly.

Piezas insertas Uno de los varios téminos que puede aplicarse a las placas de cambio de velocidades encontradas en un conjunto sincronizador.

Insert springs Round wire springs that hold the inserts or shift plates in contact with the synchronizer sleeve. Located around the synchronizer hub.

Muelles insertos Muelles redondos de alambre que mantienen el contacto entre las piezas insertas o placas de cambio de velocidades y el manguito sincronizador. Ubicados alrededor del cubo sincronizador.

Inspection cover A removable cover that permits entrance for inspection and service work.

Cubierta de inspección Cubierta desmontable que permite la entrada para inspección y reparación.

Insulator A material that does not allow electrons to flow easily through it.

Aislador Sustancia que no permite el flujo fácil de los electrones.

Integral Built into, as part of the whole.

Integral Pieza incorporada, como parte del todo.

Integral axle housing A rear axle housing-type in which the parts are serviced through an inspection cover and adjusted within and relative to the axle housing.

Puente trasero integral Tipo de puente trasero donde se reparan las piezas a través de una cubierta de inspección, y se ajustan dentro y con relación al puente trasero.

Integrated circuit (IC) An electrical circuit composed of many diodes and transistors. The IC is the basis for computerized systems.

Circuito integrado (IC) Un circuito eléctrico compuesto de muchos diodos y transitores. El IC es básico en los sistemas computarizados.

Integrated full-time 4WD Systems that use computer controls to enhance full-time operation, adjusting the torque split depending on which wheels have traction.

Tracción a las cuatro ruedas a tiempo completo integral Sistemas que utilizan controles de computadoras para mejorar el funcionamiento a tiempo completo, ajustando el reparto de torsión dependiendo de cuáles de las ruedas tienen tracción.

Interaxle differential The center differential in some 4WD systems.

Diferencial entre ejes Diferencial central en algunos sistemas de tracción a las cuatro ruedas.

Interlock mechanism A mechanism in the transmission shift linkage that prevents the selection of two gears at one time.

Mecanismo de enganche Mecanismo en la biela motriz del cambio de velocidades de la transmisión que impide la selección de dos velocidades a la vez.

Intermediate drive shaft Located between the left and right drive shafts, it equalizes drive shaft length.

Árbol de mando intermedio Ubicado entre los árboles de mando izquierdo y derecho, compensa la longitud del árbol de mando.

Internal gear A gear with teeth pointing inward, toward the hollow center of the gear.

Engranaje interno Engranaje con los dientes orientados hacia adentro, hacia el centro hueco del engranaje.

IRS A common abbreviation for independent rear suspension.

IRS Abreviatura común para suspensión trasera independiente.

Jam nut A second nut tightened against a primary nut to prevent it from working loose. Used on inner and outer tie-rod adjustment nuts and on many pinion bearing adjustment nuts.

Contratuerca Segunda tuerca apretada contra una tuerca principal para evitar que ésta se suelte. Utilizada en las tuercas de las barras de acoplamiento interiores y exteriores y en muchas tuercas de ajuste del cojinete del piñón.

Jitter cycle A common term for a duty cycle.

Ciclo jitter Término común para un ciclo de trabajo.

Joint angle The angle formed by the input and output shafts of CV joints. Outer joints can typically operate at angles up to 45 degrees, whereas inner joints have more restricted angles.

Ángulo de las juntas Ángulo formado por el árbol impulsor y el árbol de rendimiento de las juntas de velocidad constante. Las juntas exteriores pueden funcionar típicamente en ángulos de hasta 45 grados, mientras que las juntas interiores funcionan dentro de ángulos más limitados.

Jounce The up and down movement of the car in response to road surfaces.

Sacudida Movimiento de ascenso y descenso del vehículo debido a las superficies de la carretera.

Journal The area on a shaft that rides on the bearing.

Gorrón La sección en un árbol que se mueve sobre el cojinete.

Key A small block inserted between the shaft and hub to prevent circumferential movement.

Chaveta Pasador pequeño insertado entre el árbol y el cubo para impedir el movimiento circunferencial.

Keyway A groove or slot cut to permit the insertion of a key.

Chavetero Ranura o hendidura cortada para que pueda insertarse una chaveta.

Kinetic energy Energy in motion.

Energía cinética Energía que se mueve.

Knock A heavy metallic sound usually caused by a loose or worn bearing.

Golpeteo Sonido metálico pesado normalmente causado por un cojinete suelto o desgastado.

Knuckle The part of the suspension that supports the wheel hub and serves as the steering pivot. The bottom of the knuckle is attached to the lower control arm with a ball joint, and the upper portion is usually bolted to the strut.

Muñón Parte de la suspensión que apoya al cubo de la rueda, y que sirve como pivote de dirección. La parte inferior del muñón se fija al brazo de mando inferior por medio de una junta esférica, y la superior normalmente se emperna al montante.

Knurl To indent or roughen a finished surface.

Estriar Endentar o poner áspera una superficie acabada.

Lapping The process of fitting one surface to another by rubbing them together with an abrasive material between the two surfaces.

Pulido Proceso de ajustar una superficie contra otra rozando la una contra la otra con un material abrasivo colocado entre las dos superficies.

Lash The amount of free motion in a gear train, between gears, or in a mechanical assembly, such as the lash in a valve train.

Juego Cantidad de movimiento libre en un tren de engranajes, entre engranajes o en un conjunto mecánico, como por ejemplo, el juego en un tren de válvulas.

Latent heat The heat required to change a mass's state of matter.

Calor latente Calor necesario para cambiar la condición de material de una masa.

Leaf spring A spring made up of a single flat steel plate or of several plates of graduated lengths assembled one on top of another; used on vehicles to absorb road shocks by bending or flexing.

Muelle de láminas Muelle compuesto de una sola placa de acero plana o de varias placas de longitudes graduadas montadas una sobre la otra; utilizado en vehículos para absorber la aspereza de la carretera a través de la flexión.

Lever   A device made up of a bar turning about a fixed pivot point, called the fulcrum, that uses a force applied at one point to move a mass on the other end of the bar.

Palanca   Mecanismo construído de una percha que gira por una fuerza aplicada a una punta fija para mover una masa al otro lado de la percha.

Limited-slip differential   A differential designed so that when one wheel is slipping, a major portion of the drive torque is supplied to the wheel with the better traction; also called a nonslip differential.

Diferencial de deslizamiento limitado   Diferencial diseñado para que cuando una rueda se deslice, una mayor parte del par de torsión de mando llegue a la rueda que tiene mejor tracción. Llamado también diferencial antideslizante.

Linkage   Any series of rods, yokes, and levers, and so forth, used to transmit motion from one unit to another.

Biela motriz   Cualquier serie de varillas, yugos y palancas, etc., utilizada para transmitir movimiento de una unidad a otra.

Live axle   A shaft that transmits power from the differential to the wheels.

Eje motor   Árbol que transmite fuerza motriz del diferencial a las ruedas.

Load   A term normally used to describe an electrical device that is operating in a circuit. Load also can be used to describe the relative amount of work a driveline must do.

Carga   Un término que normalmente describe un dispositivo eléctrico operando en un circuito. La carga también puede describir la cantidad relativa del trabajo que debe efectuar una flecha motriz.

Lock pin   Used in some ball sockets (inner tie-rod end) to keep the connecting nuts from working loose. Also used on some lower ball joints to hold the tapered stud in the steering knuckle.

Pasador de cierre   Utilizado en algunas rótulas para bolas (extremo interior de la barra de acoplamiento) para que no se suelten las tuercas de conexión. Utilizado también en algunas juntas esféricas inferiores para mantener el espárrago cónico en el muñón de dirección.

Locked differential   A differential with the side and pinion gears locked together.

Diferencial trabado   Diferencial en el que el engranaje lateral y el de piñón están sujetos entre sí.

Locknut   A second nut turned down on a holding nut to prevent loosening.

Contratuerca   Una segunda tuerca montada boca abajo sobre una tuerca de ensamble para evitar que se suelte.

Lockplates   Metal tabs bent around nuts or bolt heads.

Placa de cierre   Orejetas metálicas dobladas alrededor de tuercas o cabezas de perno.

Lockwasher   A type of washer that, when placed under the head of a bolt or nut, prevents the bolt or nut from working loose.

Arandela de muelle   Tipo de arandela que cuando se coloca debajo de la cabeza de un perno o de una tuerca, evita que se suelte el perno o la tuerca.

Low gears   First and second gears in a typical transmission.

Engranajes de baja velocidad   En una transmisión típica son los engranajes de primera y segunda velocidad.

Low speed   The gearing that produces the highest torque and lowest speed of the wheels.

Baja velocidad   Engranajes que producen el par de torsión más alto y la velocidad más baja de las ruedas.

Lubricant   Any material, usually a petroleum product such as grease or oil, that is placed between two moving parts to reduce friction.

Lubricante   Cualquier material, normalmente un derivado del petróleo, como la grasa o el aceite, que se coloca entre dos piezas móviles para disminuir la fricción.

Lug nut   The nuts that fasten the wheels to the axle hub or brake rotor. Missing lug nuts should always be replaced. Overtightening can cause warpage of the brake rotor in some cases.

Tuerca de orejetas   Tuercas que sujetan las ruedas al cubo del eje o el rotor de freno. Se deben reemplazar siempre las tuercas de orejetas que se hayan perdido. En algunos casos apretarlas demasiado puede causar el torcimiento del rotor de freno.

MR (Magneto-Rheological) fluid   A synthetic hydrocarbon fluid containing soft particles that changes its viscosity when introduced to a magnetic field.

Fluído de imán reológica   Un fluido sintético de hidrocarburo que contiene partículas blandas y que cambia de viscosidad durante la introducción al magnetismo.

Mass   The amount of matter in an object.

Masa   Cantidad de materia en un objeto.

Master cylinder   The liquid-filled cylinder in the hydraulic brake system or clutch, in which hydraulic pressure is developed when the driver depresses a foot pedal.

Cilindro primario   Cilindro lleno de líquido en el sistema de freno hidráulico o en el embrague, donde se produce presión hidráulica cuando el conductor oprime el pedal.

Matched gear set code   Identification marks on two gears that indicate they are matched. They should not be mismatched with another gear set and placed into operation.

Código del juego de engranaje emparejado   Señales de identificación en dos engranajes que indican que ambos hacen pareja. No deben emparejarse con otro juego de engranaje y ponerse en funcionamiento.

Meshing   The mating, or engaging, of the teeth of two gears.

Engranar   Emparejar, o endentar, los dientes de dos engranajes.

Micrometer   A precision measuring device used to measure small bores, diameters, and thicknesses. Also called a mike.

Micrómetro   Instrumento de precisión utilizado para medir calibres, espesores y diámetros pequeños. Llamado también mic.

Misalignment   When bearings are not on the same center line.

Mal alineamiento   Cuando los cojinetes no se encuentran en la misma línea central.

Molecule   The smallest particle of an element or compound that can exist in the free state and still retain the characteristics of the element or compound.

Molécula   Partícula más pequeña de un elemento o compuesto que pueda existir en estado libre, que pueda retener los atributos de dicho elemento o compuesto.

Momentum   A type of mechanical energy that is the product of an object's weight times its speed.

Momento   Tipo de energía que se deriva del peso de un producto por su velocidad.

Mounts   Made of rubber to insulate vibrations and noise while they support a powertrain part, such as engine or transmission mounts.

Monturas   Hechas de caucho para aislar vibraciones y ruido mientras apoyan una pieza del tren transmisor de potencia, como por ejemplo, las monturas del motor o de la transmisión.

Multiple disc   A clutch with a number of driving and driven discs as compared to a single plate clutch.

Disco múltiple   Embrague con muchos discos de accionamiento y accionados, comparado con un embrague de placa simple.

Needle bearing   An antifriction bearing using a great number of long, small diameter rollers. Also known as a quill bearing.

Cojinete de agujas   Cojinete de antifricción que utiliza una gran cantidad de rodillos largos con diámetros pequeños. Llamado también cojinete de manguito.

Neoprene   A synthetic rubber not affected by the various chemicals that are harmful to natural rubber.

Neopreno   Caucho sintético que no lo afectan las distintas sustancias químicas nocivas para el caucho natural.

Neutral   In a transmission, the setting in which all gears are disengaged and the output shaft is disconnected from the drive wheels.

Neutral   En una transmisión, la regulación a la cual se desengranan todos los engranajes y se desconecta el árbol de rendimiento de las ruedas motrices.

Neutral start switch   A switch wired into the ignition switch to prevent engine cranking unless the transmission shift lever is in neutral or the clutch pedal is depressed.

Interruptor de encendido neutral   Interruptor conectado al botón de encendido para impedir el arranque del motor a menos que la palanca de cambios esté en neutro o se oprima el pedal de embrague.

Newton-meter (N•m)   Metric measurement of torque or twisting force equal to foot-pounds multiplied by 1.355.

Metro-Newton (N•m)   Medida métrica de par de torsión o fuerza de torsión equivalente a libras-pies multiplicado por 1,355.

Nominal shim   A shim with a designated thickness.

Laminillas nominales   Laminillas de un espesor específico.

Nonhardening   A gasket sealer that never hardens.

Antiendurecedor   Junta de estanqueidad de la guarnición que nunca se endurece.

Nut   A removable fastener used with a bolt to lock pieces together; made by threading a hole through the center of a piece of metal that has been shaped to a standard size.

Tuerca   Herramienta de fijación desmontable utilizada con un perno para que las piezas queden sujetas entre sí; se hace abriendo un hueco en el centro de una pieza de metal conformada a un tamaño estándar.

O-ring   A type of sealing ring, usually made of rubber or a rubberlike material. In use, the O-ring is compressed into a groove to provide the sealing action.

Anillo-O   Tipo de anillo de estanqueidad normalmente hecho de caucho o de un material parecido al caucho. Cuando se le utiliza, el anillo-O se comprime en una ranura para proveer estanqueidad.

OD   Outside diameter.

OD   Diámetro exterior.

Ohm   A unit of measurement of electrical resistance.

Ohmio   Medida de la resistencia eléctrica.

Ohm's law   A statement that describes the characteristics of electricity as it flows in a circuit.

Ley de Ohm   Declaración que describe las características de la electricidad al fluir dentro de un circuito.

Ohmmeter   The instrument used to measure electrical resistance.

Ohmímetro   Instrumento que sirve para medir la resistencia eléctrica.

Oil seal   A seal placed around a rotating shaft or other moving part to prevent leakage of oil.

Junta de aceite   Junta de estanqueidad colocada alrededor de un árbol giratorio u otra pieza móvil para evitar fugas de aceite.

On-demand 4WD   Systems that power a second axle only after the first begins to slip.

Tracción a las cuatro ruedas por demanda   Sistemas que proveen fuerza motriz a un segundo eje solamente cuando el primero comienza a patinar.

Open differential   A standard-type differential.

Diferencial abierto   Diferencial de tipo estándar.

Operating angle   The difference between the drive shaft and transmission installation angles is the operating angle.

Ángulo de funcionamiento   Ángulo de funcionamiento es la diferencia entre los ángulos del montaje del árbol de mando y de la transmisión.

Outboard constant velocity joint   The outer CV joint, or the one closest to the wheels. The outer joint is a fixed joint.

Junta de velocidad constante fuera de borde   Junta de velocidad constante exterior, o la que está más cerca de las ruedas. La junta exterior es una junta fija.

Outer bearing race   The outer part of a bearing assembly on which the balls or rollers rotate.

Anillo de cojinete exterior   Parte exterior de un conjunto de cojinetes sobre la cual giran las bolas o los rodillos.

Out-of-round   Wear of a round hole or shaft that when viewed from an end will appear egg shaped.

Con defecto de circularidad   Desgaste de un agujero o árbol redondo que, vistos desde uno de los extremos, parecen ovalados.

Output driver   The part of a circuit that controls an actuator.

Impulsor de salida   La parte de un circuito que controla al actuador.

Output shaft   The shaft or gear that delivers the power from a device such as a transmission.

Árbol de rendimiento   Árbol o engranje que transmite la fuerza motriz desde un mecanismo, como por ejemplo, la transmisión.

Overcenter spring   A heavy coil spring arrangement in the clutch linkage to assist the driver with disengaging the clutch and returning the clutch linkage to the full engagement position.

Muelle sobrecentro   Distribución de muelles helicoidales gruesos en la biela motriz del embrague para ayudar al conductor a desengranar el embrague, y devolver la biela motriz del embrague a la posición de enganche total.

Overall ratio   The product of the transmission gear ratio multiplied by the final drive or rear axle ratio.

Relación total   Producto de la relación del engranaje de la transmisión multiplicado por la relación de la transmisión final o del eje trasero.

Overdrive   A gear ratio whereas the output shaft of the transmission rotates faster than the input shaft. Any arrangement of gearing that produces more revolutions of the driven shaft than of the driving shaft.

Sobremultiplicación   Relación de engranajes donde el árbol de rendimiento de la transmisión gira de manera más rápida que el árbol impulsor. Cualquier distribución de engranajes que produce más revoluciones del árbol accionado que del árbol de accionamiento.

Overdrive ratio   Identified by the decimal point indicating less than one driving input revolution compared to one output revolution of a shaft.

Relación de sobremultiplicación Identificada por el punto decimal, lo que indica menos de una revolución impulsora de mando comparada con una revolución de rendimiento de un árbol.

Overrun coupling A free-wheeling device to permit rotation in one direction but not in the other.

Acoplamiento de giro libre Mecanismo de marcha en rueda libre que permite que se lleve a cabo la rotación en una dirección pero no en la otra.

Overrunning clutch A device consisting of a shaft or housing linked together by rollers or sprags operating between movable and fixed races. As the shaft rotates, the rollers or sprags jam between the movable and fixed races. This jamming action locks together the shaft and housing. If the fixed race should be driven at a speed greater than the movable race, the rollers or sprags will disconnect the shaft.

Embrague de rueda libre Mecanismo que consiste de un árbol o un alojamiento conectados por rodillos u horquillas que funcionan entre anillos móviles o fijos. Mientras el árbol gira, los rodillos o las horquillas se acuñan entre los anillos móviles y los anillos fijos. Este acuñamiento sujeta al árbol y al alojamiento entre sí. Si el anillo fijo debe accionarse a una velocidad más alta que el anillo móvil, los rodillos o las horquillas desconectarán el árbol.

Oxidation Burning or combustion; the combining of a material with oxygen. Rusting is slow oxidation and combustion is rapid oxidation.

Oxidación Quema o combustión; la combinación de oxígeno con otro elemento. La corrosión es una oxidación lenta, mientras que la combustión es una oxidación rápida.

Parallel The quality of two items being the same distance from each other at all points; usually applied to lines and, in automotive work, to machined surfaces.

Paralelo Calidad de dos objetos que se encuentran a la misma distancia el uno del otro en todos los puntos; normalmente se aplica a líneas y en la reparación de automóviles, a superficies maquinadas.

Part-time 4WD Systems that can be shifted in and out of 4WD.

Tracción a las cuatro ruedas a tiempo parcial Sistemas que pueden cambiarse a/o de tracción a las cuatro ruedas según sea necesario.

Pawl A lever that pivots on a shaft. When lifted, it swings freely and when lowered, it locates in a detent or notch to hold a mechanism stationary.

Trinquete Palanca que gira sobre un árbol. Cuando se la levanta, se mueve libremente y cuando se la baja, se acuña en un retén o en una muesca para bloquear el movimiento de un mecanismo.

Pedal play The distance the clutch pedal and release bearing assembly move from the fully engaged position to the point at which the release bearing contacts the pressure plate release levers.

Holgura del pedal Distancia a la que el pedal del embrague y el conjunto del cojinete de desembrague se mueven de una posición enteramente engranada a un punto donde el cojinete de desembrague entra en contacto con las palancas de desembrague de la placa de presión.

Peen To stretch or clinch over by pounding with the rounded end of a hammer.

Granallar Estirar o remachar golpeando con el extremo redondo de un martillo.

Permeable Materials that absorb fluids.

Penetración Materiales que absorben los fluídos.

Phasing The rotational position of the U-joints on the drive shaft.

Fasaje Posición de rotación de las juntas universales sobre el árbol de mando.

Pilot bearing A small bearing, such as in the center of the flywheel end of the crankshaft, which carries the forward end of the clutch shaft.

Cojinete piloto Cojinete pequeño, como por ejemplo, el ubicado en el centro del extremo del volante del cigüeñal, que soporta el extremo delantero del árbol del embrague.

Pilot bushing A plain bearing fitted in the end of a crankshaft. The primary purpose is to support the input shaft of the transmission.

Buje piloto Cojinete sencillo insertado en el extremo de un cigüeñal. El propósito principal es apoyar el árbol impulsor de la transmisión.

Pilot shaft A shaft used to align parts that is removed before final installation of the parts; a dummy shaft.

Árbol piloto Árbol que se utiliza para alinear piezas y que se remueve antes del montaje final de las mismas; árbol falso.

Pinion gear The smaller of two meshing gears.

Engranaje de piñón El más pequeño de los dos engranajes de engrane.

Pinion carrier The mounting or bracket that retains the bearings supporting a pinion shaft.

Portador de piñón Montaje o soporte que sujeta los cojinetes que apoyan un árbol del piñón.

Pitch The number of threads per inch on any threaded part.

Paso Número de filetes de un tornillo por pulgada en cualquier pieza fileteada.

Pivot A pin or shaft on which another part rests or turns.

Pivote Chaveta o árbol sobre el cual se apoya o gira otra pieza.

Planet carrier In a planetary gear system, the carrier or bracket in a planetary system that contains the shafts on which the pinions or planet gears turn.

Portador planetario En un sistema de engranaje planetario, portador o soporte en un sistema planetario que contiene los árboles sobre los cuales giran los piñones o los engranajes planetarios.

Planet gears The gears in a planetary gear set that connect the sun gear to the ring gear.

Engranajes planetarios Engranajes en un tren de engranaje planetario que conectan el engranaje principal a la corona.

Planet pinions In a planetary gear system, the gears that mesh with, and revolve about, the sun gear; they also mesh with the ring gear.

Piñones planetarios En un sistema de engranaje planetario, los engranajes con que se engranan y giran entorno del engranaje principal; se engranan también con la corona.

Planetary gear set A system of gearing modeled after the solar system. A pinion is surrounded by an internal ring gear, and planet gears are in mesh between the ring gear and pinion around which all revolves.

Tren de engranaje planetario Sistema de engranaje inspirado en el sistema solar. Un piñón está rodeado por una corona interna, y los engranajes planetarios se engranan entre la corona y el piñón, alrededor de los cuales todo gira.

Plate loading Force developed by the pressure plate assembly to hold the driven disc against the flywheel.

Carga de placa Fuerza producida por el conjunto de la placa de presión para sujetar el disco accionado contra el volante.

Plunging action Telescoping action of an inner front-wheel-drive U-joint.

Acción sumergible Acción telescópica de una junta universal interior de tracción delantera.

Plunging constant velocity joint Usually the inner CV joint. The joint is designed so that it can telescope slightly to compensate for suspension motions.

Junta de velocidad constante sumergible Normalmente junta de velocidad constante interior. La junta está diseñada para que pueda extenderse ligeramente, y así compensar el movimiento de la suspensión.

Potential energy Stored energy.

Energía potencial Acumulación de energía.

Potentiometer A variable resistor with three connections that is typically used to change voltage.

Potenciómetro Un resistor variable con tres conexiones, usado para cambiar el voltaje.

Pound-foot See foot-pound.

Libra-pie Vea pie-libra (foot-pound).

Power The measure of work being done.

Fuerza Medida de trabajo en accion.

Powertrain The mechanisms that carry the power from the engine crankshaft to the drive wheels; these include the clutch, transmission, driveline, differential, and axles.

Tren transmisor de potencia Mecanismos que transmiten la potencia desde el cigüeñal del motor hasta las ruedas motrices; éstos incluyen el embrague, la transmisión, la línea de transmisión, el diferencial y los ejes.

Preload A fixed amount of pressure constantly applied to a component. Preload on bearings eliminates looseness. Preload on limited-slip differential clutches also provides torque transfer to the driven wheel with the least traction.

Carga previa Cantidad fija de presión aplicada continuamente a un componente. La carga previa sobre los cojinetes elimina el juego. La carga previa sobre los embragues del diferencial de deslizamiento limitado provee también transferencia de par de torsión a la rueda accionada de menor tracción.

Press-fit Forcing a part into an opening that is slightly smaller than the part itself to make a solid fit.

Ajuste en prensa Forzar una pieza dentro de una apertura un poco más pequeña que la pieza misma para lograr un ajuste sólido.

Pressure Force per unit area, or force divided by area. Usually measured in pounds per square inch (psi) or in kilopascals (kPa) in the metric system.

Presión Fuerza por unidad de área, o fuerza dividida por área. Normalmente se mide en libras por pulgada cuadrada (lpc) o en kilopascales (kPa) en el sistema métrico.

Pressure plate That part of the clutch that exerts force against the friction disc; it is mounted on and rotates with the flywheel. A heavy steel ring pressed against the clutch disc by spring pressure.

Placa de presión Pieza del embrague que ejerce fuerza contra el disco de fricción; se monta encima y gira con el volante, anillo pesado de acero, comprimido contra el disco de embrague, mediante presión elástica.

Propeller shaft A common term for a drive shaft.

Árbol transmisor Término común para árbol de mando.

Prussian blue A blue pigment; in solution, useful in determining the area of contact between two surfaces.

Azul de Prusia Pigmento azul; en una solución, sirve para determinar el área de contacto entre dos superficies.

psi Abbreviation for pounds per square inch, a measurement of pressure.

psi Abreviatura de libras por pulgada cuadrada; medida de presión.

Puller Generally, a shop tool used to separate two closely fitted parts without damage. Often contains a screw, or several screws, which can be turned to apply a gradual force.

Tirador Gencralmente la herramienta de taller utilizada para separar dos piezas fuertemente apretadas sin averiarlas. A menudo contiene uno o varios tornillos a los que se les puede dar vuelta para aplicar una fuerza gradual.

Pulley A wheel with a grooved rim in which a rope, belt, or chain runs to raise something by pulling on the other end of the rope, belt, or chain.

Polispasto Una rueda con llanta estriada en que corre la cuerda, la correa, o la cadena para levanter un objeto al tirar del otro lado de la cuerda, de la correa, o de la cadena.

Pulsation To move or beat with rhythmic impulses.

Pulsación Mover o golpear con impulsos rítmicos.

Quadrant A section of a gear. A term sometimes used to identify the shift-lever selector mounted on the steering column.

Cuadrante Sección de un engranaje. Término utilizado en algunas ocasiones para identificar el selector de la palanca de cambio de velocidades montado sobre la columna de dirección.

Quill shaft The hollow shaft on the front of the front bearing retainer.

Árbol de manguito Árbol hueco en la parte frontal del retenedor del cojinete delantero.

Race A channel in the inner or outer ring of an antifriction bearing in which the balls or rollers roll.

Anillo Canal en el anillo interior o exterior de un cojinete de antifricción en el cual giran las bolas o los rodillos.

Raceway A groove or track designed into the races of a bearing or U-joint housing to guide and control the action of the balls or trunnions.

Anillo de rodadura Ranura o canal construido en el interior de los anillos de un cojinete o de un alojamiento de junta universal para guiar y controlar el movimiento de las bolas o de las muñequillas.

Radial The direction moving straight out from the center of a circle; perpendicular to the shaft or bearing bore.

Radial Dirección que sale directamente del centro de un círculo perpendicular al árbol o al calibre de cojinete.

Radial clearance (radial displacement) Clearance within the bearing and between balls and races perpendicular to the shaft.

Espacio libre radial (desplazamiento radial) Dentro del cojinete y entre las bolas y los anillos, espacio libre perpendicular al árbol.

Radial load A force perpendicular to the axis of rotation. Loads applied at 90 degrees to an axis of rotation.

Carga radial Fuerza perpendicular al pivote de rotación. Cargas aplicadas a un pivote de rotación a 90°.

Ramp-type differential A design of limited-slip differential.

Diferencial tipo rampa Un diseño de diferencial con deslizamiento limitado.

Ratcheting mechanism Uses a pawl and gear arrangement to transmit motion or to lock a particular mechanism by having the pawl drop between gear teeth.

Mecanismo de trinquete Utiliza un conjunto de retén y engranaje para transmitir movimiento o para bloquear un mecanismo específico haciendo que el retén caiga entre los dientes del engranaje.

Ratio The relation or proportion that one number bears to another.

Relación Razón o proporción que existe entre un número y otro.

Reactivity A statement of how easily a substance can cause or be part of a chemical reaction.

Reactividad Enunciado que expresa cuán fácil una substancia puede provocar o ser parte de una reacción química.

Reamer A round metal cutting tool with a series of sharp cutting edges; enlarges a hole when turned inside it.

Escariador Herramienta redonda metálica de corte que tiene una serie de aristas agudas; ensancha un agujero cuando se le da vuelta dentro de éste.

Rear axle torque The torque received and multiplied by the rear driving axle assembly.

Torsión del eje trasero Par de torsión recibido y multiplicado por el conjunto del eje motor trasero.

Rear-wheel drive (RWD) A term associated with a vehicle in which the engine is mounted at the front and the driving axle and driving wheels are mounted at the rear of the vehicle.

Tracción trasera Término relacionado a un vehículo donde el motor se monta en la parte delantera, y el eje motor y las ruedas motrices en la parte trasera.

Rebound The movement of the suspension system as it attempts to bring the car back to normal heights after jounce.

Rebote Movimiento que se obsrva en la suspensión mientras intenta estabilizar el funcionamiento normal del vehículo después de una sacudida.

Reference voltage sensor A type of sensor that changes a reference voltage in response to chemical or mechanical changes.

Sensor de voltaje de referencia Un tipo de sensor que cambia el voltaje de referencia según la respuesta a los cambios químicos o mecánicos.

Relay An electrical device that allows a low current circuit to control a high current circuit.

Relé Un dispositivo eléctrico que permite que un circuito de baja corriente controle un circuito de alta corriente.

Release bearing A ball-type bearing moved by the clutch pedal linkage to contact the pressure plate release levers to either engage or disengage the driven disc with the clutch driving members.

Cojinete de desembrague Cojinete de tipo bola accionado por la biela motriz del pedal del embrague para entrar en contacto con las palancas de desembrague de la placa de presión o para engranar o desengranar el disco accionado con los mecanismos de accionamiento del embrague.

Release levers In the clutch, levers that are moved by throwout-bearing movement, causing clutch spring force to be relieved so that the clutch is disengaged, or uncoupled, from the flywheel.

Palancas de desembrague En el embrague las palancas accionadas por el movimiento del cojinete de desembrague, que hacen disminuir la fuerza del muelle del embrague para que el embrague se desengrane, o se desacople del volante.

Release plate Plate designed to release the clutch pressure plate's loading on the clutch driven disc.

Placa de desembrague Placa diseñada para desembragar la carga de la placa de presión del embrague en el disco accionado del embrague.

Removable carrier housing A type of rear axle housing from which the axle carrier assembly can be removed for parts service and adjustment.

Alojamiento portador desmontable Tipo de puente trasero del cual se puede desmontar el conjunto del portador del eje para la reparación y el ajuste de las piezas.

Retaining ring A removable fastener used as a shoulder to retain and position a round bearing in a hole.

Anillo de retención Aparato fijador desmontable utilizado como punto de apoyo para sujetar y colocar un cojinete redondo en un agujero.

Retractor clips Spring steel clips that connect the diaphragm's flexing action to the pressure plate.

Grapas retractoras Grapas de acero para muelles que conectan el movimiento flexible del diafragma a la placa de presión.

Reverse idler gear In a transmission, an additional gear that must be meshed to obtain reverse gear; a gear used only in reverse that does not transmit power when the transmission is in any other position.

Piñón de marcha atrás En una transmisión, el engranaje adicional que debe engranarse para obtener un engranaje de marcha atrás; el engranaje utilizado solamente durante la inversión de marcha, que no transmite fuerza cuando la transmisión se encuentra en cualquier otra posición.

Rheostat A variable resistor with two connections that is used to change current flow through a circuit.

Reóstato Un resistor variable de dos conexiones que sirve para cambiar el flujo de corriente por un circuito.

Ring gear A gear that surrounds or rings the sun and planet gears in a planetary system. Also the name given to the spiral bevel gear in a differential.

Corona Engranaje que rodea los engranajes planetario y el principal en un sistema planetario. También es el nombre que se le da al engranaje cónico con dentado espiral en un diferencial.

Rivet A headed pin used for uniting two or more pieces by passing the shank through a hole in each piece, and securing it by forming a head on the opposite end.

Remanche Chaveta de cabeza utilizada para unir dos o más piezas insertando la espinilla en cada una de las piezas a través de un agujero. La espinilla se asegura formando una cabeza en el extremo opuesto.

Roller bearing An inner and outer race on which hardened steel rollers operate.

Cojinete de rodillos Anillo interior y exterior sobre el cual funcionan unos rodillos de acero templado.

Rollers Round steel bearings that can be used as the locking element in an overrunning clutch or as the rolling element in an antifriction bearing.

Rodillos Cojinetes redondos de acero que pueden utilizarse como el elemento de bloqueo en un embrague de rueda libre, o como el elemento rodante en un cojinete de antifricción.

rpm Abbreviation for revolutions per minute, a measure of rotational speed.

rpm Abreviatura de revoluciones por minuto; medida de velocidad de rotación.

RTV sealer Room Temperature Vulcanizing gasket material, which cures at room temperature; a plastic paste squeezed from a tube to form a gasket of any shape.

Junta de estanqueidad VTA Material vulcanizador de guarnición a temperatura ambiente, que se conserva a temperatura ambiente; pasta plástica que viene en tubo, utilizada para formar una guarnición de cualquier tamaño.

Rubber coupling Rubber-based disc used as a U-joint between the driving and driven shafts.

Acoplamiento de caucho Disco con base de caucho; utilizado como junta universal entre el árbol de accionamiento y el árbol accionado.

Runout Deviation of the specified normal travel of an object. The amount of deviation or wobble a shaft or wheel has as it rotates. Runout is measured with a dial indicator.

Desviación Desalineación del movimiento normal indicado de un objeto. Cantidad de desalineación o bamboleo que tiene un árbol o una rueda mientras gira. La desviación se mide con un indicador de cuadrante.

Rzeppa constant velocity joint The name given to the ball-type CV joint (as opposed to the tripod-type CV joint). Rzeppa joints are usually the outer joints on most FWD cars. Named after its inventor, Alfred Rzeppa, a Ford engineer.

Junta de velocidad constante Rzeppa Nombre que se le da a la junta de velocidad constante de tipo bola (en contraste con la junta de velocidad constante de tipo trípode). Las juntas Rzeppa normalmente son las juntas exteriores en la mayoría de automóviles de tracción delantera. Nombrada por su creador, Alfred Rzeppa, ingeniero de la Ford.

SAE Society of Automotive Engineers.

SAE Sociedad de Ingenieros Automotrices.

Score A scratch, ridge, or groove marring a finished surface.

Muesca Rayado, rotura o ranura que estropea una superficie acabada.

Scuffing A type of wear in which there is a transfer of material between parts moving against each other; shows up as pits or grooves in the mating surfaces.

Frotamiento Tipo de desgaste en el cual se transfiere material entre piezas que se mueven la una contra la otra; aparece en forma de hendiduras o ranuras en las superifices emparejadas.

Seal A material, shaped around a shaft, used to close off the operating compartment of the shaft, preventing oil leakage.

Junta de estanqueidad Material conformado alrededor de un árbol, que se utiliza para sellar el compartimiento de funcionamiento del árbol, y así evitar la fuga de aceite.

Sealer A thick, tacky compound, usually spread with a brush, which may be used as a gasket or sealant to seal small openings or surface irregularities.

Líquido de estanqueidad Compuesto grueso y viscoso, normalmente esparcido con una brocha, que puede emplearse como guarnición o compucsto obturador para rellenar pequeñas aperturas o irregularidades en la superficie.

Seat A surface, usually machined, on which another part rests or seats; for example, the surface on which a valve face rests.

Asiento Superficie, normalmente maquinada, sobre la cual se coloca o se sienta otra pieza; por ejemplo, la superficie sobre la cual se coloca una cara de válvula.

Self-adjusting clutch linkage Monitors clutch pedal play through a clutch control cable and ratcheting mechanism to automatically adjust clutch pedal play.

Biela motriz del embrague de ajuste automático Controla el juego del pedal del embrague mediante un cable de mando del embrague y un mecanimso de trinquete para ajustar automáticamente el juego del pedal del embrague.

Semicentrifugal pressure plate The release levers of this pressure plate are weighted to take advantage of centrifugal force to increase plate loading resulting in reduced driven disc slip.

Placa de presión semicentrífuga A las palancas de desembrague de esta placa de presión se les añade peso para aprovechar la fuerza centrífuga, y hacer que ésta aumente la carga de la placa. El resultado será un deslizamiento menor del disco accionado.

Semifloating rear axle An axle that supports the weight of the vehicle on the axle shaft in addition to transmitting driving forces to the rear wheels.

Eje trasero semi-flotante Eje que apoya cl peso del vehículo sobre el árbol motor, además de transmitir las fuerzas motrices a las ruedas traseras.

Separators A component in an antifriction bearing that keeps the rolling components apart.

Separadores Componente en un cojinete de antifricción que mantiene separados los componentes de rodamiento.

Shank The portion of the shoe that protects the ball of your foot.

Arqueo Parte del zapato que protege el hueso de la planta del pie.

Shift-on-the-fly 4WD A system that can be shifted from two- to four-wheel drive while the vehicle is moving.

Tracción a las cuatro ruedas shift-on-the-fly Sistema que puede cambiarse de marcha a tracción a las dos ruedas a marcha a tracción a las cuatro ruedas mientras el vehículo está en movimiento.

Shift forks Mechanisms attached to shift rails that fit into the synchronizer hub for change of gears.

Horquillas de cambio de velocidades Las ranuras en el anillo sincronizador del embrague cónico deben ser afiladas para lograr la sincronización.

Shift lever The lever used to change gears in a transmission. Also, the lever on the starting motor that moves the drive pinion into or out of mesh with the flywheel teeth.

Palanca de cambio de velocidades Palanca utilizada para cambiar las velocidades en una transmisión. También es la palanca en el motor de arranque que engrana o desengrana el piñón de mando con los dientes del volante.

Shift rails Rods placed within the transmission housing that are a part of the transmission gearshift linkage.

Barras de cambio de velocidades Varillas ubicadas dentro del alojamiento de transmisión que forman parte de la biela motriz del cambio de velocidades de la transmisión.

Shifter A common term for the shift lever of a transmission.

Cambiador Término común para la palanca de cambio de velocidades de una transmisión.

Shim Thin sheets used as spacers between two parts, such as the two halves of a journal bearing.

Chapa de relleno Láminas delgadas utilizadas como espaciadores entre dos piezas, como por ejemplo, las dos mitades de un cojinete liso.

Shim stock Sheets of metal of accurately known thickness that can be cut into strips and used to measure or correct clearances.

Material de chapa de relleno Láminas de metal de espesor preciso que puede cortarse en tiras, y utilizarse para medir el espacio libre correcto.

Side clearance The clearance between the sides of moving parts when the sides do not serve as load-carrying surfaces.

Despojo lateral Espacio libre entre los dos lados de piezas móviles cuando éstos no sirven como superficies de carga.

Side gears Gears that are meshed with the differential pinions and splined to the axle shafts (RWD) or drive shafts (FWD).

Engranajes laterales Engranajes que se engranan con los piñones del diferencial, y son ranurados a los árboles motores en vehículos de tracción trasera o a los árboles de mando en vehículos de tracción delantera.

Side thrust Longitudinal movement of two gears.

Empuje lateral Movimiento longitudinal de dos engranajes.

Slave cylinder   Located at a lower part of the clutch housing. Receives fluid pressure from the master cylinder to engage or disengage the clutch.

Cilindro secundario   Ubicado en la parte inferior del alojamiento del embrague. Recibe presión de fluido del cilindro primario para engranar o desengranar el embrague.

Sliding fit   Where sufficient clearance has been allowed between the shaft and journal to allow free-running without overheating.

Ajuste deslizante   Donde se ha permitido espacio suficiente entre el árbol y el gorrón para permitir un funcionamiento libre sin ocasionar un recalentimiento.

Sliding yoke   Slides on internal and external splines to compensate for driveline length changes.

Yugo deslizante   Se desliza sobre las lengüetas internas y externas para compensar los cambios de longitud de la línea de transmisión.

Sliding gear transmission   A transmission in which gears are moved on their shafts to change gear ratios.

Transmisión por engranaje desplazable   Transmisión en la cual los engranajes se mueven sobre sus árboles para cambiar la relación de los engranajes.

Slip fit   Running or sliding fit.

Ajuste corredizo   Ajuste deslizante o de marcha.

Slip joint   In the powertrain, a variable length connection that permits the drive shaft to change its effective length.

Junta corrediza   En el tren transmisor de potencia, una conexión de longitud variable que le permite al árbol de mando cambiar su longitud eficaz.

Snapring   Split spring-type ring located in an internal or external groove to retain a part.

Anillo de resorte   Anillo hendido de tipo muelle ubicado en una ranura interna o externa para sujetar una pieza en su lugar.

Solenoid   An electromagnet with a movable core. The core is used to complete an electrical circuit or to cause a mechanical action.

Solenoide   Un electroimán de núcleo móvil. El núcleo sirve para completar un circuito eléctrico o para causar una acción mecánica.

Solid axle   A rear axle design that places the final drive, axles, bearings, and hubs into one housing.

Eje sólido   Diseño del eje trasero que coloca la transmisión final, los ejes, los cojinetes y los cubos dentro de un solo alojamiento.

Solution   Formed when a solid dissolves into a liquid, its particles break away from this structure and mix evenly in the liquid.

Solución   Una solución se forma cuando se disuelve un sólido en líquido; sus partículas se rompen de esta estructura y se combinan equitativamente en el líquido.

Spalling   A condition in which the material of a bearing surface breaks away from the base metal.

Esquirla   Condición en la que el material de la superficie de un cojinete se separa del metal de base.

Speed   The distance an object travels in a set amount of time.

Velocidad   La distancia que se mueve un objeto en duración fija.

Speed gears   Driven gears located on the transmission output shaft. This term differentiates between the gears of the counter gear and cluster assemblies and gears on the transmission output shaft.

Engranajes de velocidades   Engranajes accionados ubicados en el árbol de rendimiento de la transmisión. Este término distingue entre los engranajes de los conjuntos del mecanismo contador y de los engranajes desplazables y los engranajes sobre el árbol de rendimiento de la transmisión.

Spindle   The shaft on which the wheels and wheel bearings mount.

Huso   Árbol sobre el cual se montan las ruedas y los cojinetes de rueda.

Spiral bevel gear   A ring gear and pinion wherein the mating teeth are curved and placed at an angle with the pinion shaft.

Engranaje cónico con dentado espiral   Corona y piñón cuyos dientes emparejados son curvos, y están montados en ángulo con el árbol de piñón.

Spiral gear   A gear with teeth cut according to a mathematical curve on a cone. Spiral bevel gears that are not parallel have center lines that intersect.

Engranaje helocoidal   Engranaje con dientes cortados en un cono según una curva matemática. Los engranajes cónicos con dentado espiral que no son paralelos tienen líneas centrales que se cruzan.

Spline   A slot or groove cut in a shaft or bore; a splined shaft onto which a hub, wheel, gear, and so forth, with matching splines in its bore is assembled so that the two must turn together.

Lengüeta   Hendidura o ranura excavada en un árbol o un calibre; árbol ranurado sobre el que se montan un cubo, una rueda, un engranaje, etc., con lengüetas que hacen pareja en su calibre, para que ambos giren juntos.

Splined hub   Several keys placed radially around the inside diameter of a circular part, such as a wheel or driven disc.

Cubo ranurado   Varias chavetas ubicadas de manera radial alrededor del diámetro interior de una pieza circular, como por ejemplo, una rueda o un disco accionado.

Split lip seal   Typically, a rope seal, sometimes used to denote any two-part oil seal.

Junta de estanqueidad de reborde hendido   Típicamente una junta de estanqueidad de cable utilizada en algunas ocasiones para denominar cualquier junta de aceite de dos partes.

Split pin   A round split-spring steel tubular pin used for locking purposes; for example, locking a gear to a shaft.

Pasador hendido   Pasador tubular redondo hendido de acero para muelles utilizado para sujetar; por ejemplo, para asegurar un engranaje a un árbol.

Spring   A device that changes shape when it is stretched or compressed, but returns to its original shape when the force is removed; the component of the automotive suspension system that absorbs road shocks by flexing and twisting.

Muelle   Pieza que cambia de forma cuando se estira o comprime, pero que recobra su forma original cuando se detiene la fuerza; componente del sistema de suspensión automotriz que absorbe las sacudidas de la carretera doblándose y torciéndose.

Spring retainer   A steel plate designed to hold a coil or several coil springs in place.

Retenedor de muelle   Placa de acero diseñada para sujetar uno o varios muelles helicoidales en su lugar.

Spur gear   Gears cut on a cylinder with teeth that are straight and parallel to the axis.

Engranaje recto   Engranajes cortados en un cilindro que tienen dientes rectos y paralelos al pivote.

Squeak   A high-pitched noise of short duration.

Rechinamiento   Sonido agudo de corta duración.

Squeal   A continuous high-pitched noise.

Chirrido   Sonido agudo continuo.

Standard shift   A common name for a manual transmission.

Cambio de velocidades estándar   Nombre común para la transmisión manual.

Stick-shift   A common name for a manual transmission.
Palanca de marcha   Nombre común para la transmisión manual.

Stress   The force to which a material, mechanism, or component is subjected.
Esfuerzo   Fuerza a la que se somete un material, mecanismo o componente.

Strut assembly   Refers to all the strut components, including the strut tube, shock absorber, coil spring, and upper bearing assembly.
Conjunto de montante   Se refiere a todas las piezas del montante, inclusive al tubo de montante, al amortigador, al muelle helicoidal y al conjunto del cojinete superior.

Stub axle   A common name for the spindle shaft of an axle.
Muñón corto   Nombre común para el árbol huso de un eje.

Stub shaft   A very short shaft.
Árbol corto   Árbol sumamente corto.

Sun gear   The central gear in a planetary gear system around which the rest of the gears rotate. The innermost gear of the planetary gear set.
Engranaje principal   Engranaje central en un sistema de engranaje planetario alrededor del cual giran los demás engranajes. Es el engranaje más interior del tren de engranaje planetario.

Synchromesh transmission   Transmission gearing that aids the meshing of two gears or shift collars by matching their speed before engaging them.
Transmisión de engranaje sincronizado   Engranaje transmisor que facilita el engrane de dos engranajes o collares de cambio de velocidades al igualar la velocidad de éstos antes de engranarlos.

Synchronize   To cause two events to occur at the same time; for example, to bring two gears to the same speed before they are meshed to prevent gear clash.
Sincronizar   Hacer que dos sucesos ocurran al mismo tiempo; por ejemplo, hacer que dos engranajes giren a la misma velocidad antes de que se engranen para evitar el choque de engranajes.

Synchronizer assemblies   A device that uses cone clutches to bring two parts rotating at two speeds to the same speed. A synchronizer assembly operates between two gears: first and second gear, and third and fourth gear.
Conjuntos sincronizadores   Mecanismo que utiliza embragues cónicos para hacer que dos piezas que giran a dos velocidades giren a una misma velocidad. Un conjunto sincronizador funciona entre dos engranajes; engranajes de primera y segunda velocidad, y engranajes de tercera y cuarta velocidad.

Synchronizer blocker ring   Usually a brass ring that acts as a clutch and causes driving and driven units to turn at the same speed before final engagement.
Anillo de bloque sincronizador   Normalmente un anillo de latón que sirve de embrague, y hace que las piezas de accionamiento y las accionadas giren a la misma velocidad antes del acoplamiento final.

Synchronizer hub   Center part of the synchronizer assembly that is splined to the synchronizer sleeve and transmission output shaft.
Cubo sincronizador   Pieza central del conjunto sincronizador ranurada al manguito sincronizador y al árbol de rendimiento de la transmisión.

Synchronizer sleeve   The sliding sleeve that fits over the complete synchronizer assembly.
Manguito sincronizador   Manguito deslizante que cubre todo el conjunto sincronizador.

Tap   To cut threads in a hole with a tapered, fluted, threaded tool.
Aterrajar   Cortar filetes de tornillo en un agujero con una herramienta cónica, estriada y fileteada.

Temper   To change the physical characteristics of a metal by applying heat.
Templar   Cambiar las características físicas de un metal aplicándole calor.

Tension   Effort that elongates or "stretches" a material.
Tensión   Fuerza que alarga o estira un material.

Thickness gauge   Strips of metal made to an exact thickness, used to measure clearances between parts.
Calibrador de espesor   Tiras de metal hechas a un espesor exacto, utilizadas para medir el espacio libre entre las piezas.

Thread chaser   A device, similar to a die, that is used to clean threads.
Fileteadora de tornillo   Utensilio, parecido a un troquel, utilizado para limpiar tornillos.

Threaded insert   A threaded coil that is used to restore the original thread size to a hole with damaged threads.
Piezas insertas fileteadas   Espiral fileteado que se utiliza para devolver el tamaño original del tornillo a un agujero que tiene tornillos averiados.

Three-quarter floating axle   An axle in which the axle housing carries the weight of the vehicle while the bearings support the wheels on the outer ends of the axle housing tubes.
Eje flotante de tres cuartos   Puente trasero que soporta el peso del vehículo mientras los cojinetes soportan las ruedas en los extremos exteriores de los tubos del puente trasero.

Throw-out bearing   In the clutch, the bearing that can be moved inward to the release levers by clutch-pedal action to cause declutching, which disengages the engine crankshaft from the transmission. A common name for a clutch release bearing.
Cojinete de desembrague   En el embrague, el cojinete que puede moverse hacia adentro hasta las palancas de desembrague, por medio de la acción del pedal del embrague, para lograr el desembrague. Esta acción desengrana el cigüeñal del motor de la transmisión. Nombre común para cojinete de desembrague del embrague.

Thrust load   A load that pushes or reacts through the bearing in a direction parallel to the shaft.
Carga de empuje   Carga que empuja o reacciona mediante el cojinete en una dirección paralela al árbol.

Thrust washer   A washer designed to take up end thrust and prevent excessive end play.
Arandela de empuje   Arandela diseñada para asegurar el empuje longitudinal, y prevenir un juego longitudinal excesivo.

Tie-rod   The linkage between the steering rack and the steering knuckle arm. The tie-rod is threaded into a tie-rod end or has a threaded split member for making toe adjustments.
Barra de acoplamiento   Biel mortiz entre la cremallera y el brazo del muñón de dirección, la barra de acoplamiento se filetea en un extremo de la barra de acoplamiento o tiene una pieza fileteada hendida para ajustar el tope.

Tie-rod end   The fittings on the ends of the tie-rods. The outer tie-rod end connects to the steering arm, and the inner one connects to the steering rack. Both ends include ball sockets to allow pivotal action, as well as up and down flexing.
Extremo de la barra de acoplamiento   Conexiones en los extremos de las barras de acoplamiento. El extremo exterior de la barra de acoplamiento se conecta al brazo de dirección, y el extremo interior se conecta a la cremallera. Ambos extremos incluyen juntas de rótula para permitir tanto el movimiento giratorio como el de ascenso y descenso.

Tolerance A permissible variation between the two extremes of a specification or dimension.

Tolerancia Variación permisible entre los dos extremos de una especificación o dimensión.

Torque A twisting motion, usually measured in ft.-lbs. (N•m).

Par de torsión Fuerza de torsión, normalmente medida en libras-pies (N•m).

Torque converter A turbine device utilizing a rotary pump, one or more reactors (stators), and a driven circular turbine or vane whereby power is transmitted from a driving to a driven member by hydraulic action. It provides varying drive ratios; with a speed reduction, it increases torque.

Convertidor de par de torsión Turbina que utiliza una bomba giratoria, uno o más reactores (estátores) y una turbina circular accionada o paleta; la fuerza se transmite del mecanismo de accionamiento al mecanismo accionado mediante una acción hidráulica. Provee relaciones de accionamiento variadas; con una reducción de velocidad, aumenta el par de torsión.

Torque curve A line plotted on a chart to illustrate the torque personality of an engine. When the engine operates on its torque curve it is producing the most torque for the quantity of fuel being burned.

Curva de torsión Línea trazada en una gráfica para ilustrar las características de torsión de un motor. Cuando un motor funciona según su curva de torsión, produce mayor par de torsión por cantidad de combustible quemado.

Torque multiplication The result of meshing a small driving gear and a large driven gear to reduce speed and increase output torque.

Multiplicación de par de torsión Resultado de engranar un engranaje de accionamiento pequeño y un engranaje accionado grande para reducir la velocidad y aumentar el par de torsión de rendimiento.

Torque steer A self-induced steering condition in which the axles twist unevenly under engine torque. An action felt in the steering wheel as the result of increased torque.

Dirección de torsión Condición automática de dirección en la cual los ejes giran de manera irregular bajo el par de torsión del motor. Acción que se advierte en el volante de dirección como resultado de un aumento en el par de torsión.

Torque tube A fixed tube over the drive shaft on some cars. It helps locate the rear axle and takes torque reaction loads from the drive axle so the drive shaft will not sense them.

Tubo de eje cardán Tubo fijo sobre el árbol de mando en algunos automóviles. Ayuda a colocar el eje trasero y remueve las cargas de reacción de torsión del eje de mando para que no las reciba el árbol de mando.

Torsional springs Round, stiff coil springs placed in the driven disc to absorb the torsional disturbances between the driving flywheel and pressure plate and the driven transmission input shaft.

Muelles de torsión Muelles helicoidales redondos y rígidos ubicados en el disco accionado para absorber las alteraciones de torsión entre el volante motor y la placa de presión, y el árbol impulsor accionado de la transmisión.

Total travel Distance the clutch pedal and release bearing move from the fully engaged position until the clutch is fully disengaged.

Avance total Distancia a la que el pedal del embrague y el cojinete de desembrague se mueven de la posición enteramente engranada hasta que el embrague se desengrane por completo.

Toxicity A statement of how poisonous a substance is.

Toxicidad Enunciado que expresa la cualidad tóxica de una sustancia.

Traction The gripping action between the tire tread and the road's surface.

Tracción Agarrotamiento entre la banda de la llanta y la superficie de la carretera.

Transaxle A type of construction in which the transmission and differential are combined in one unit.

Transeje Tipo de construcción en la que la transmisión y el diferencial se combinan en una sola unidad.

Transaxle assembly A compact housing most often used in front-wheel-drive vehicles that houses the manual transmission, final drive gears, and differential assembly.

Conjunto del transeje Alojamiento compacto que normalmente se utiliza en vehículos de tracción delantera, y que aloja a la transmisión manual a los engranajes de transmisión final y al conjunto del diferencial.

Transfer case An auxiliary transmission mounted behind the main transmission used to divide engine power and transfer it to both front and rear differentials, either full time or part time.

Caja de transferencia Transmisión secundaria montada detrás de la transmisión principal. Se utiliza para separar la energía del motor y transferirla a los diferenciales delantero y trasero, a tiempo completo o a tiempo parcial.

Transmission The device in the powertrain that provides different gear ratios between the engine and drive wheels as well as reverse.

Transmisión Mecanismo en el tren transmisor de potencia que provee diferentes relaciones de engranajes tanto entre el motor y las ruedas motrices como en la marcha atrás.

Transmission case An aluminum or iron casting that encloses the manual transmission parts.

Caja de transmisión Pieza fundida en aluminio o en hierro que encubre las piezas de la transmisión manual.

Transverse Powertrain layout in a front-wheel-drive automobile extending from side to side.

Transversal Distribución del tren transmisor de potencia en un vehículo de tracción delantera, que se extiende de un lado al otro.

Tripod (also called tripot) A three-prong bearing that is the major component in tripod CV joints. It has three arms (or trunnions) with needle bearings and rollers that ride in the grooves or yokes of a tulip assembly.

Trípode Cojinete de tres puntas; componente principal en juntas trípode de velocidad constante. Tiene tres brazos (o muñequillas) con cojinetes de agujas y rodillos que van montados sobre las ranuras o los yugos de un conjunto tulipán.

Tripod Universal joints A U-joint consisting of a hub with three arm and roller assemblies that fit inside a casting called a tulip.

Juntas universales de trípode Juntas universales que consisten de un cubo con conjuntos de tres brazos y de rodillos que se insertan dentro de una pieza fundida llamada tulipán.

Trunnion One of the projecting arms on a tripod or on the cross of a four-point U-joint. Each trunnion has a bearing surface that allows it to pivot within a joint or slide within a tulip assembly.

Muñequilla Uno de los brazos salientes en un trípode o en la cruz de una junta universal de cuatro puntas. La superficie del cojinete de cada muñequilla permite que ésta gire dentro de una junta o se deslice dentro de un conjunto tulipán.

Tulip assembly The outer housing containing grooves or yokes in which trunnion bearings move in a tripod CV joint.

Conjunto tulipán Alojamiento exterior que contiene ranuras o yugos en los cuales se mueven los cojinetes de muñequilla en una junta trípode de velocidad constante.

Two-disc clutch A clutch with two friction discs for additional holding power; used in heavy duty equipment.

Embrague de dos discos Embrague con dos discos de fricción para proporcionar más fuerza de retención; utilizado en equipos de gran potencia.

U-bolt An iron rod with threads on both ends, bent into the shape of a U and fitted with a nut at each end.

Perno en U Varilla de hierro con filetes de tornillo en los dos extermos, acodada en forma de U, y provista de una tuerca a cada extremo.

U-joint A four-point cross connected to two U-shaped yokes that serves as a flexible coupling between shafts.

Junta cardánica Cruz con cuatro puntas fijadas a dos yugos en forma de U, que sirve de acoplamiento flexible entre árboles.

Universal joint A mechanical device that transmits rotary motion from one shaft to another shaft at varying angles.

Junta universal Dispositivo mecánico que transmite movimiento giratorio de un árbol a otro a ángulos cambiantes.

Universal joint operating angle The difference in degrees between the drive shaft and transmission installation angles.

Ángulo de funcionamiento de la junta universal Diferencia en grados entre los ángulos del árbol de mando y del montaje de la transmisión.

Upshift To shift a transmission into a higher gear.

Cambio de velocidades ascendente Acción de cambiar la transmisión a un engranaje de alta multiplicación.

Vehicle identification number (VIN) The number assigned to each vehicle by its manufacturer, primarily for registration and identification purposes.

Número de identificación del vehículo Número asignado por el fabricante a cada vehículo, principalmente para su registro e identificación.

Velocity The speed of an object in a particular direction.

Velocidad La rapidez de un objeto que va en una dirección particular.

Vibration A quivering, trembling motion felt in the vehicle at different speed ranges.

Vibración Estremecimiento y temblor que se advierte en el vehículo a diferentes gamas de velocidades.

Viscosity The resistance to flow exhibited by a liquid. A thick oil has greater viscosity than a thin oil.

Viscosidad Resistencia de un fluido al movimiento relativo. Un aceite pesado tiene mayor viscosidad que un aceite liviano.

Viscous Thick, tending to resist flowing.

Viscoso Espeso, que tiende a resistir el movimiento.

Viscous friction The friction between layers of a liquid.

Fricción viscosa Fricción entre las capas de un fluido.

Volatility A statement of how easily the substance vaporizes or explodes.

Volatilidad Enunciado que expresa la facilidad de una substancia de evaporizarse o explotar.

Voltage Electrical pressure that causes current to flow.

Voltaje Presión eléctrica que causa que fluya la corriente.

Voltage generating device A type of sensor that generates voltage in response to the movement of a shaft or other device.

Dispositivo para generar el voltaje Tipo de sensor que genera el voltaje como respuesta al movimiento de un eje u otro dispositivo.

Voltmeter The instrument used to measure electrical pressure or potential.

Voltímetro Instrumento para medir la presión eléctrica o la energía potencial.

Volts The unit of measurement for electrical pressure or EMF.

Voltio Unidad de medida de la presión eléctrica o EMF (la fuerza electromagnética).

Weight A force on a mass by the gravitational force.

Peso Fuerza ejercida sobre una masa por la fuerza de atracción.

Wet-disc clutch A clutch in which the friction disc (or discs) is operated in a bath of oil.

Embrague de disco húmedo Embrague en el que el disco de fricción (o discos) funciona bañado en aceite.

Wheel A disc or spokes with a hub at the center that revolves around an axle, and a rim around the outside on which the tire is mounted.

Rueda Disco o rayos con un cubo en el centro que gira alrededor de un eje y una llanta alrededor del exterior sobre la cual se monta la rueda.

Work What is accomplished when a force moves a certain mass a specific distance.

Tarabajo Lo que se logra cuando una fuerza mueve una masa particular a una distancia especifica.

Worm gear A gear with teeth that resemble a thread on a bolt. It is meshed with a gear that has teeth similar to a helical tooth except that it is dished to allow more contact.

Engranaje sinfín Engranaje con dientes parecidos a los filetes de tornillo en un perno. Se engrana con un engranaje cuyos dientes son parecidos a un diente helicoidal, pero se comba para permitir un mejor contacto.

Yoke In a U-joint, the driveable torque-and-motion input and output member, attached to a shaft or tube.

Yugo En una junta universal, el mecanismo accionable impulsor y de rendimiento de torsión y movimiento, que se fija a un árbol o un tubo.

Yoke bearing A U-shaped, spring-loaded bearing in the rack-and-pinion steering assembly that presses the pinion gear against the rack.

Cojinete de yugo En el conjunto de dirección de cremallera y piñón, cojinete en forma de U, con cierre automático, que sujeta el piñón contra la cremallera.

Zerk fitting A very small check valve that allows grease to be injected into a part but keeps the grease from squirting out again.

Conexión Zerk Válvula de retención sumamente pequeña que permite inyectar la grasa, y que a la vez impide que esa grasa se derrame nuevamente.

INDEX

Note: Page numbers followed by "f" indicate material in a figure.